Civil Engineering: Structures and Materials

Civil Engineering: Structures and Materials

Edited by **Seth Royal**

LANRYE INTERNATIONAL

New Jersey

Published by Clanrye International,
55 Van Reypen Street,
Jersey City, NJ 07306, USA
www.clanryeinternational.com

Civil Engineering: Structures and Materials
Edited by Seth Royal

International Standard Book Number: 978-1-63240-527-2 (Hardback)

Contents

Preface

Civil engineering is a discipline that dates back to the beginning of civilization. It is a diverse field that takes into consideration structural engineering, earthquake engineering, transportation engineering, architectural engineering, environmental engineering, municipal or urban engineering, water resources engineering, etc. It deals with the construction and maintenance of roads, bridges, dams, canals, power plants and residential buildings. This book traces the progress of this field and highlights some of its key concepts and applications. From theories to research to practical applications, case studies related to all contemporary topics of relevance to this area of study have been included in it. The aim of this book is to present researches that have transformed this discipline and aided its advancement. Students, researchers, experts and all associated with civil engineering will benefit alike from this book.

This book is a result of research of several months to collate the most relevant data in the field.

When I was approached with the idea of this book and the proposal to edit it, I was overwhelmed. It gave me an opportunity to reach out to all those who share a common interest with me in this field. I had 3 main parameters for editing this text:

1. Accuracy – The data and information provided in this book should be up-to-date and valuable to the readers.
2. Structure – The data must be presented in a structured format for easy understanding and better grasping of the readers.
3. Universal Approach – This book not only targets students but also experts and innovators in the field, thus my aim was to present topics which are of use to all.

Thus, it took me a couple of months to finish the editing of this book.

I would like to make a special mention of my publisher who considered me worthy of this opportunity and also supported me throughout the editing process. I would also like to thank the editing team at the back-end who extended their help whenever required.

Editor

Strength and Durability Evaluation of Recycled Aggregate Concrete

Sherif Yehia*, Kareem Helal, Anaam Abusharkh, Amani Zaher, and Hiba Istaitiyeh

Abstract: This paper discusses the suitability of producing concrete with 100 % recycled aggregate to meet durability and strength requirements for different applications. Aggregate strength, gradation, absorption, specific gravity, shape and texture are some of the physical and mechanical characteristics that contribute to the strength and durability of concrete. In general, the quality of recycled aggregate depends on the loading and exposure conditions of the demolished structures. Therefore, the experimental program was focused on the evaluation of physical and mechanical properties of the recycled aggregate over a period of 6 months. In addition, concrete properties produced with fine and coarse recycled aggregate were evaluated. Several concrete mixes were prepared with 100 % recycled aggregates and the results were compared to that of a control mix. SEM was conducted to examine the microstructure of selected mixes. The results showed that concrete with acceptable strength and durability could be produced if high packing density is achieved.

Keywords: recycled aggregate, concrete properties, physical properties, mechanical properties.

1. Introduction

Utilizing recycled aggregate is certainly an important step towards sustainable development in the concrete industry and management of construction waste. Recycled aggregate (RA) is a viable alternative to natural aggregate, which helps in the preservation of the environment. One of the critical parameters that affect the use of recycled aggregate is variability of the aggregate properties. Quality of the recycled aggregate is influenced by the quality of materials being collected and delivered to the recycling plants. Therefore, production of recycled aggregate at an acceptable price rate and quality is difficult to achieve due the current limitations on the recycling plants. These issues concern the clients about the stability of production and variability in aggregate properties. The main goal of the current research project is to investigate variability of aggregate properties and their impact on concrete production. Aggregate strength, gradation, absorption, moisture content, specific gravity, shape, and texture are some of the physical and mechanical characteristics that contribute to the strength and durability of concrete. Therefore, it is necessary to evaluate these properties before utilizing the aggregate. In this paper, properties of recycled aggregate from an unknown source collected over a period of 6 months from a recycling plant were evaluated. In addition, properties of concrete produced with 100 % recycled aggregates were investigated.

Department of Civil Engineering, American University of Sharjah, Sharjah, United Arab Emirates.
*Corresponding Author; E-mail: syehia@aus.edu

2. Background

2.1 Economical and Environmental Impact

The evolution in the construction industry introduces several concerns regarding availability of natural aggregate resources, as they are being rapidly depleted. Recent statistics showed the increasing demand of construction aggregate to reach 48.3 billion metric tons by the year 2015 with the highest consumption being in Asia and Pacific as shown in Fig. 1 (The Freedonia Group 2012). This increasing demand is accompanied by an increase of construction waste. For example, construction waste from European Union countries represents about 31 % of the total waste generation per year (Marinkovic et al. 2010; Ministry of Natural Resources 2010). Similarly, in Hong Kong, the waste production was nearly 20 million tons in the year 2011, which constitutes about 50 % of the global waste generation (Tam and Tam 2007; Lu and Tam 2013; Ann et al. 2013). Disposal in landfills is the common method to manage the construction waste, which creates large deposits of construction and demolition waste sites (Marinkovic et al. 2010; Tam and Tam 2007; Naik and Moriconi 2005). Efforts to limit this practice and to encourage recycling of construction and demolition waste in different construction applications led to utilizing up to 10 % of the recycled aggregate in different construction applications (Marinkovic et al. 2010; Ministry of Natural Resources 2010; Naik and Moriconi 2005; European Aggregate Association 2010; Cement, Concrete, and Aggregates 2008; Tepordei 1999). Therefore, recycling has the potential to reduce the amount of waste materials disposed of in landfills and to preserve natural resources (Sonawane and Pimplikar 2013; Llatas 2011; Lu and Yuan 2011; Braunschweig et al. 2011; Marinkovic et al. 2010; Gupta 2009; Rao et al. 2010; Tam 2008; Topcu and Guncan 1995).

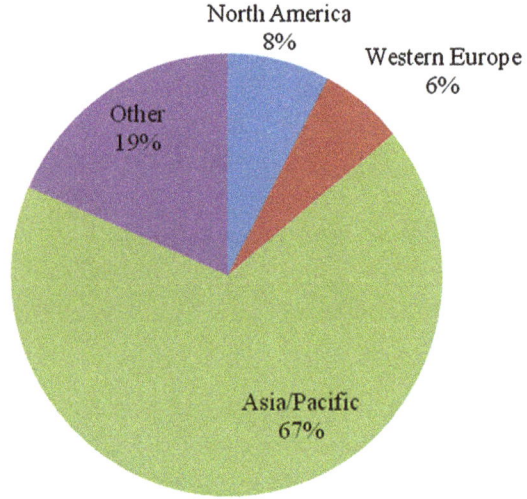

Fig. 1 Demand on construction aggregates worldwide (The Freedonia Group 2012).

2.2 Properties of Recycled Aggregate Concrete

Durability and other concrete properties are affected by the use of recycled aggregate in concrete mixes. Research efforts to introduce RA into the construction industry and to address their effects on properties could be classified to the following categories:

(1) Policies, cost and benefits: the goals are to standardize the use of RA in concrete, highlight the cost of capital investments and to emphasize environmental and economic benefits. Land protection and preservation of natural resources are the main benefits of utilizing recycled materials in the construction industry (Hansen 1986; Kartam et al. 2004; FHWA 2004; Oikonomou 2005; Tam and Tam 2007; EU Directive 2008/98/EC; Ministry of Natural Resources 2010; Marinkovic et al. 2010; Ann et al. 2013; Silva et al. 2014; Lu and Tam 2013; Bodet 2014).

(2) Evaluation of physical and mechanical properties of RA: absorption, aggregate texture (type of crushers, number of crushing stages), aggregate size and gradation, specific gravity, density, mortar content, percentage and type of contamination, aggregate strength and abrasion resistance are the main properties that affect utilizing RA in concrete production. Variation in the RA properties due to loading, different environmental conditions in addition to the crushing process, contamination and impurities such as wood and plastic pieces, affect concrete properties produced using RA. Mortar adhered to RA lead to lower density, high absorption, and high L.A. abrasion loss. In addition, sulphate and alkali contents cause expansive reactions which can be controlled if the maximum sulphate is in the range of 0.8–1.0 % by mass and alkali content below 3.5 kg/m^3 (Tam et al. 2008; De Juan and Gutiérrez 2009; McNeil and Kang 2013; De Brito and Saikia 2013; Akbarnezhad et al. 2013; Silva et al. 2014).

(3) Mix design and proportioning: direct volume replacement, weight replacement and equivalent mortar

replacement are some of the approaches that could be followed to design mixtures with RA. In addition, the mixing process can affect overall concrete properties. Both volume replacements and pre-soaking approaches showed improved properties of concrete produced with RA (Tam et al. 2007a, b; Cabral et al. 2010; Fathifazl et al. 2009; Knaack and Kurama 2013; Wardeh et al. 2014).

(4) Evaluation of fresh and hardened concrete made with RA: there are numerous efforts to evaluate fresh and hardened properties of concrete with RA. Optimizations to determine the percent of RA that could be used without affecting the short and long term performance were also investigated. Design equations based on data collected from many publications were also proposed. In general, the use of recycled aggregate led to reduction in all mechanical properties, in addition to influencing the fresh stage properties and concrete durability due to high absorption and porosity (Xiao et al. 2006; Yang et al. 2008; Kwan et al. 2012; Manzi et al. 2013; Akbarnezhad et al. 2013; Ulloa et al. 2013; Xiao et al. 2014; McNeil and Kang 2013; Silva et al. 2014).

(5) Improving durability of RA concrete: concerns about durability and the long-term performance of concrete with RA are hurdles that limit utilization of RA in many applications. Chloride conductivity, oxygen and water permeability, carbonation depth, alkaline aggregate reaction, sulphate resistance, shrinkage and creep performance, abrasion resistance and freeze resistance are some of the parameters that could be used as durability and long-term performance indicators of concrete material. In general, concrete made with RA showed less durability due to high pore volume which led to high permeability and water absorption. High water absorption is due to cement paste adhered on the aggregate surface. However, this can be countered by achieving saturated surface dry (SSD) conditions before mixing. This might not be practical in some cases of mass production. Therefore, aggregate absorption can be accounted for during the mix design stage by adjusting the mixing water that will be absorbed by the recycled aggregate. Surface coating was another approach to control absorption and improve properties (Olorunsogo and Padayachee 2002; Zaharieva et al. 2003; Levy and Helene 2004; Ann et al. 2008; Yang et al. 2008; Abbas et al. 2009; Thomas et al. 2013; Lederle and Hiller 2013; Fathifazl and Razaqpur 2013; Xiao et al. 2014; Ryou and Lee 2014). In addition, many research efforts showed that the use of supplementary cementitious materials (SCM) as a replacement for cement or addition by weight can improve concrete durability due to improvement of pore structure and reduction of the volume of macro pores. Fly ash (25–35 %), silica fume (10 %) and ground-granulated blast-furnace slag (up to 65 %) are the most commonly SCM which are used to improve concrete strength and durability properties (Berndt

2009; Kou and Poon 2012; Amorim et al. 2012; Eisa 2014).

(6) Microstructure, interfacial transition zone (ITZ) and bond characteristics: close inspection of the interfacial transition zone (ITZ) showed porous microstructure which can be attributed to high porosity and high absorption capacity of the recycled aggregate. In addition, possible cracking due to crunching and processing and exposure to several chemicals and depositions of harmful substances on the surface of aggregate can lead to cracks in concrete and reduction in the bond between the cement and aggregate. The mixing process, less w/c ratio and addition of SCM can improve the ITZ and bond characteristics of recycled aggregate concrete (Otsuki et al. 2003; Poon et al. 2004; Tam et al. 2005; Evangelista and Brito 2007; Tabsh and Abdelfatah 2009; Xiao et al. 2012a)

Table 1 summarizes some of the findings, limitations and potential challenges in using recycled aggregate in concrete applications.

3. Aggregates Used in the Study

Quality and availability of recycled aggregate are the main factors towards stable use and introduction of recycled aggregate concrete to the construction industry. The crushed stone aggregate used in the study was obtained from a recycling plant which was established and directed towards reducing waste produced from the construction industry to provide an efficient alternative for the reuse of recycled aggregate. The waste is received and processed to produce several products; however, the main product is aggregate. The process involves crushing, separation of metals by a magnet, manual removal of other impurities (plastic, wood, etc..), and classification of aggregate to different grades based on particle size. The facility produces 5 grades that vary from fine aggregate (grade 5) to 63 mm particle size (grade 3). The percentage produced from each grade depends on the materials delivered to the facility; however, grades 1, 2, 4 and 5 represent about 80 % of the plant production that ensure availability of these grades for the use in the construction industry.

4. Experimental Program

The main objectives of the experimental program were to (i) investigate variability of recycled aggregate properties and their impact on concrete production and (ii) evaluate properties of concrete prepared with 100 % recycled aggregate. Therefore, the experimental program was divided into two phases; Phase 1 deals with evaluation of the aggregate properties and Phase 2 focuses on the evaluation of concrete mixtures utilizing 100 % recycled aggregates. Figure 2 summarizes the experimental program and list of physical and mechanical properties included in the

investigation. All results were compared to that of a control mix prepared with virgin aggregate (crushed lime stone). In addition, Scanning Electron Microscopy (SEM) was conducted to examine the microstructure of some samples to provide an idea about the bond strength between cement and aggregate and identify potential weak points within the mix.

4.1 Phase 1: Evaluation of Aggregate Properties

The recycling facility was the source of the recycled aggregate (RA) used in the investigation. Aggregate was collected at different time intervals to evaluate the effect of consistency and variability in the quality on concrete properties. Only four grades were included in the investigation, grade 1 (maximum size of 10 mm), 2 (maximum size of 25 mm), 4 (mixture of course and fine aggregate along with impurities) and 5 (fine sand). Grade 3 was excluded because of the particle size (63 mm). In this phase, several physical and mechanical properties of aggregate that are directly related to concrete properties were evaluated, as shown in Fig. 2.

4.1.1 Results of Aggregate Evaluation

Results of the physical and mechanical tests conducted on RA showed expected variations from virgin aggregate mainly due to the presence of mortar adhered on the aggregate which is reflected in the high absorption capacity of the aggregate. Figure 3 shows sample of different aggregate grades used in the study. Small percentage of impurities (wood and plastic chips) was found in the aggregate, such impurities are expected due to the recycling process.

Sieve analysis Four batches of RA were obtained from the recycling facility between December 2012 and April 2013. All batches went through the same evaluation to investigate any variability in production. Figure 4 shows the sieve analysis results of the RA and virgin aggregate (control) compared to the upper and lower limits specified by (ASTM C33/C33 M 2013a, ASTM C136 2011a). Although the gradation varies from that of the control and did not meet any ASTM grading requirements, there was a clear similarity in the gradation of the last 3 batches of each grade which indicates a consistent RA production. Additionally, the authors decided to use the RA to produce concrete without any alteration of the gradations already obtained from the plant. The reasons for the decision are to avoid additional costs and to utilize available gradations to achieve acceptable particle distribution.

Aggregate crushing value (ACV) provides an indication of the aggregate strength. Aggregate with lower ACV is recommended to ensure that the aggregate will be able to resist applied loads. The test was conducted on coarse aggregate of different grades. The ACV is calculated as the ratio between the weight passing sieve 2.36 and the original weight. Values were in the range of 20–30, as shown in Fig. 5a.

Abrasion resistance is an indication of the aggregates' toughness. The Los Angeles (LA) test was conducted according to (ASTM C131 2006) and the test results are shown in Fig. 5b. The coarse aggregates in grade 4 had a higher

Table 1 Effect of recycled aggregate on concrete properties.

Durability	Durability of Recycled Aggregate (RA) can be influenced by coarse aggregate replacement ratio, concrete age, w/c ratio, and moisture content; generally, a lower w/c ratio generates a more durable concrete mix. RA concrete is less durable due to high porosity of recycled aggregate. However, lower resistance to ingress of certain agents might be compensated by the combination of recycled aggregate with CO_2 and chlorides which reduces their penetration rates. SCM are used to improve strength and durability of RA concrete	Thomas et al. (2013), Fathifazl and Razaqpur (2013), Kou and Poon (2012), Chen and Ying (2011), Corinaldesi and Moriconi (2009), Gonclaves et al. (2004)
Compressive strength	50 to 100 % replacement of virgin aggregates with recycled aggregate decreases the compressive strength by 5 to 25 %. However, it was found that up to 30 % virgin aggregate can be substituted with RCA without any effects on concrete strength. Strength gain for RCA concrete is lower than normal aggregate concrete (NAC) for the first 7 days. On the other hand, fine RA has a more detrimental effect on compressive strength than coarse RA	Malešev et al. (2010), Rahal (2007), Yehia et al. (2008), Limbachiya et al. (2004), Xiao et al. (2012b), Corinaldesi (2010), Rahal (2007), Garg et al. (2013), Sim and Park (2011)
Fresh concrete Properties: Workability Moisture Content	More water is needed to achieve similar workability to that of NAC due to higher absorption capacity of recycled aggregate which can be attributed to the presence of impurities and attached cement hydrates. As the RA content increase in the mix, the workability reduces especially at lower w/c ratio in their study found that the entrapped air content was similar when compared to normal concrete mix having a range of 2.4 ± 0.2 %. In fact, there is no significant effect regarding the air content up to 25 % replacements	Xiao et al. (2012b), Sagoe-Crentsil et al. (2001), Tabsh and Abdelfatah (2009), Medina et al. (2014), Qasrawi and Marie (2013), Sagoe-Crentsil et al. (2001)
Flexural strength	Recycled aggregate has marginal influence on flexural strength, some studies showed that flexural strength reduction is limited to 10 % in RA concrete. Others indicated that RA concrete has very similar flexural behavior with virgin aggregate concrete	Malešev et al. (2010), Xiao et al. (2012b), Chen et al. (2010), Limbachiya et al. (2004)
Modulus of elasticity	Modulus of elasticity is greatly reduced by the use of recycled aggregate; it can reach 45 % of the modulus of elasticity of corresponding conventional concrete. This percentage reduction varies based on the percentage substitution. The 45 % reduction was found at 100 % substitution, while up to 15 % reduction was observed at 30 % substitution	Vyas and Bhatt (2013), Xiao et al. (2012b), Corinaldesi (2010)
Split tensile strength	A reduction of up to 10 % in split tensile strength was observed when virgin aggregate was substituted with recycled aggregate. Studies suggest that split tensile strength is more dependent on the binder quality rather than the aggregate type	Malešev et al. (2010), Thomas et al. (2013), Sagoe-Crentsil et al. (2001)

Table 1 continued

Specific gravity and bulk density	Padmini et al. (2009) found that the specific gravity and bulk density are relatively low for recycled aggregates when compared to fresh granite aggregate (FGA). This is mainly due to the high water absorption of the RA, as mortar has higher porosity than aggregates; hence RA absorbs more water than FGA	Padmini et al. (2009).
Aggregate size	Padmini et al. (2009) found that as the maximum size of the RA increases, the achieved strength increases	Padmini et al. (2009).
Shrinkage and creep	Shrinkage and creep deformation of RA concrete are higher than those of conventional concrete, 25 and 35 % higher, respectively. Percentage of substitution, size and source of parent aggregate, mixing procedure, curing, SCM and chemical admixture affect shrinkage and creep of the RA concrete. Recent studies showed improved behavior could be achieved by mix proportioning, low w/c ratio and curing	Silva et al. (2015), Fathifazl and Razaqpur (2013), Fathifazl et al. (2011), Henschen et al. (2012), Domingo-Cabo et al. (2009), Xiao et al. (2014).

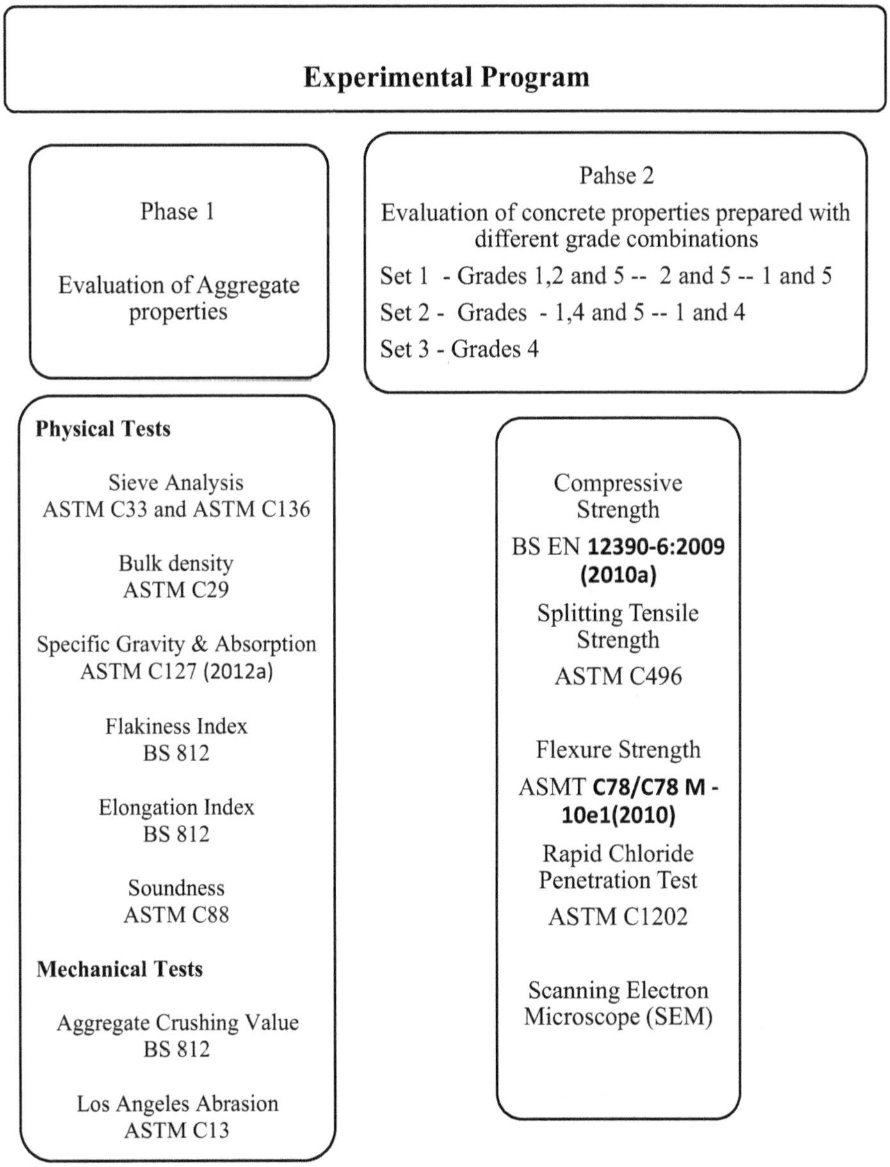

Experimental Program

Phase 1

Evaluation of Aggregate properties

Pahse 2

Evaluation of concrete properties prepared with different grade combinations

Set 1 - Grades 1,2 and 5 -- 2 and 5 -- 1 and 5

Set 2 - Grades - 1,4 and 5 -- 1 and 4

Set 3 - Grades 4

Physical Tests

Sieve Analysis
ASTM C33 and ASTM C136

Bulk density
ASTM C29

Specific Gravity & Absorption
ASTM C127 (2012a)

Flakiness Index
BS 812

Elongation Index
BS 812

Soundness
ASTM C88

Mechanical Tests

Aggregate Crushing Value
BS 812

Los Angeles Abrasion
ASTM C13

Compressive Strength
BS EN **12390-6:2009 (2010a)**

Splitting Tensile Strength
ASTM C496

Flexure Strength
ASMT **C78/C78 M - 10e1(2010)**

Rapid Chloride Penetration Test
ASTM C1202

Scanning Electron Microscope (SEM)

Fig. 2 Summary of the experimental program conducted in the investigation.

Close inspection of the aggregate

(a) Grade 1 - maximum size of 10mm

Aggregate Texture

(b) Grade 2 - maximum size of 25mm

(c) Grade 5 - Fine sand

(d) Grade 4 - Mixture of course and fine   Small size aggregate connected by mortar
aggregate along with impurities such as
wood and plastic pieces

Fig. 3 Different grades of recycled aggregates produced by the recycling facility.

percentage of weight loss, close inspection showed weak aggregate (small-sized aggregate covered with mortar, Fig. 3d).

Absorption grades 1 and 4 showed high absorption capacity (up to 8 %) while it was in the range of 3 % for grade 2 and 5. These values indicate high porosity which will require special considerations during mixing to achieve workability and to control water demand.

Soundness Soundness test was conducted according to (ASTM C88. 2013b) using Sodium Sulphate salt. Coarse aggregates from Grades 2 and 4 were sieved to different sizes and the retained on each sieve was exposed to four cycles of soaking in the solution and drying in air. Figure 5d shows percentage of the weight loss in size 9.5 mm. There

was about 20 % weight loss in grade 2; however, the loss in grade 4 was in the range of 20 to 40 %. The reasons for this high loss in volume from exposure to deicing agents are weak strength and high porosity of the recycled aggregate as indicated by high absorption.

4.1.2 Comparison Between Properties of the Virgin Aggregate and RA

Table 2 shows a sample of the results obtained from the physical and mechanical tests of recycled aggregate in December 2012 and April 2013 respectively. The last three batches indicated a similar trend with slight variations in properties, while aggregate gradation and particle sizes were maintained. However, there was an increase in the specific

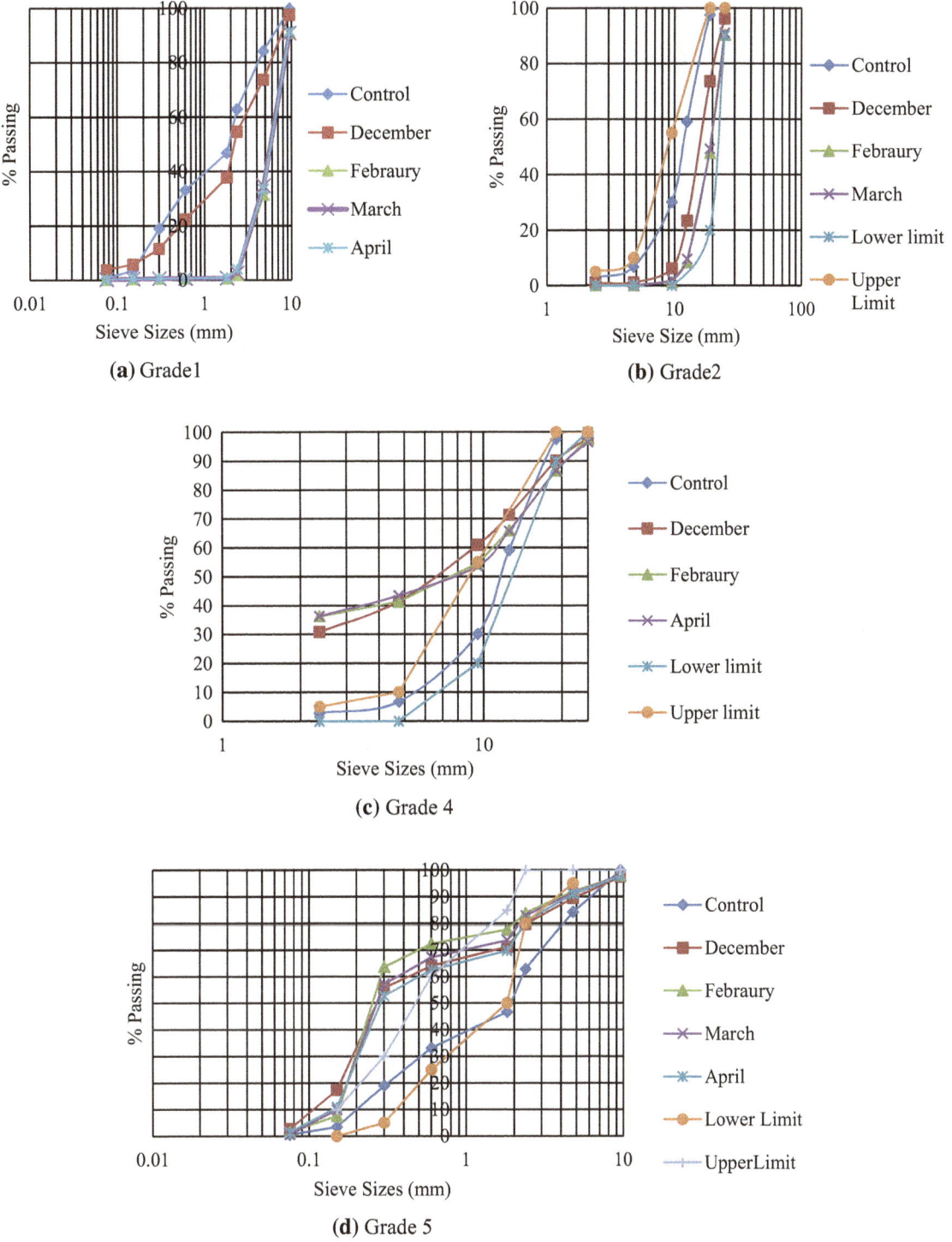

Fig. 4 Sieve analysis of RA and virgin (control) aggregate.

gravity values of Grade 5, which may have resulted from the addition of asphalt to increase its selling value.

Values obtained from the evaluation of the physical and mechanical properties of RA were compared against the values obtained from the same evaluation process conducted on virgin aggregate, as shown in Table 3. The results showed that RA has higher absorption capacity due to the mortar adhered on the surface, higher abrasion loss, high crushing value, and soundness loss which could be attributed to previous exposure to weathering and loading.

4.2 Phase 2: Evaluation of Concrete Properties Prepared with Different Grade Combinations

In this phase, extensive evaluation was conducted to select the grade combinations as-delivered that could be used in concrete production to meet the target strength and durability requirements for different applications. Compressive strength, splitting tensile strength, flexural strength, and modulus of elasticity tests were performed to determine suitability of these mixes to different applications. Additionally, the rapid chloride penetration tests (RCPT) (Kwan et al. 2012) for all mixes and scanning electron microscopy (SEM) scans to examine the micro-structural features for selected samples were conducted to provide information about the long-term durability.

Materials Grades 1, 2, 4, and 5 as fine and coarse aggregates, in addition, type I cement were used in all mixes. No supplementary cementitious materials were used in the mixes; only high range water reducer admixture was used to achieve the target workability.

Control mix the mix proportioning is based on the absolute volume method to produce self-consolidated concrete (SCC). The main reason for selecting a SCC mix that issues related to workability and aggregate gradation could be emphasized with a SCC mix. In addition, if recycled

aggregate (RA) could be used to produce SCC; hence, RA could be used for other mixes with target slump. The following volumetric ratios of 14 % cement, 17.6 % water (w/c = 0.4) and 68.4 % aggregate. The aggregate percentage (68.4 %) was divided into 37.6 % coarse aggregate (crushed lime stone) and 30.8 % fine aggregate based on the optimization of packing density of normal weight fine and coarse aggregates used for the control mix. The target cube compressive strength was 50 MPa (7000 psi) and total slump flow was 500 mm (20 in.) spread.

Packed density of RA based on the volumetric ratios, the weight of grade 1, grade 2 and grade 5 were proportioned and collected in a measuring cylinder has a volume of

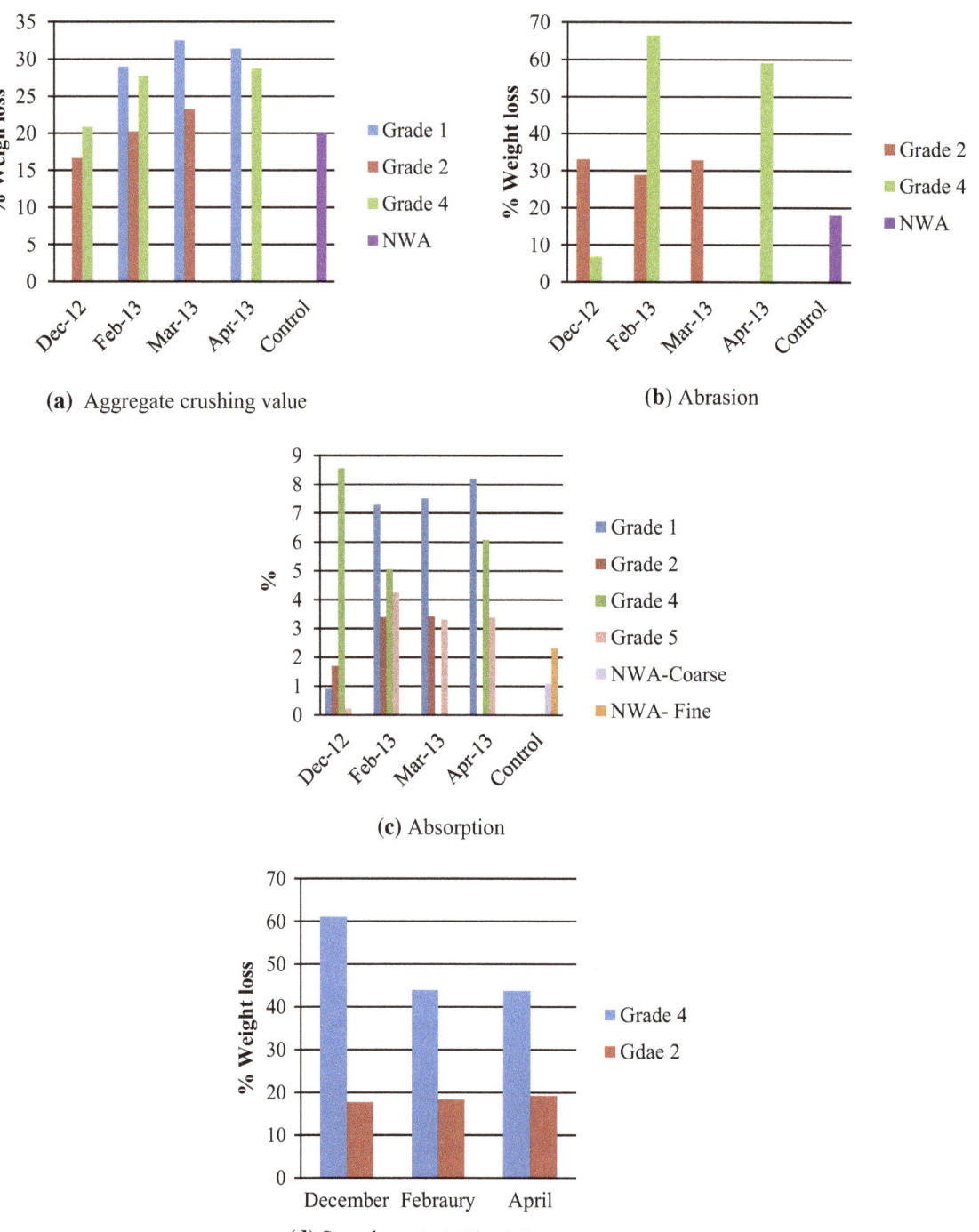

(a) Aggregate crushing value

(b) Abrasion

(c) Absorption

(d) Soundness test - Size 9.5 mm

Fig. 5 Evaluation of physical and mechanical properties of RA.

Table 2 Summary of results of physical and mechanical tests.

Grades	Flakiness index BSI 812-105.2 (1990)	Elongation index BSI 812-105.2 (1990)	Bulk density (kg/m³) ASTM C29/29 M (2009)		Absorption %	Specific gravity			Mechanical	
						Bulk dry	Bulk	Apparent specific gravity	ACV BSI EN 1097-2:2010 (2010b)	Abrasion ASTM C131 (2006)
	%	%	Loose	Compacted		Specific gravity	Specific gravity		% (Weight loss)	% (Weight loss)
December 2012										
1	–	–	1385.8	1575.5	0.9	1.88	1.9	1.91	–	–
2	10.6	15.34	1161.9	1228.2	1.69	3.40	3.46	3.61	16.63	33
4	–	–	1281.8	1297.25	8.56	3.05	3.31	4.12	20.8	6.8
5	–	–	1549.3	1704.3	0.23	1.69	2.08	2.78	–	–
April 2013										
1	–	–	1182.3	1276.7	8.2	2.97	3.21	3.94	31.4	–
4	12.3	6.7	1758.1	1821.5	6.08	2.32	2.46	2.7	28.71	59.1
5	–	–	1552	1639.7	3.38	4.57	4.73	5.41	–	–
Control										
Coarse	15	8.7	1411.5	1512.65	1.1	2.62	2.65	2.7	20	18
Fine	–	–	1585.2	1728.8	2.32	2.51	2.57	2.59	–	–

Table 3 Summary of the tests results: Virgin and recycled aggregate properties.

Property	Virgin aggregate	RA
Absorption	1–2.5 %	1–8.5 %
Specific gravity	2.4–2.7	2–4.8
Crushing value	15–20 %	20–35 %
L.A abrasion	15–30 %	25–65 %
Sodium sulfate soundness (mass loss)	7–21 %	5–36 %

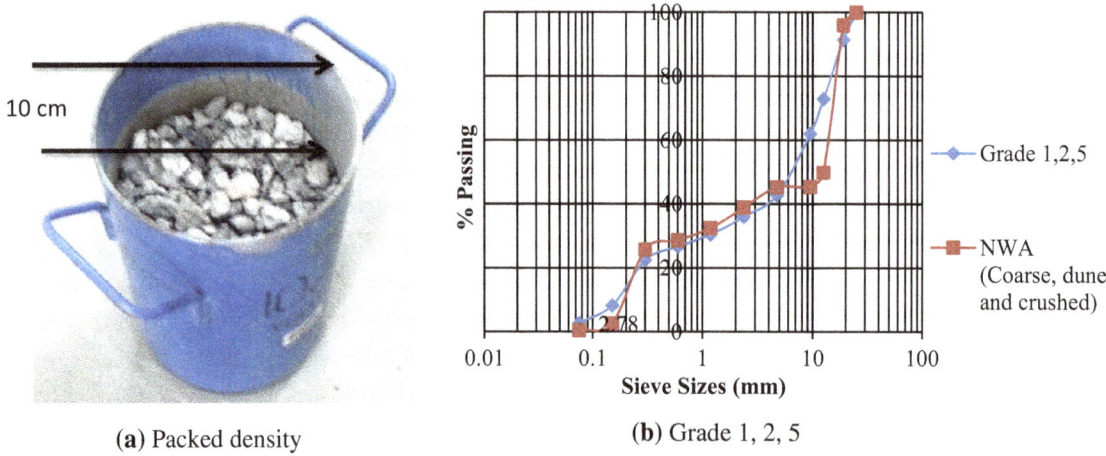

(a) Packed density   (b) Grade 1, 2, 5

Fig. 6 Evaluation of grades 1,2, 5 combined.

10 dm^3 (cubic decimeters), which is equivalent to 10 l. This cylinder is used in determining loose and compacted bulk density of aggregates according to ASTM C29/29 M (2009). The sum of the design volumes of these materials is 68.4 % of the total volume; however, when the dry materials were placed and tamped in three layers, as shown in Fig. 6a, the materials occupied 68 % of the volume. This indicates that the mix proportioning utilizing grades 1, 2 and 5 leads to a dense matrix, which in turn should reflect on strength and durability performance.

Mix proportioning for recycled aggregate concrete the same volumetric ratios of the control mix was adopted for the recycled aggregate, however, since different grades of the recycled aggregate with different particle sizes were available, the following approach was considered in the current study: (i) in case of mixes contain grades 1 and 2, percentage of the coarse aggregate was divided to 50–50 %, (ii) mixes with grades 1, 4, and 5, 37.6 % of grade 4, 15 % of grade 1 and 15.8 % of grade 5 were used. These ratios were verified according to the packed density as discussed before.

Water and moisture adjustment mixing water of different mixes was adjusted during the mix design stage according to the moisture content and percentage absorption of each grade included in a specific mix. In addition, the decision was to use the same quantity of the admixture used for the control mix and monitor the slump/flow for the mixes with recycled aggregate. The concrete mixes had the same water to cement ratio (*w/c*) and cement content.

Several mixes were prepared utilizing four grades, grades 1, 2, 4, and 5 of the recycled aggregate. Mixes were identified according to the grades used in each mix, for example, Mix 1,2,5 indicates that grade 1, grade 2 and grade 5 were used in that mix. Six mixes from the four grades were prepared in addition to the control mix.

4.2.1 Fresh Stage Evaluation

Table 4 summarizes the results of slump, air content, and unit weight, which were recorded immediately after every mix. All mixes achieved the target flow except Mix 1,5 because of the particle size and distribution. Figure 7 shows slump test for Mixes 1,2,5 and 1,5. Mix 1,4 produced the least unit weight, which could be attributed to the existence of mortar attached to the aggregate as shown in Fig. 3d. Air content varied between 0.8 and 2.4 % for mixes with recycled aggregate, which indicates variation in aggregate gradation, particle size and distribution.

4.2.2 Hardened Stage Evaluation: Mechanical and Microstructure Evaluation

Table 4 summarizes the test results of splitting tensile strength and flexural strength for all mixes compared to the compressive strength. Results of split tensile and modulus of rupture from the current study were compared to corresponding equations from BSI EN 1097-2:2010 (2010b) and proposed equations by (Xiao et al. 2006). In addition, Table 5 shows typical failure modes of several samples from different mixes.

Table 4 Summary of the test results—Phase II.

Mix	Unit weight kg/m³	Slump/flow (cm)	Air content %	f'_c	Split tensile f_{ct}				f_r (Mpa)	Modulus of rupture			RCPT	
					f_{ct} (Mpa)	f_{ct}/f'_c (%)	Equation 1	Equation 2		$\frac{f_r}{(f'_c)^{0.5}}$	Equation 3	Equation 4	Coulombs	Class
2,5	2187	52 Flow	3.3	47.68	2.8	5.90	3.11	2.96	5.61	0.81	4.83	5.17	3965	Moderate
1,2,5	2317	59 Flow	3.9	46.90	3.5	7.4	3.08	2.93	6.9975	1.02	4.79	5.14	2214	Low
4	2085	52 Flow	3.3	40.89	3.0	7.46	2.88	2.67	5.67	0.89	4.48	4.80	4508	High
1,4	2194	49 Flow	2.3	51.89	2.6	5.02	3.24	3.13	5.4225	0.75	5.04	5.40	3436	Moderate
1,5	2172	49 Flow	4.3	50.80	2.34	4.61	3.21	3.08	6.345	0.89	4.99	5.35	5573	High
1,4,5	2143	14 Slump	3.5	42.24	3.2	7.59	2.92	2.73	4.455	0.69	4.55	4.87	4385	High
Control	2338	58 Flow	3.1	51.85	3.0	5.78	3.24	3.12	6.57	0.91	5.04	5.40	2007	Low

Equation 1 Split tensile—$f_{ct} = 0.45(f'_c)^{0.5}$, f'_c = cube strength in MPa BSI EN 1097-2:2010 (2010b)
Equation 2 Split tensile—$f_{ct} = 0.24(f'c)^{0.65}$, f'_c = cube strength in MPa (Xiao et al. 2006)
Equation 3 Modulus of rupture $f_r = 0.7(f'_c)^{0.5}$, f'_c = cube strength in MPaBSI EN 1097-2:2010 (2010b)
Equation 4 Modulus of rupture $f_r = 0.75(f'_c)^{0.5}$, f'_c = cube strength in MPa (Xiao et al. 2006)

(a) Mix 1,2,5   (b) Mix 1,5

Fig. 7 Fresh stage evaluation—Workability.

Compressive strength Cubes (150 mm × 150 mm × 150 mm) were tested for compressive strength according to (ASTM 2011a) at 3, 7, 14, 21, and 28 days, strength development with time is shown in Fig. 8. Compressive strength of concrete produced with the recycled aggregate was in the range of 41 to 52 MPa. Mix 4 had the lowest compressive strength. This was expected due to the nature of grade 4, which has poor particle distribution and contains different impurities. Mix 1,4 and Mix 1,2,5 showed similar compressive strength to that of the control. Mix 1,4 consisted of grade 1 (10 mm) as coarse aggregate in addition to grade 4, which has different particle sizes varying from 20 mm and different distribution of fine aggregate. This aggregate gradation provided a dense matrix, which reduces the amount of voids within the mix leading to higher compressive strength. In Mix 1,2,5, grades 1, 2 and 5 provided good distribution of fine and coarse aggregate, which led to higher compressive strength and unit weight similar to that of the control mix. This was also supported by the sieve analysis and packed density as shown in Fig. 6. On the other hand, Mix 4 had the lowest strength out of all mixes due to the gap-gradation that shows an absence of an appropriate distribution of the coarse aggregate. Most of the aggregate sizes are either 20 mm coarse or fine aggregate. In addition, failure modes were observed during testing as shown in Table 5. All failure modes were similar to that of the control. Plane of failures did not go through the coarse aggregates, instead the failure was in the mortar or aggregates were pulled out during the flexural tests, as indicated in Table 5.

Splitting tensile strength Splitting tensile tests were conducted according to ASTM C496/C496 M (2011b) to determine indirect tensile strength of concrete. Mix 1,2,5 had the highest splitting tensile strength while Mix 1,5 showed

the least splitting tensile strength at 28 days. The test results did not show a clear trend, which might be attributed to the aggregate distribution and particle size. However, values in Table 4 were in the range of 4.6–7.46 % of the cube compressive strength, which is close to the range predicted by Eq. 1 (6–7 %). Split tensile results calculated using Eq. 2 were different from those of the current study and Eq. 1. The predicted values are scattered and not close to the test data.

Flexural strength Third-point loading was applied on simple concrete prisms to determine the flexural strength for all mixes. Mix 1,2,5 and Mix 1,5 showed flexural strength higher or similar to that of the control mix. This could be attributed to the improved mechanical interlocking due to better bond between crushed coarse aggregate and cement paste. This was observed from the failure modes and cracking of aggregate as shown in Table 5. In addition, results in Table 4 showed that all mixes achieved flexural strength similar or higher than that predicted using Eq. 3. The average ratio of $f_r/\sqrt{f_c'}$ is 0.85 which is higher than the 0.7 used in Eq. 3; however, it is closer to that proposed by Eq. 4.

Modulus of elasticity Several samples from each mix were tested to evaluate the stress–strain relationship and to calculate the modulus of elasticity values. The modulus of elasticity values were in the range of 25–28 GPa. This variation could be attributed to low aggregate strength and the variation of the volumetric ratio of the course aggregate (some grades have coarse aggregate within their distribution).

Rapid chloride penetration test (RCPT) Ability of concrete to resist chloride ion penetration at 60 voltage direct current (VDC) and 6 h of testing is taken as an indicator of the concrete durability. The results in Coulombs are summarized in

Table 5 Failure modes at 28-day testing—Phase II.

	Compression	Flexure	Split tension
Mix 1,2,5			
Mix 1,4			
1,4,5			
Control			

Table 4 and categorized according to ASTM C1202 (2012b). All mixes except Mix 1,2,5 had high or close to the upper boundary of moderate permeability which could be attributed to the poor aggregate distribution. On the other hand, Mix 1,2,5 produced similar results to that of the control. The use of two different course aggregate distributions along with the fine aggregate led to a dense mix with less voids and better resistance to the chloride ion penetration.

5. SEM Scan

The SEM scans were conducted on samples of two mixes, which had high and low chloride ion permeability according to the RCPT classifications. Figure 9a shows a SEM scan for Mix 1,2,5 (low permeability mix), a good bond and no sign of the wall effect at the Interfacial Transition Zone (ITZ) between the cement paste and the recycled aggregate was

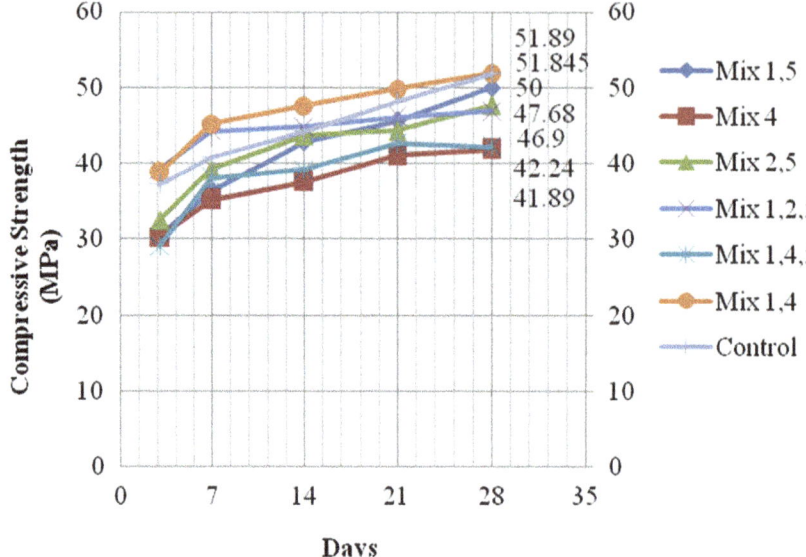

Fig. 8 Development of compressive strength with time—Phase II.

observed. On the hand, a close inspection to the SEM scans for Mix 1,5 in Fig. 9b (high permeability mix) shows that a porous layer exists between the aggregate and cement paste, which confirm the formation of the wall effect at ITZ in this mix. This layer, which could be cement hydrates, adhered to the coarse aggregate, in addition to different contaminants and voids contributed to the higher absorption and higher chloride ion permeability in this mix category. In both mixes, micro cracks (not due to sample preparation) were found in the cement paste; this type of cracks usually occur due to shrinkage and difference in modulus of elasticity between the paste and the coarse aggregate particles (Neville 1995).

6. Discussion

Results of Phase II evaluations showed that Mix 1,2,5 achieved acceptable compressive, flexural, and splitting tensile strength. In addition, it had the best performance in RCPT which was confirmed with the microstructure evaluation as shown in the SEM scans. The main reason of this performance was achieving high packing density by utilizing different grades. The high packing density provided solution for limitations in particle distribution and aggregate strength. This led to reduction in total pore volume which in turn improved the strength and durability of the mix. This also is in agreement with that reported by (Levy and Helene 2004; McNeil and Kang 2013). In addition, the absolute volume method used in the current study took into consideration variability in specific gravity

of the RA during mix proportioning which led to improved properties. This is also in agreement with the findings by (Knaack and Kurama 2013). Examination of the SEM and crack propagation, Fig. 10, showed that cracks are initiated at the interface between the aggregates and mortar. Figure 10 shows that regardless of the sample shape the cracks started at the pours mortar adhered to the recycled aggregate. This indicates, in this case, a weakness of the old mortar which led to reduced bond between the old and new mortar. Similar behavior was discussed by (Tam et al. 2007a, b; Xiao et al. 2012a).

Table 6 provides a summary of the results from several investigations found in the literature compared to that of Mix 1,2,5. The results included in Table 6 are only those of concrete mixes with 100 % RA or from full replacement of coarse aggregate. No results of partial replacement of natural aggregate are included. Although the testing environment, aggregate source, and w/c ratios are different, there is a good agreement in all the mechanical properties. This summary emphasizes that concrete with similar results could be produced with recycled aggregate regardless the source of the aggregate. In addition, the following could be observed from over all the results in Table 6, (1) RA with high absorption capacity and low specific gravity lead to concrete with less compressive strength compared to target strength; (2) 7 to 15 % reduction in compressive strength compared to target strength when w/cm ratio is maintained in the range of 0.4 to 0.45; (3) flexural and splitting strength varied based on the w/c ratio and aggregate source and (4) reduction of about 10 to 15 % in the modulus of elasticity.

(a) Aggregate – cement paste interaction– Mix 1,2,5

(b) Aggregate – cement paste interaction– Mix 1,5

Fig. 9 SEM features of concrete with recycled aggregate.

6.1 Recommendations from the Current Study

The following recommendations could be drawn from the study:

- For every batch of recycled aggregate:
  - Particle size and distribution should be evaluated every batch

- Absorption capacity, abrasion resistance, and soundness are important properties that need to be evaluated.

- Mixture design method based on direct volume replacement and high packing density is the key to achieve strength and durable concrete.

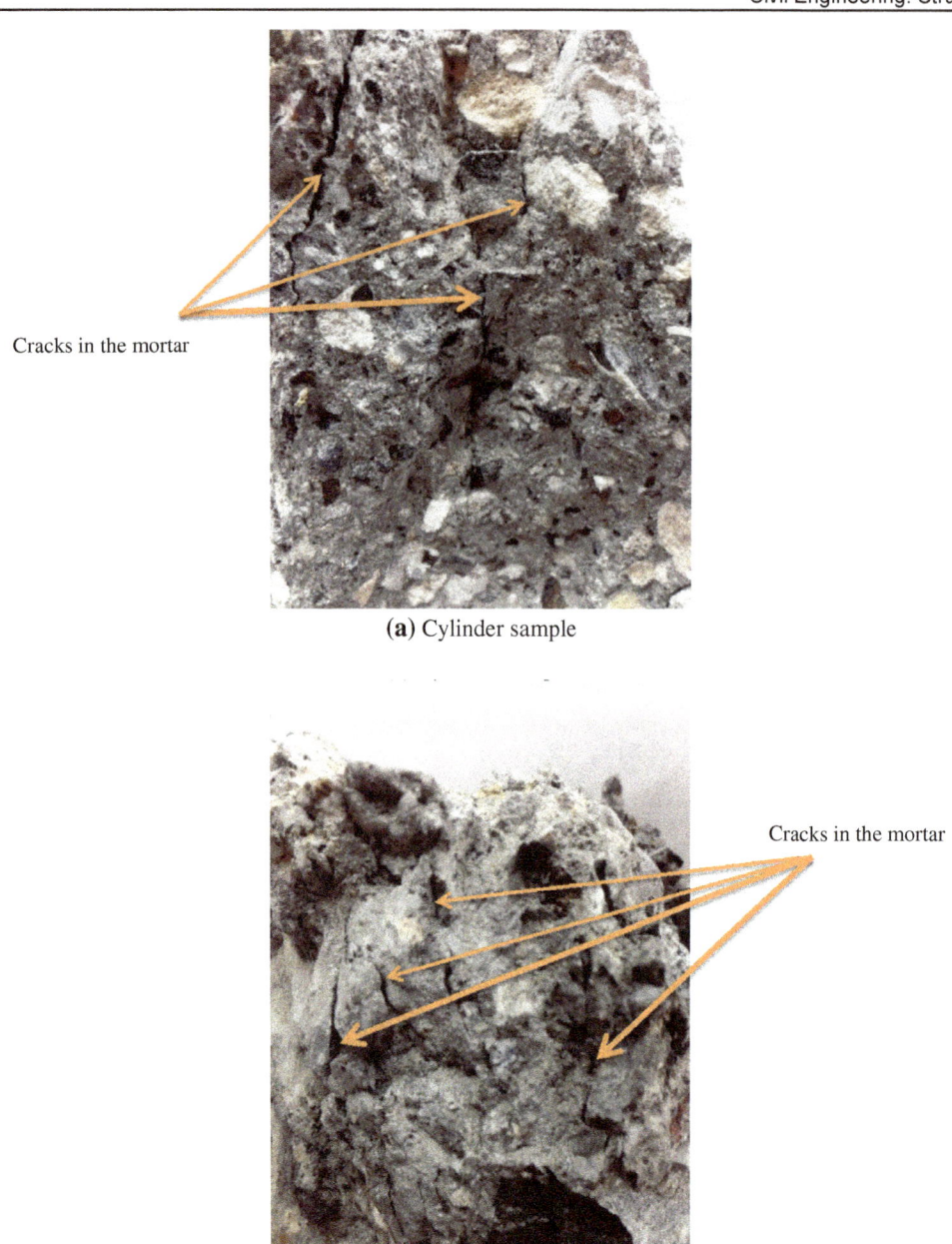

(a) Cylinder sample

(b) Cube sample

Fig. 10 Crack initiation and propagation in RA concrete.

- w/c ratio ≤ 0.4 is preferred to improve strength and durability of concrete with RA
- Effect of SCM and high packing density on strength and durability of concrete with RA need to be investigated.

7. Conclusions

The work presented in this paper evaluates the effect of recycled aggregate quality on the properties of concrete. Evaluation of the aggregate physical and mechanical properties showed an acceptable variation in properties when samples were collected and evaluated from unknown source over 6 months. However, limitations in gradation requirements; high absorption and aggregate strength could be resolved during the proportioning stage and by achieving high packing density. Furthermore, concrete produced utilizing different combination of coarse and fine aggregate without alteration in particle size or distribution showed that comparable compressive, flexural, splitting strength, and modulus of elasticity could be achieved. All mixes except Mix 1,2,5 did not show acceptable performance in the RCPT because of the high porosity supported by the examination of the microstructure of the hardened concrete. High concrete porosity and permeability might be attributed to the variability in aggregate gradation and existence of contamination. It is also important to monitor the long-term performance and volume change (creep and shrinkage) to have better assessment of the concrete produced with recycled aggregate.

Table 6 Comparison with available data from literature.

Reference	% of target compressive strength (MPa)	Flexural strength (MPa)	Split tensile (MPa)	Elasticity (GPa)	w/c ratio	Aggregate source
Mix 1,2,5 Current study[b]	93.8 (50)	6.99	3.48	27	0.40	Recycling facility
De Brito and Saikia (2013)[b]	88 (N/A)	5.0	3.3	26.7	0.50	C and D Waste
Vivian A. Ulloa et al. (2013)[a]						C and D Waste
	–(31.4)		X		0.51	6.1 % Abs-Demolition of old concrete structure
	–(26)				0.61	
	–(36.7)		X		0.51	5.8 % Abs
	–(29.5)				0.62	
	–(42.9)		X		0.45	3.9 % Abs
	–(37.7)				0.54	
	–(38.7)		X		0.4	4.5 % Abs
	–(31.4)				0.5	
	–(37)		X		0.43	4.7 % Abs
	–(31.2)				0.56	
Abdelfatah et al. (2011)[b]	85.7 (42)	X	X	X	0.40	Old concrete with known strength
Malešev et al. (2010)[a]	91.3 (50)	5.2	2.78	29.1	0.513	Crushed laboratory test cubes
Tabsh and Abdelfatah (2009)[b]	92 (50)	X	4	X	0.40	Old concrete with known strength
Corinaldesi and Moriconi (2009)[b]	89 (28)	X	1.45	27	0.4	Rubble Recycling Plant
Yang et al. (2008)[a]—G1	90 (36.0)	3.84	3.49	29.22	0.42	Old concrete with unknown strength
Yang et al. (2008)[a]—G3	73.75 (29.5)	3.20	2.56	23.72	0.42	G1—SG 2.53—1.9 % Abs G3—SG 2.4—6.2 % Abs
Rahal (2007)[a]	93 (50)	X	X	29.5	0.6	Field demolished concrete
Etxeberria et al. (2007)[b]	93.3 (30)	X	2.72	27.76	0.52	Selected and processed for the study
Etxeberria et al. (2007)[a]	93.3 (28)	X	2.72	27.76	0.50	C and D Waste
	78.3 (47)				0.50	C and D Waste
	85(51)				0.43	
	93.3(56)				0.40	
	93.3(56)				0.40	
	66.7(40)				0.52	
Limbachiya et al. (2004)[a]	94 (35)	4.5	X	25	0.6	C and D Waste

Table 6 continued

Reference	% of target compressive strength (MPa)	Flexural strength (MPa)	Split tensile (MPa)	Elasticity (GPa)	w/c ratio	Aggregate source
Katz (2003)[b]	77.46 (26.8)	5.4	3.1	11.3	0.60	Old concrete with known strength

[a] Coarse aggregate replacement.

[b] Full replacement.

– Target strength is not available.

SG Specific gravity, *Abs* Absorption capacity.

References

Abbas, A., Fathifazl, G., Isgor, O. B., Razaqpur, A. G., Fournier, B., & Foo, S. (2009). Durability of recycled aggregate concrete designed with equivalent mortar volume method. *Cement & Concrete Composites, 31*(8), 555–563.

Abdelfatah, A., Tabsh, S., & Yehia, S. (2011). Utlilization of recycled coarse aggregate in concrete mixes. *Journal of Civil Engineering and Architecture, 5*(6), 562–566.

Akbarnezhad, A., Ong, K. C. G., Tam, C. T., & Zhang, M. H. (2013). Effects of the parent concrete properties and crushing procedure on the properties of coarse recycled concrete aggregates. *Journal of Materials in Civil Engineering, 25*(12), 1795–1802.

Amorim, P., De Brito, J., & Evangelista, L. (2012). Concrete made with coarse concrete aggregate: Influence of curing on durability. *ACI Materials Journal, 109*(2), 195.

Ann, K. Y., Moon, H. Y., Kim, Y. B., & Ryou, J. (2008). Durability of recycled aggregate concrete using pozzolanic materials. *Waste Management, 28*(6), 993–999.

Ann, T. W., Poon, C. S., Wong, A., Yip, R., & Jaillon, L. (2013). Impact of construction waste disposal charging scheme on work practices at construction sites in Hong Kong. *Waste Management, 33*(1), 138–146.

ASTM C131. (2006). Standard test method for resistance to degradation of small-size coarse aggregate by abrasion and impact in the Los Angeles machine. West Conshohocken, PA: ASTM International.

ASTM C29/29 M. (2009). Standard test method for bulk density (unit weight) and voids in aggregate. West Conshohocken, PA: ASTM International.

ASTM C78/C78 M-10e1. (2010). Standard test method for flexural strength of concrete (using simple beam with third-point loading). West Conshohocken, PA: ASTM International.

ASTM C136. (2011a). Standard test method for sieve analysis of fine and coarse aggregates ASTM West Conshohocken, PA: ASTM International.

ASTM C496/C496 M. (2011b). Standard test method for splitting tensile strength of cylindrical concrete specimens. West Conshohocken, PA: ASTM International.

ASTM C127. (2012a). Standard test method for density, relative density (specific gravity), and absorption of coarse aggregate. West Conshohocken, PA: ASTM International.

ASTM C1202. (2012b). Standard test method for electrical indication of concrete's ability to resist chloride ion penetration. West Conshohocken, PA: ASTM International.

ASTM C33/C33. M (2013a). Standard specification for concrete aggregates. West Conshohocken, PA: ASTM International.

ASTM C88. (2013b). Standard test method for soundness of aggregates by use of sodium sulfate or magnesium sulfate. West Conshohocken, PA: ASTM International.

Berndt, M. L. (2009). Properties of sustainable concrete containing fly ash, slag and recycled concrete aggregate. *Construction and Building Materials, 23*(7), 2606–2613.

Bodet, R. (2014). Review French/European standards and regulations. 08/07/2014 http://navier.enpc.fr/IMG/pdf/Standards_and_regulation_R-BODET.pdf.

Braunschweig, A., Kytzia, S., & Bischof, S. (2011). Recycled concrete: Environmentally beneficial over virgin concrete?, www.lcm2011.org.

BSI EN 12390-6:2009. (2010a). Testing hardened concrete Compressive strength of test specimens. British Standard Institute.

BSI EN 1097-2:2010. (2010b). Tests for mechanical and physical properties of aggregates: Methods for the determination of resistance to fragmentation. London, UK: British Standards Institute.

BSI 812-105.2. (1990). Method for determination of particle size: Elongation index of coarse aggregate. London, UK: British Standard Institute.

Cabral, A. E. B., Schalch, V., Dal Molin, D. C. C., & Ribeiro, J. L. D. (2010). Mechanical properties modeling of recycled aggregate concrete. *Construction and Building Materials, 24*(4), 421–430.

Cement, Concrete, and Aggregates (2008). Use of recycled aggregate. Australia: Hong Kong Housing Authority.

Chen, H.-G., & Ying, J.-W. (2011). *Analysis of factors influencing durability of recycled aggregate: A review.* Paper presented at the electric technology and civil engineering, Lushan.

Chen, Z.-P., Huang, K.-W., Zhang, X.-G., & Xue, J.-Y. (2010). *Experimental research on the flexural strength of recycled coarse aggregate concrete.* Paper presented at the 2010 international conference on mechanic automation and control engineering (MACE), Wuhan, China.

Corinaldesi, V. (2010). Mechanical and elastic behaviour of concretes made of recycled-concrete coarse aggregates. *Construction and Building Materials, 24*(9), 1616–1620. doi:10.1016/j.conbuildmat.2010.02.031.

Corinaldesi, V., & Moriconi, G. (2009). Influence of mineral additions on the performance of 100% recycled aggregate concrete. *Construction and Building Materials, 23*(8), 2869–2876.

De Brito, J., & Saikia, N. (2013). Recycled aggregate in concrete: Use of industrial, construction and demolition waste. 445 p., London, UK: Springer.

de Juan, M. S., & Gutiérrez, P. A. (2009). Study on the influence of attached mortar content on the properties of recycled concrete aggregate. *Construction and Building Materials, 23*(2), 872–877.

EU. Directive 2008/98/EC of the European Parliament and the Council of 19 November 2008 on waste and repealing certain Directives. European Union.

Domingo-Cabo, A., Lázaro, C., López-Gayarre, F., Serrano-López, M. A., Serna, P., & Castaño-Tabares, J. O. (2009). Creep and shrinkage of recycled aggregate concrete. *Construction and Building Materials, 23*(7), 2545–2553.

Eisa, A. (2014). Properties of concrete incorporating recycled post-consumer environmental wastes. *International Journal of Concrete Structures and Materials, 8*(3), 251–258.

Etxeberria, M., Vazques, E., Mari, A., & Barra, M. (2007). Influence of amount of recycled coarse aggregates and production process on properties of recycled aggregate concrete. *Cement and Concrete Research, 37*(5), 735–742.

European Aggregate Association. (2010). *Planning policies and permitting procedures to ensure the sustainable supply of aggregates in Europe.* Austria: University of Leoben.

Evangelista, L., & Brito, J. D. (2007). Mechanical behaviour of concrete made with fine recycled concrete aggregates. *Cement & Concrete Composites, 29*(5), 397–401. doi: 10.1016/j.cemconcomp.2006.12.004.

Fathifazl, G., & Razaqpur, A. G. (2013). Creep rheological models for recycled aggregate concrete. *ACI Materials Journal, 110*(2), 115–126.

Fathifazl, G., Abbas, A., Razaqpur, A. G., Isgor, O. B., Fournier, B., & Foo, S. (2009). New mixture proportioning method for concrete made with coarse recycled concrete aggregate. *Journal of Materials in Civil Engineering, 21*(10), 601–611.

Fathifazl, G., Razaqpur, A. G., Isgor, O. B., Abbas, A., Fournier, B., & Foo, S. (2011). Creep and drying shrinkage characteristics of concrete produced with coarse recycled concrete aggregate. *Cement & Concrete Composites, 33*(10), 1026–1037.

Garg, P., Singh, H., & Walia, B. S. (2013). Optimum Size of Recycled Aggregate. *GE-International Journal of Engineering Research.* pp. 35–41, ISSN:2321-1717

Gonclaves, A., Esteves, A., & Vieira, M. (2004). *Influence of recycled concrete aggregate on concrete durability.* Paper presented at the National Laboratory of Civil Engineering, Portugal.

Gupta, Y. P. (2009). *Use of recycled aggregate in concrete construction: A need for sustainable environment.* Paper presented at the our world in concrete and structures, Singapore.

Hansen, T. C. (1986). Recycled aggregates and recycled aggregate concrete second state-of-the-art report developments 1945–1985. *Materials and Structures, 19*(3), 201–246.

Henschen, J., Teramoto, A., & Lange, D. A. (2012, January). Shrinkage and creep performance of recycled aggregate concrete. In 7th RILEM international conference on cracking in pavements (pp. 1333–1340). Springer Netherlands.

Kartam, N., Al-Mutairi, N., Al-Ghusain, I., & Al-Humoud, J. (2004). Environmental management of construction and demolition waste in Kuwait. *Waste Management, 24*(10), 1049–1059.

Katz, A. (2003). Properties of concrete made with recycled aggregate from partially hydrated old concrete. *Cement and Concrete Research, 33*(5), 703–711.

Knaack, A. M., & Kurama, Y. C. (2013). Design of concrete mixtures with recycled concrete aggregates. *ACI Materials Journal, 110*(5), 483–493.

Kou, S. C., & Poon, C. S. (2012). Enhancing the durability properties of concrete prepared with coarse recycled aggregate. *Construction and Building Materials, 35*, 69–76.

Kwan, W. H., Ramli, M., Kam, K. J., & Sulieman, M. Z. (2012). Influence of the amount of recycled coarse aggregate in concrete design and durability properties. *Construction and Building Materials, 26*(1), 565–573.

Lederle, R. E., & Hiller, J. E. (2013). Reversible shrinkage of concrete made with recycled concrete aggregate and other aggregate types. *ACI Materials Journal, 110*(4), 423.

Levy, S. M., & Helene, P. (2004). Durability of recycled aggregates concrete: A safe way to sustainable development. *Cement and Concrete Research, 34*(11), 1975–1980.

Limbachiya, M. C., Koulouris, A., Roberts, J. J., & Fried, A. N. (2004). *Performance of Recycled Concrete Aggregate.* Paper presented at the RILEM international symposium on environmental-conscious materials and systems for sustainable development

Llatas, C. (2011). A model for quantifying construction waste in projects according to the European waste list. *Waste Management, 31*(6), 1261–1276.

Lu, W., & Tam, V. W. (2013). Construction waste management policies and their effectiveness in Hong Kong: A longitudinal review. *Renewable and Sustainable Energy Reviews, 23*, 214–223.

Lu, W., & Yuan, H. (2011). A framework for understanding waste management studies in construction. *Waste Management, 31*(6), 1252–1260.

Malešev, M., Radonjanin, V., & Marinković, S. (2010). Recycled concrete as aggregate for structural concrete production. *Sustainability, 2*(5), 1204–1225. doi:10.3390/su2051204.

Manzi, S., Mazzotti, C., & Bignozzi, M. C. (2013). Short and long-term behavior of structural concrete with recycled concrete aggregate. *Cement & Concrete Composites, 37,* 312–318.

Marinkovic, S., Radonjanin, V., Malesev, M., & Ignjatovic, I. (2010). Comparative environmental assessment of natural and recycled aggregate concrete. *Waste Management, Elsevier, 30*(11), 2255–2264.

McNeil, K., & Kang, T. H.-K. (2013). Recycled concrete aggregates: A review. *International Journal of Concrete Structures and Materials, 7*(1), 61–69.

Medina, C., Zhu, W., Howind, T., Sanchez de Rojas, M. I., & Frias, M. (2014). Influence of mixed recycled aggregate on the physical-mechanical properties of recycled concrete. *Journal of Cleaner Production, 68*(1), 216–225.

Ministry of Natural Resources (2010). State of the aggregate resource in Ontario study. Toronto, Canada: Queen's Printer for Ontario.

Naik, T. R., & Moriconi, G. (2005) Environmental-friendly durable concrete made with recycled materials for sustainable concrete construction CANMET/ACI international symposium on sustainable development of cement and concrete, Toronto, Canada, October 2005 (pp. 1–13).

Neville, A. M. (1995). *Properties of concrete.* Harlow, NY: Pearson.

Oikonomou, N. D. (2005). Recycled concrete aggregates. *Cement & Concrete Composites, 27*(2), 315–318.

Olorunsogo, F. T., & Padayachee, N. (2002). Performance of recycled aggregate concrete monitored by durability indexes. *Cement and Concrete Research, 32*(2), 179–185.

Otsuki, N., Miyazato, S. I., & Yodsudjai, W. (2003). Influence of recycled aggregate on interfacial transition zone, strength, chloride penetration and carbonation of concrete. *Journal of Materials in Civil Engineering, 15*(5), 443–451.

Padmini, A. K., Ramamurthy, K., & Mathews, M. S. (2009). Influence of parent concrete on the properties of recycled aggregate concrete. *Construction and Building Materials, 23*(2), 829–836. doi:10.1016/j.conbuildmat.2008.03.006.

Poon, C. S., Shui, Z. H., & Lam, L. (2004). Effect of microstructure of ITZ on compressive strength of concrete prepared with recycled aggregates. *Construction and Building Materials, 18*(6), 461–468.

Qasrawi, H., & Marie, I. (2013). Towards better understanding of concrete containing recycled concrete aggregate. *Advances in Materials and Science Engineering.* doi: 10.1155/2013/636034.

Rahal, K. (2007). Mechanical properties of concrete with recycled coarse aggregate. *Building and Environment, 42*(1), 407–415. doi:10.1016/j.buildenv.2005.07.033.

Rao, M. C., Bhattacharyya, S. K., & Barai, S. V. (2010). *Recycled aggregate concrete: A sustainable built environment.* Paper presented at the ICSBE: International conference on sustainable built environment.

Ryou, J. S., & Lee, Y. S. (2014). Characterization of recycled coarse aggregate (RCA) via a surface coating method.

International Journal of Concrete Structures and Materials, 8(2), 165–172.

Sagoe-Crentsil, K., Brown, T., & Taylor, A. H. (2001). Performance of concrete made with commercially produced coarse recycled concrete aggregate. *Cement and Concrete Research, 31*(5), 707–712. doi:10.1016/S0008-8846(00)00476-2.

Silva, R. V., De Brito, J., & Dhir, R. K. (2014). Properties and composition of recycled aggregates from construction and demolition waste suitable for concrete production. *Construction and Building Materials, 65,* 201–217.

Silva, R. V., de Brito, J., & Dhir, R. K. (2015). Prediction of the shrinkage behavior of recycled aggregate concrete: A review. *Construction and Building Materials, 77,* 327–339.

Sim, J., & Park, C. (2011). Compressive strength and resistance to chloride ion penetration and carbonation of recycled aggregate concrete with varying amount of fly ash and fine recycled aggregate. *Waste Management, 31*(11), 2352–2360. doi:10.1016/j.wasman.2011.06.014.

Sonawane, T. R., & Pimplikar, S. S. (2013). Use of recycled aggregate in concrete. *International Journal of Engineering Research and Technology (IJERT), 2*(1), 1–9.

Tabsh, S. W., & Abdelfatah, A. S. (2009). Influence of recycled concrete aggregates on strength properties of concrete. *Construction and Building Materials, 23*(2), 1163–1167. doi:10.1016/j.conbuildmat.2008.06.007.

Tam, V. W. Y. (2008). Economic comparison of concrete recycling: A case study approach. *Elsevier, 52*(5), 821–828.

Tam, V. W., Gao, X. F., & Tam, C. M. (2005). Microstructural analysis of recycled aggregate concrete produced from two-stage mixing approach. *Cement and Concrete Research, 35*(6), 1195–1203.

Tam, V. W., Gao, X. F., Tam, C. M., & Chan, C. H. (2008). New approach in measuring water absorption of recycled aggregates. *Construction and Building Materials, 22*(3), 364–369.

Tam, V. W. Y., & Tam, C. M. (2007). Economic comparison of recycling over-ordered fresh concrete: A case study approach. *Resources, Conservation and Recycling, 52*(2), 208–218. doi:10.1016/j.resconrec.2006.12.005.

Tam, V. W., Tam, C. M., & Le, K. N. (2007a). Removal of cement mortar remains from recycled aggregate using pre-soaking approaches. *Resources, Conservation and Recycling, 50*(1), 82–101.

Tam, V. W., Tam, C. M., & Wang, Y. (2007b). Optimization on proportion for recycled aggregate in concrete using two-stage mixing approach. *Construction and Building Materials, 21*(10), 1928–1939.

Tepordei, V. V. (1999). Natural aggregates—Foundation of America's future. *USGS: Science for a changing world.*

The Freedonia Group. (2012). World Construction Aggregates.

Thomas, C., Setien, J., Polanco, J. A., Alaejos, P., & Sanchez de Juan, M. (2013). Durability of recycled concrete aggregate. *Construction and Building Materials, 40,* 1054–1065.

Topcu, I. B., & Guncan, N. F. (1995). Using waste concrete as aggregate. *Cement and Concrete Research, 25*(7), 1385–1390. doi:10.1016/0008-8846(95)00131-U.

Transportation Applications of Recycled Concrete Aggregate—FHWA State of the Practice National Review 2004; U.S.

Department of Transportation Federal Highway Administration: Washington, DC, 2004; pp. 1–47.

Ulloa, V. A., García-Taengua, E., Pelufo, M. J., Domingo, A., & Serna, P. (2013). New views on effect of recycled aggregates on concrete compressive strength. *ACI Materials Journal, 110*(6), 1–10.

Vyas, C. M., & Bhatt, D. R. (2013). Evaluation of modulus of elasticity for recycled coarse aggregate concrete. *International Journal of Engineering Science and Innovative Technology, 2*(1), 26.

Wardeh, G., Ghorbel, E., & Gomart, H. (2014). Mix design and properties of recycled aggregate concretes: Applicability of Eurocode 2. *International Journal of Concrete Structures and Materials, 9*(1), 1–20.

Xiao, J., Li, W., Sun, Z., & Shah, S. P. (2012a). Crack propagation in recycled aggregate concrete under uniaxial compressive loading. *ACI Materials Journal, 109*(4), 451–462.

Xiao, J., Fana, L. Y., & Huang, X. (2012b). An overview of study on recycled aggregate concrete in China (1996–2011). *Construction and Building Materials, 31*, 364–383. doi: 10.1016/j.conbuildmat.2011.12.074.

Xiao, J., Li, L., Tam, V. W., & Li, H. (2014). The state of the art regarding the long-term properties of recycled aggregate concrete. *Structural Concrete, 15*(1), 3–12.

Xiao, J. Z., Li, J. B., & Zhang, C. (2006). On relationships between the mechanical properties of recycled aggregate concrete: An overview. *Materials and Structures, 39*(6), 655–664.

Yang, K. H., Chung, H. S., & Ashour, A. F. (2008). Influence of type and replacement level of recycled aggregates on concrete properties. *ACI Materials Journal, 105*(3), 289–296.

Yehia, S., Khan, S., & Abudayyeh, O. (2008). Evaluation of mechanical properties of recycled aggregate for structural applications. *HBRC Journal, 4*(3), 7–16.

Zaharieva, R., Buyle-Bodin, F., Skoczylas, F., & Wirquin E. (2003). Assessment of the surface permeation properties of recycled aggregate concrete. *Cement and Concrete Composites, 25*(2), 223–232. doi:10.1016/S0958-9465(02)00010-0

2

Physical and Mechanical Properties of Cementitious Specimens Exposed to an Electrochemically Derived Accelerated Leaching of Calcium

Arezou Babaahmadi[1],* ⓘ, Luping Tang[1], Zareen Abbas[2], and Per Mårtensson[3]

Abstract: Simulating natural leaching process for cementitious materials is essential to perform long-term safety assessments of repositories for nuclear waste. However, the current test methods in literature are time consuming, limited to crushed material and often produce small size samples which are not suitable for further testing. This paper presents the results from the study of the physical (gas permeability as well as chloride diffusion coefficient) and mechanical properties (tensile and compressive strength and elastic modulus) of solid cementitious specimens which have been depleted in calcium by the use of a newly developed method for accelerated calcium leaching of solid specimens of flexible size. The results show that up to 4 times increase in capillary water absorption, 10 times higher gas permeability and at least 3 times higher chloride diffusion rate, is expected due to complete leaching of the Portlandite. This coincides with a 70 % decrease in mechanical strength and more than 40 % decrease in elastic modulus.

Keywords: nuclear waste management, service life, concrete, mechanical properties, leaching, acceleration.

1. Introduction

In repositories for nuclear waste, concrete and other cementitious materials are extensively used in both the structural components such as the barrier construction and to fill the voids between the waste containers inside the barrier construction. During the very long periods of time considered in an analysis of the long term safety of such a repository exchange of ions between the cementitious materials and the surrounding groundwater (Berner 1992; Reardon 1992) due to concentration differences will occur. This will result in dissolution or precipitation of minerals, and consequently in alteration of the microstructure as well as the chemical and mineralogical composition of the cementitious materials. There are a lot of concerns in normal constructions such as chloride corrosion of the reinforcement, carbonation

or sulfate attack (Morga and Marano 2015; Park and Choi 2012; Pham and Prince 2014; Pritzl et al. 2014), however, here the most important process is the decalcification of the material through the dissolution of the calcium hydrates (Portlandite $Ca(OH)_2$ and the Calcium silicate hydrates (CSH) phases, which constitute the major portion of hydrated cementitious material (Hinsenveld 1992).

Safety assessments of repositories for low and intermediate level radioactive waste (LILW), require prediction of changes in the properties of the cementitious barriers over a very long period of time, up to 100,000 years. In order to improve the accuracy and reduce the uncertainties of the assessments a detailed understanding of the processes occurring in the repository and their effect on the properties of the cementitious materials is of great importance.

The effect of degradation on the properties of cementitious materials has been reported in several studies in the literature (Adenot and Buil 1992; Carde et al. 1997; Carde and François 1997; Carde et al. 1996; Faucon et al. 1996, 1998; Haga et al. 2005; Maltais et al. 2004). In these studies leaching of calcium from cementitious materials have been accomplished both through immersion of the solid cementitious specimens in water (Faucon et al. 1996; Haga et al. 2005; Mainguy et al. 2000) and through immersion of the specimens in chemical agents in order to accelerate the leaching process (Adenot and Buil 1992; Carde and François 1997; Faucon et al. 1996, 1998; Haga et al. 2005; Heukamp et al. 2001; Mainguy et al. 2000; Maltais et al. 2004; Revertegat et al. 1992; Ryu et al. 2002; Saito et al. 1992; Wittmann 1997). The results in these studies indicate that a layered structure is developed in the leached samples comprising an unaltered core delineated by total

[1])Division of Building Technology (Building Materials), Chalmers University of Technology, Gothenburg, Sweden.
*Corresponding Author;
E-mail: arezou.babaahmadi@chalmers.se

[2])Department of Chemistry, University of Gothenburg, Gothenburg, Sweden.

[3])Division of Low and Intermediate Level Nuclear Waste, Swedish Nuclear Fuel and Waste Management Company, Stockholm, Sweden.

dissolution of Portlandite followed by different zones separated by dissolution/precipitation fronts and progressive decalcification of the CSH gel (Adenot and Buil 1992). Moreover, it is concluded that depletion in calcium changes the bulk density and the pore structure of the hydrated cement paste (Haga et al. 2005; Mainguy et al. 2000) and the changes in pore volume also alters the mechanical properties of the cementitious materials (Carde and François 1997; Heukamp et al. 2001; Saito and Deguchi 2000).

However, although some important conclusions have been drawn from these studies regarding in particular the chemical properties of the Ca-depleted materials, they have been limited to the use of crushed materials or small solid samples. This has limited the possibilities to study the mechanical and physical properties, e.g. compressive strength and diffusivity, which require the use of larger samples. In addition it should be noted that there are not many studies reported in the literature with implication of concrete specimens (Choi and Yang 2013; Marinoni et al. 2008; Nguyen et al. 2007; Sellier et al. 2011) of proper size but instead paste specimens or powder samples have been used (Adenot and Buil 1992; Carde and François 1997; Carde et al. 1996; Faucon et al. 1996, 1998; Haga et al. 2005; Heukamp et al. 2001; Maltais et al. 2004; Revertegat et al. 1992; Ryu et al. 2002; Saito and Deguchi 2000; Ulm et al. 2003; Wittmann 1997).

Finally, although the common feature for both natural and accelerated leaching scenarios will be a total dissolution of Portlandite and a significant decalcification of the CSH phases, other effects of the aging processes may differ considerably between specimens aged by different acceleration methods and comparably natural leaching methods. This emphasizes the importance of reproducing accelerating tests and characterizing the aged samples to account for the comparability of the ageing function of different methods in order to demonstrate properties of degraded cementitious materials.

All this implies that effective acceleration methods with comprehensible kinetics, simulating the natural calcium leaching process for concrete specimens with a size suitable for further evaluation of the mechanical and physical properties of the specimens are needed. In order to comply with this requirement a new method for accelerated leaching of cementitious materials is developed and utilized in this study.

This paper presents the results from the study of the changes in mechanical and physical properties of solid cementitious specimens which have been depleted in calcium. The following properties have been studied: tensile strength, elastic modulus, permeability and water absorption. In addition, the chloride diffusion coefficient of concrete specimens has been studied in order to give an indication of the transport properties of the specimens.

2. Materials and Methods

2.1 Specimen Preparation

The paste specimens were cast from a mixture of Swedish structural Portland cement for civil engineering

(CEM I 42.5N MH/SR3/LA) and deionized water at a water-cement ratio of 0.5. The chemical composition of the cement is listed in Table 1. Fresh cement paste was cast in acrylic cylinders with an internal diameter of 50 mm and a length of 250 mm. The cylinder's ends were sealed with silicone rubber stops. The cylinders containing fresh paste were rotated longitudinally at a rate of 12–14 rpm for the first 18–24 h of hydration in order to produce specimens with a homogeneous composition and structure. Afterwards the rubber stops were removed and the ends of the cylinders were sealed with plastic tape. The specimens were stored for over 6 months in a moist plastic box and then cut to cylinders with the size of Ø50 × 75 mm for use as specimens in the experiments. In order to prevent carbonation, saturated lime water was used at the bottom of the plastic box as absorbent for carbon dioxides during the storage of specimens. To further ensure that the specimens used in the leaching experiments were not carbonated, the paste portions about 10–20 mm near the ends of the cylinders was cut off prior to the specimen cutting.

The mortar specimens at water- cement ratio of 0.5 and a cement: sand ratio of 1:2, were cast from mixtures of Swedish structural Portland cement for civil engineering (Table 1), deionized water and a siliceous natural sand with maximum particle size of 1 mm. Similar casting procedure as of paste specimens was followed.

The concrete specimens were cast from mixtures of Swedish structural Portland cement for civil engineering, natural sand (the sand was a type of siliceous gravel with size of 0–8 mm and fineness modulus of 3.82, in accordance with European standard EN 933-1 (EN 933-1 Tests for geometrical properties of aggregates)) and crushed coarse aggregate with maximum size of 16 mm. 65 % of total aggregate content was fine aggregate (0–8 mm) and the rest was the course aggregate (8–16 mm). The course aggregate was an equal bland of (50 %), 8–12 mm and 12–16 mm of crushed aggregates. The specimens were cast in cylinders in two different dimensions of Ø100 × 200 mm and Ø50 × 250 mm with two different water cement W/C-ratios (according to the properties of the concrete used in the Final Repository for Short-lived Radioactive Waste, SFR, in Sweden (Emborg et al. 2007; Höglund 2001)). The observations from the slump test prior to casting was 25 mm for the concrete with W/C = 0.48 and 35 mm for W/C = 0.62. The specimens were demolded 24 h after casting and then cured in the saturated lime water for more than 3 months after which they were stored for over 3 months in a moist plastic box and then cut to cylinders with the dimensions of Ø50 × 75 and Ø100 × 50 mm to be depleted in Ca in the leaching experiments.

The specifications of all the specimens such as mix proportions as well as cast sizes and cut sizes are presented in Table 2. It should be noted that the W/C-ratios for paste and mortar specimens were chosen as to obtain better homogenized mixes. However, as noted the concrete specimens were cast according to the used W/C-ratios used in repository of nuclear waste in Sweden, SFL.

Table 1 Chemical characteristics of Swedish CEM I 42.5N MH/SR3/LA.

Chemical formulation	CaO	SiO$_2$	Al$_2$O$_3$	Fe$_2$O$_3$	MgO	Na$_2$O	K$_2$O	SO$_3$	Cl
Percentage	64	22.2	3.6	4.4	0.94	0.07	0.72	2.2	0.01

Table 2 Cast specimen's specifications.

Specimens	Cement type	W/C	Aggregate fraction	Cement content (kg/m^3)	Size of cast specimens (mm)	Size of specimens after cutting (mm)
Paste	CEM I 42.5 N MH/SR3/LA	0.5	–	1225	Ø50 × 250	Ø50 × 75
Mortar		0.5	0.5	635		
Concrete		0.48	0.7	350	Ø50 × 250 and Ø100 × 200	Ø50 × 75 and Ø100 × 50
		0.62	0.7	300		

2.2 Electrochemical Acceleration Method

Concrete specimens were depleted in calcium by the use of a newly developed method for accelerated calcium leaching of solid specimens of flexible size (Babaahmadi et al. 2015). As illustrated in Fig. 1, a cementitious specimen acting as a porous medium for ion migration is placed between two electrolyte solutions. The electrical migration enables faster transport of calcium ions, whilst application of ammonium nitrate as catholyte enables higher rate of dissolution of calcium hydroxides. As of using lithium hydroxide solution as anolyte, the pH of the anolyte was always kept at 13.5–14 and as of using ammonium nitrate solution as catholyte, the pH of catholyte was around a pH level of around 9. The alkaline characteristic of anolyte solution prevented acid attach due to production of H$^+$ ions at the anode.

As it is concluded by Babaahmadi et al. (Babaahmadi et al. 2015), with a current density of 125–130 A/m^2 an approximately 53 days of experimental time is predicted to reach to complete leaching of Portlandite for a paste specimen of Ø50 × 75 mm and W/C-ratio of 0.5. In this study similar experimental time (53 days) was chosen for the specimens. It should be noted that depending on the W/C-ratio the Portlandite content varies in the specimens. In this study the concrete specimens have the same volume of aggregate, implying the same volume of CSH, Portlandite and capillary pores. Which means with a higher W/C- ratio, the capillary pore volume is larger, implying the CSH and Portlandite volume is smaller. Accordingly under the same degree of hydration, Portlandite in the concrete with higher W/C is less. As a consequence the experimental time for concrete specimens with W/C-ratio of 0.62 to reach to

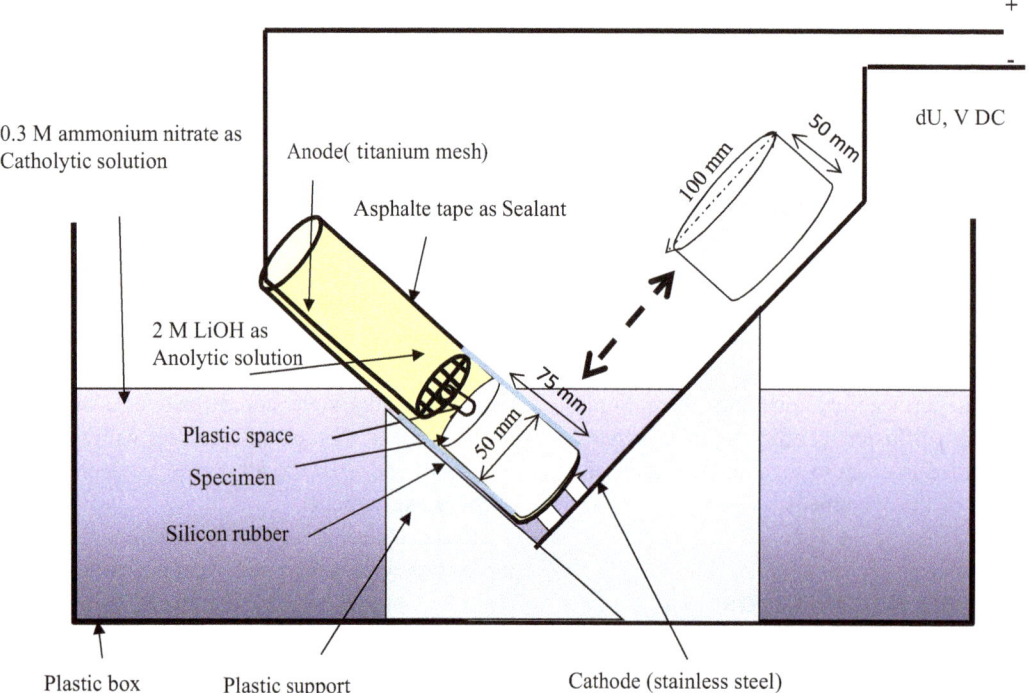

Fig. 1 Electrochemical migration set-up design.

complete leaching of Portlandite should be shorter than that of specimens with W/C-ratio of 0.48. However, the prolongation of experimental time after leaching of Portlandite only affects the phase changes in CSH gel and as it was concluded by Carde et al. the changes in CSH phases due to leaching is not affecting the mechanical properties (Carde et al. 1996).

2.3 Natural Leaching Test

The natural immersion test involved the immersion of the test specimen in groundwater to account for the leaching of calcium ions under relatively natural condition. This method is chosen as most of the reference natural leaching test methods introduced in the literature involve an immersion test with frequent exchanges of the leaching solution (Adenot and Buil 1992; Carde and François 1997; Faucon et al. 1996, 1998; Gustafson 2008; Haga et al. 2005; Heukamp et al. 2001; Langton and Kosson 2009; Mainguy et al. 2000; Maltais et al. 2004; Peyronnard et al. 2009; Revertegat et al. 1992; Ryu et al. 2002; Saito et al. 1992; Wittmann 1997).

In this study the specimens of Ø46 × 100–250 mm, cured for 6 months, were coated with thick (3–4 mm) epoxy, sealing all surfaces except one for exposure. The experiment was carried out in the Äspö laboratory in Oskarshamn, Sweden. The laboratory is located 400 m deep under the ground which could facilitate access to the groundwater. The specimens were placed in a container and immersed in the groundwater. The groundwater was exchanges continuously as it was flowing into the container with a tube, Fig. 2. After 2 and 3.5 years of exposure, samples were chemically analyzed in order to obtain their leaching profiles.

2.4 Assessments of the Changes in Hydrated Phases Due to Decalcification

Line scans quantifying the axial (longitudinal) changes in Ca/Si ratios of solid samples were performed by Laser Ablation- Inductive Coupled Plasma- Mass Spectrometry (LA-ICP-MS). Laser Ablation analysis was performed using a New Wave NWR213 laser ablation system coupled to an Agilent 7500a quadrupole ICP-MS (upgraded with shield torch and a second rotary vacuum pump). A 30 μ laser spot size, beam energy density of ca. 6 J/cm^2 and repetition rate of 10 Hz was used in line scan mode. Each analysis included background measurement for 30 s, before switching on the laser. Figure 3, illustrates the cut samples used for the measurements. The surface of the cut sample was polished with a sand paper after saw cutting and later the sample was vacuum dried. An average of three line scans was used to report the results.

In order to characterize the changes in hydrated phases, especially Portlandite, in the leached specimens produced in this study thermogravimetry analysis (TGA) were performed with a Netzsch, STA 409 PC/PG. To perform the measurements powder samples were prepared out of leached specimens. For electrochemically leached specimens a longitudinal section as shown in Fig. 3 was cut from the specimen (paste, mortar and concrete) and hand crushed in a mortar and vacuum dried afterwards. However for the case

Fig. 2 Natural leaching test set-up.

of naturally leeched specimens the powders were prepared from the outermost 2 mm of the leaching front. The samples were placed in a crucible and heated in pure nitrogen (inlet flow rate of 20 ml/min) to a set temperature of 900 °C. The heating rate was a linear ramp of 10 °C/min.

2.5 Transport Properties

The gas permeability and capillary water absorption tests were performed according to state of the art report of RILEM Technical Committee 189-NEC (Torrent 2007). The measurements were performed on concrete specimen of the size Ø100 × 50 mm. The specimens were preconditioned for 2 weeks according to recommended procedures stated in RILEM technical committee 189-NEC (Torrent 2007) prior to the measurements. The step by step procedure for the preconditioning, measurements and instrument to carry out the measurements of gas permeability and capillary absorption of water is explained in 189-NEC standard.

The chloride diffusion coefficient of the pristine and leached specimens were studied by means of the rapid chloride migration test according to NT BUILD 492 described by Tang (NT BUILD 492, Concrete, Mortar and Cement-based Repair Materials: Chloride Migration Coefficient from Nonsteady-state Migration Experiments 1999; Tang 1996). For the calcium depleted specimens prediction of the duration of the experimental time was difficult due to that the increased porosity of the specimens as well as the reduced

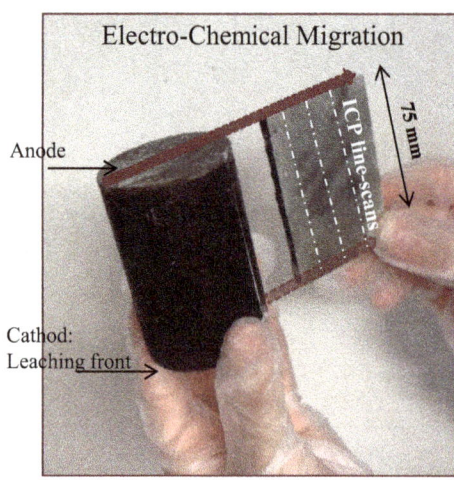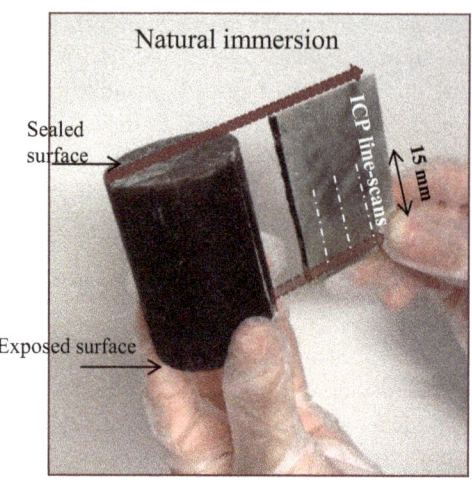

Fig. 3 Sample preparation for ICP-MS analysis.

concentration of chloride in the pore solution dramatically alters the chloride diffusion rate. Here the duration of the experiments were reduced to 15 h compared to fresh concrete specimens for which a duration of 24 h is normally used (NT BUILD 492, Concrete, Mortar and Cement-based Repair Materials: Chloride Migration Coefficient from Non-steady-state Migration Experiments 1999; Tang 1996). However, in spite of the large reduction in experimental time, chloride ions penetrated through the entire thickness of the specimens during this experiment. For that reason, no exact value of the chloride diffusion rate could be obtained but instead the minimum chloride diffusion coefficient was calculated for the calcium depleted concrete specimens.

2.6 Mechanical Properties

The tensile strength of the leached and reference concrete specimens of size Ø100 × 50 mm was measured using the splitting test on a Toni-Technik compression testing machine with a maximum capacity of 100 kN. The test procedure is in accordance with American standard ASTM C49 (ASTM C496/C496M-11, Standard Test Method for Splitting Tensile Strength of Cylindrical Concrete Specimens). The compressive strength of concrete, mortar and paste specimens of size Ø50 × 75 mm was measured with same compression testing machine and according to ASTM C39 standard (ASTM C39/C39M-14a, Standard Test Method for Compressive Strength of Cylindrical Concrete Specimens). According to this standard a correction factor of 0.96 was applied to the compressive strength results to account for the length:dimater ratio which is lower than 1.75. The calcium depleted specimens were kept in 100 % RH until being tested in order to avoid any internal cracks.

Elastic modulus of the concrete specimens of size Ø50 × 75 mm was obtained as the slope of stress–strain curves recorded by means of an ALPHA compression testing machine. The load cell had a maximum capacity of 50 kN and was loaded with a mechanical press at a constant rate of 0.01 kN/ms. The vertical strain was measured utilizing a calibrated Linear Variable Differential Transformers (LVDT) sensor. The end surfaces of the specimens, perpendicular to the longitudinal axis of specimen, were cut with a diamond saw and polished in order to

create a smooth surface. The concrete specimens of the size Ø50 × 75 mm were placed between two platens and positioned under the load cells. 4 LVDT sensors were used to measure the displacement of the bottom platen and three more sensors were employed to measure the displacement of the upper platen. The sensors were connected to a data-log system to record the gradient of strain as a function of stress.

An impact resonant apparatus was used to measure the fundamental longitudinal frequency of concrete, mortar and paste specimens of the size Ø50 × 75 mm. One end of the horizontally placed concrete cylinder was vibrated with an impactor and the other end was connected to an accelerometer. The accelerometer was connected to an amplifier and the set-up was connected to a wave form analyzer. The waveform analyzer shall have a sampling rate of at least 20 kHz and should record at least 1024 points of the wave form. By utilizing the fact that the elastic modulus is proportional to the square root of the resonant frequency the elastic modulus of each specimen can be calculated according to American standard ASTM C215. The dynamic elastic modulus was calculated according to Eq. (1).

$$E = D \times M \times (n')^2 \tag{1}$$

where E is the dynamic elastic modulus, M is the mass and n' is the fundamental longitudinal frequency and D is a factor defined according to Eq. (2) for cylindrical specimens.

$$D = 5.093 \times (L/d^2) \tag{2}$$

The values from this non-destructive test can be compared with those obtained according to stress–strain curves.

3. Results and Discussions

3.1 Characterization of Degraded Specimens

The LA-ICP-MS results, presenting the electrochemical-migration-induced longitudinal change in the Ca/Si ratio of the leached paste specimen are illustrated in Fig. 4. As can be seen, ICP-MS line scans representing the composition of leaching front in degraded specimens leached with electrochemical migration and natural immersion test methods are illustrated. It is shown that the propagation of leaching front (shape of leaching front) obtained by the two methods are relatively comparable and the longitudinal changes in the Ca/Si ratio propagates towards a homogeneous leaching state in time, indicating a complete leaching of Portlandite.

Moreover, as shown after approximately 53 days of experimental time (1.2×10^6 Coulombs), a Ca/Si ratio close to complete leaching of Portlandite is obtained in electrochemically leached specimens, while the naturally leached specimens indicate up to 10 mm of leaching front in 3.5 years of leaching. This indicates that considerable acceleration rate is gained with application of electrochemical migration method. Similar outcome can be obtained when comparing the specimen size and leaching duration in electro-chemical migration test and leaching experiments reported in the literature. Haga et al. (2005) have shown that a depletion depth of a maximum of 1.25 mm can be expected after 100 days of a natural immersion test. Faucon et al. (1998) reported that up to 60 days of natural leaching leads to 0.7 mm of dissolved thickness in exposed specimens. Lagerblad (2001) and Trägårdh and Lagerblad (1998) reported a leaching depth equal to 5–10 mm can be expected after up to 100 years. Heukamp et al. (2001) used specimens with the size of Ø11.5 × 60 mm for a leaching period of 45 days. Nguyen et al. (2007) have reported on the application of specimens of Ø32 × 100 mm and Ø110 × 220 for an experimental time of up to 547 days and Choi and Yang (2013), have used

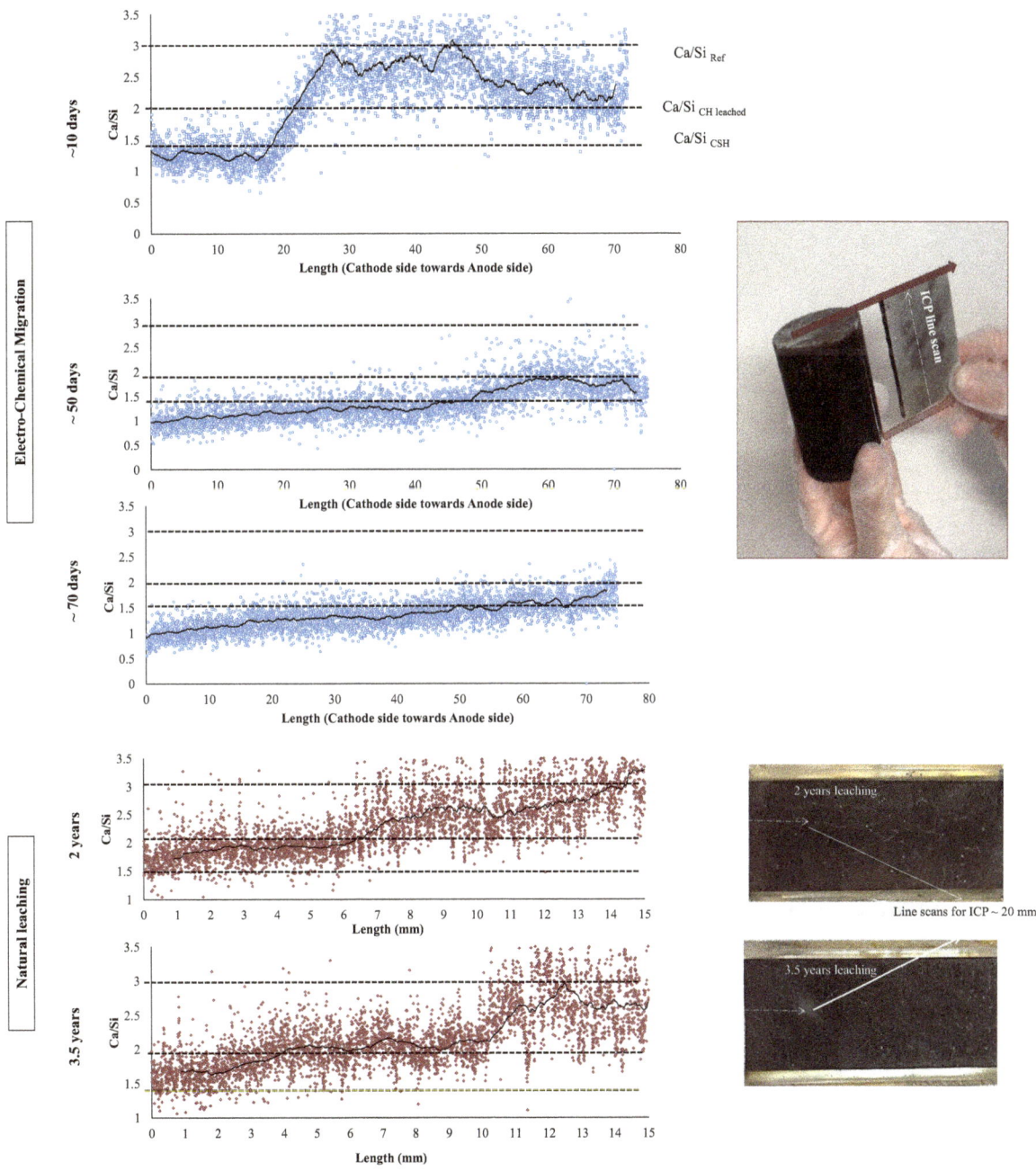

Fig. 4 LA-ICP-MS analysis for naturally and accelerated leached samples.

cylindrical concrete samples of Ø100 × 100 mm for an experimental time of up to 365 days.

Further, the TGA analyses results are presented in Fig. 5 comparing temperature-induced mass changes for reference, accelerated leached and naturally leached paste samples. It should be noted that for the case of natural leaching the samples were taken from 0 to 2 mm of the outer most part of the exposed surface. As can be seen, the reference sample exhibits a mass change at approximately 400 °C, which is the assigned peak to the Portlandite (Zhang and Ye 2012). However for all degraded samples no matter if leached naturally or with the acceleration method the Portlandite peaks are vanished. The preferential leaching of Portlandite due to decalcification has been also stated in literature

(Adenot and Buil 1992). This is also an indication that the electrochemical migration method enables leaching of cementitious specimens even when containing aggregates.

The average changes in the mass of concrete specimens representing the changes in the bulk density and eventually changes in porosity are presented in Table 3. As shown the specimens with a lower W/C-ratio are indicating higher degree of mass changes after leaching. Moreover, the changes in porosity according to total leached content of Portlandite are also presented. The results are in accordance with explanations in Sect. 2.2, stating that with a lower W/C-ratio a larger Portlandite content is expected and moreover, it is shown that the mass depletion due to leaching is mainly attributed to leaching of Portlandite. Accordingly a

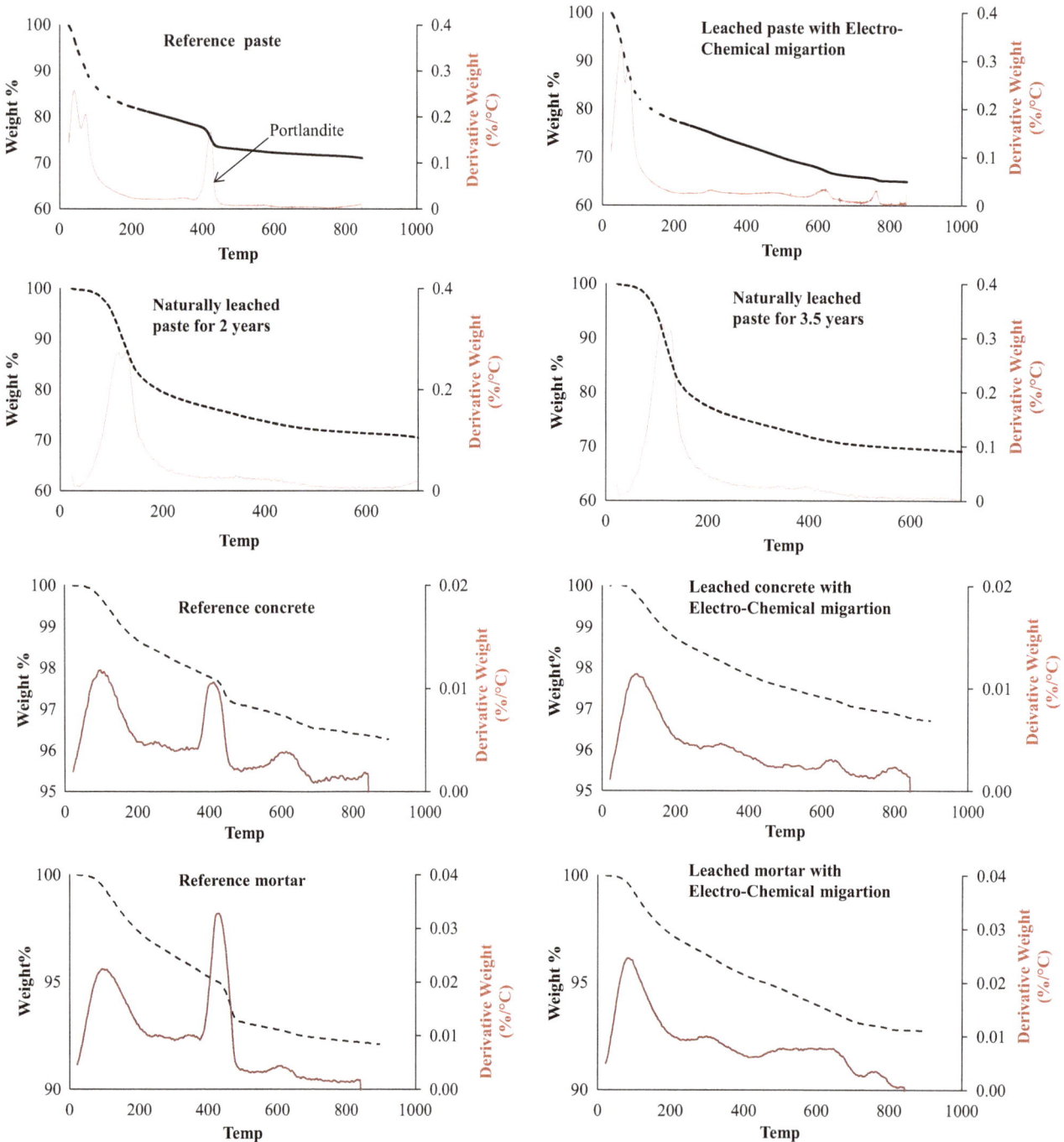

Fig. 5 TGA analysis for reference and degraded paste samples.

higher degree of mass change will be attained. It should be noted that the specimens were pre-conditioned according to the state of the art report of RILEM Technical Committee 189-NEC (Torrent 2007), before weighting to have a comparable moisture condition.

3.2 Transport Properties

The capillary water absorption results representing the changes in pore structures of the specimens are presented on the left side of Fig. 6. Considering the presented results it can be inferred that the capillary water absorption in 24 h is increased by a factor of 4 and 2 for calcium depleted specimens with W/C-ratios of 0.48 and 0.62, respectively. As can be seen that the difference in rate of capillary water absorption of aged specimens with different water cement ratios, is much less than that of reference specimens. This indicates that a very similar pore structure is expected in aged specimens no matter which initial W/C-ratio the specimens had.

Further, the change in gas permeability coefficient due to degradation is presented in the right side of Fig. 6. As illustrated the gas permeability in the calcium depleted specimens is more than 10 times higher than in the pristine material. This can be explained by the larger porosity after decalcification. Furthermore, the changes in chloride migration coefficient of the concrete specimens due to degradation are presented in Fig. 7. As it can be seen, at least three times higher chloride migration coefficient is expected after calcium depletion. It should be noted that the presented results regarding the calcium depleted specimens are the minimum chloride migration coefficient due to the full penetration through the specimen thickness as it was pointed out previously. The results are in good agreement with the results presented Choi and Yang (2013).

3.3 Mechanical Properties

The average tensile strength of concrete specimens of size Ø100 × 50 mm are presented in Fig. 8. As shown the average tensile strength for the specimens with a W/C-ratio of 0.48 is reduced by up to approximately 70 % due to calcium depletion whereas the reduction in tensile strength for the specimen with a W/C-ratio of 0.62 is about 55 %.

Further, the average compressive strength results for specimens of size Ø50 × 75 mm is presented in Fig. 9. As it can be seen the reduction in compressive strength due to leaching is 70 % for concrete specimens with W/C = 0.48, 59 % for concrete specimens with W/C = 0.62, 74 % for mortar specimens and 56 % for paste specimens. Interestingly the residual strength of the Ca-depleted concrete specimens is very similar regardless of the water cement ratios of the original specimens. This can also be seen for paste and mortar specimens. This indicates that when leaching propagates towards an approximate complete leaching of the Portlandite, comparable strength properties can be expected for samples with different original W/C-ratios. This is due to comparable pore structures that are obtained in the specimens after leaching. It has also been shown in literature that the changes in pore structure have a detrimental influence on the strength properties of cementitious materials (Carde et al. 1996; Haga et al. 2005; Mainguy et al. 2000).

The findings presented above are in good agreement with the conclusions presented by (Carde et al. 1996). In their study they showed that the loss of strength due to complete removal of Portlandite can be up to 70 %. Similar results have also been presented by (Choi and Yang 2013). Moreover, rate of change in the strain of the concrete specimens as a function of applied stress representing stress–strain curves is presented in Fig. 10. As shown the elastic modulus is represented by the slope of the stress–strain curves is reduced by up to 40 % after leaching for both types of specimens. In addition, to confirm the credibility of the results from stress–strain curves, elastic modulus was also estimated through measuring the natural frequency of the specimens. The results which are presented in Table 4 are comparable with the results presented in Fig. 10.

4. Conclusions

In this paper a study of mechanical and physical properties of concrete specimens which are depleted in calcium by a newly developed acceleration method is presented. Providing such results requires calcium depleted concrete

Table 3 Mass changes due to leaching for concrete specimens of size Ø100 × 50 mm.

	REF (g)	Aged (g)	Mass depletion	V depletion[a]	Porosity change[b]	Leached portlandite (g)[c]	Porosity change[d]
W/C = 0.48	927	881	46.00	20.81	5.3	44.4	5.12
W/C = 0.62	907	878	29.00	31.12	3.3	29.6	3.41

[a] Calculated assuming that total mass change is attributed to Portlandite depletion (V depletion = mass change/Portlandite density).

[b] Calculated according to V depletion and considering the dimension of the specimens, Ø100 × 50 mm (Porosity Change = Vdepletion/total volume of specimen).

[c] Leached Portlandite is according to measured leached calcium content in catholyte solution.

[d] Calculated according to total leached Portlandite and considering the dimension of the specimens (Ø100 × 50 mm).

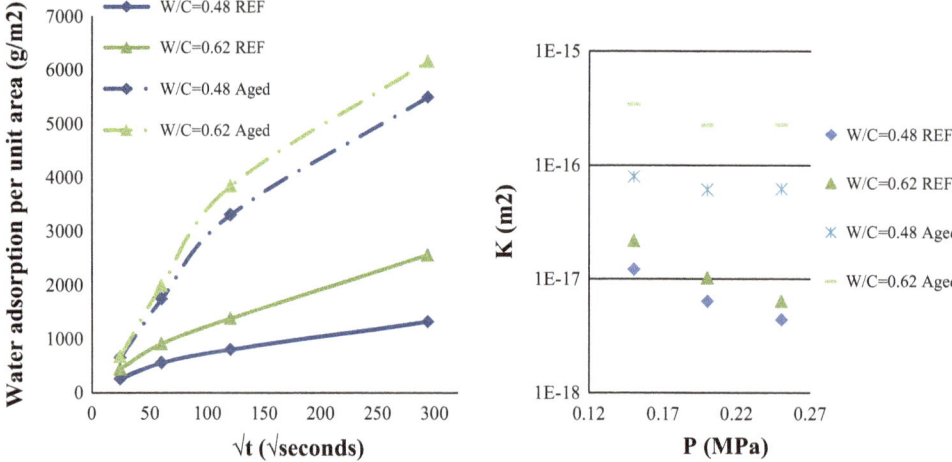

Fig. 6 Capillary water absorption for concrete specimens for W/C ratios of 0.48 and 0.62 before and after leaching of calcium (left) and Gas permeability coefficient K as function of applied absolute gas pressure P (right).

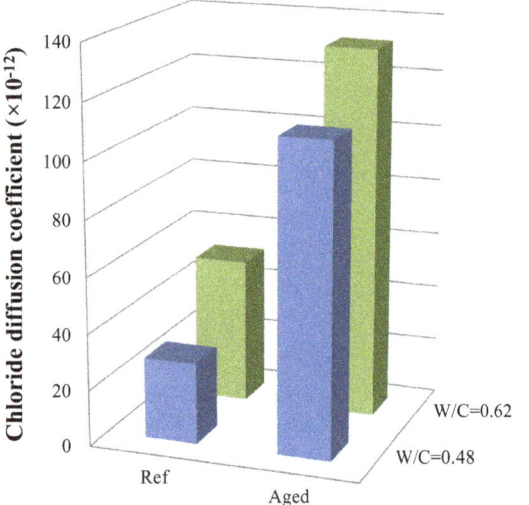

Fig. 7 Chloride diffusion coefficient for reference and aged concrete specimens. The calculated diffusion coefficient for the case of aged specimens is the least possible value. Measuring the highest possible value is not practically possible due to very high porosity of the specimens.

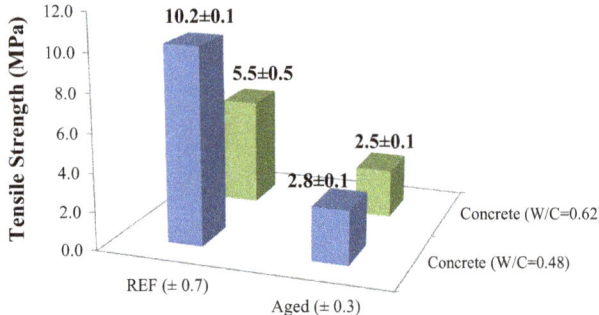

Fig. 8 Tensile strength results.

specimens of proper sizes. The presented method is shown to be effective producing decalcified concrete specimens of flexible size. The changes in properties such as tensile and

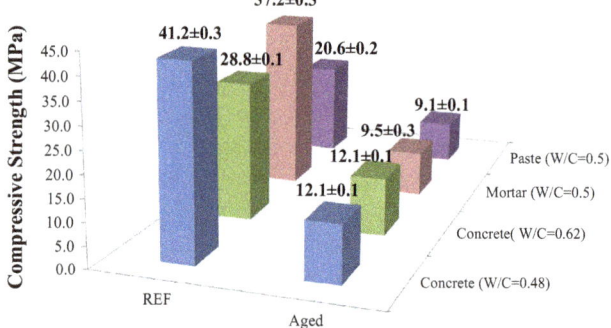

Fig. 9 Compressive strength results.

compressive strength, elastic modulus, gas permeability as well as chloride diffusion coefficient are taken into consideration. One of the direct applications of this study is to provide experimental approaches to be able to validate the risk assessment analyses regarding the functionality of engineered cementitious barriers. Based on the results presented in this study the following conclusions can be drawn:

- The electrochemically-induced leaching is not affected by the aggregate content in mortar or paste specimens.
- Leaching of Portlandite causes considerable changes in physical and mechanical properties of concrete specimen, primary due to the increase in pore volume. It was concluded that larger pore volume due to complete leaching of portlandite can be expected which would cause up to 70 % decrease in mechanical strength and 40 % decrease in elastic modulus.
- Larger pore volume after degradation causes more than 10 times increase in gas permeability and at least 3 times higher chloride diffusion rate.
- The residual strength properties of the concrete specimens after complete leaching of portlandite are shown to be relatively similar no matter which initial water cement ratios the specimens have. This is to great extent due to similar pore structures in concrete specimens after leaching.

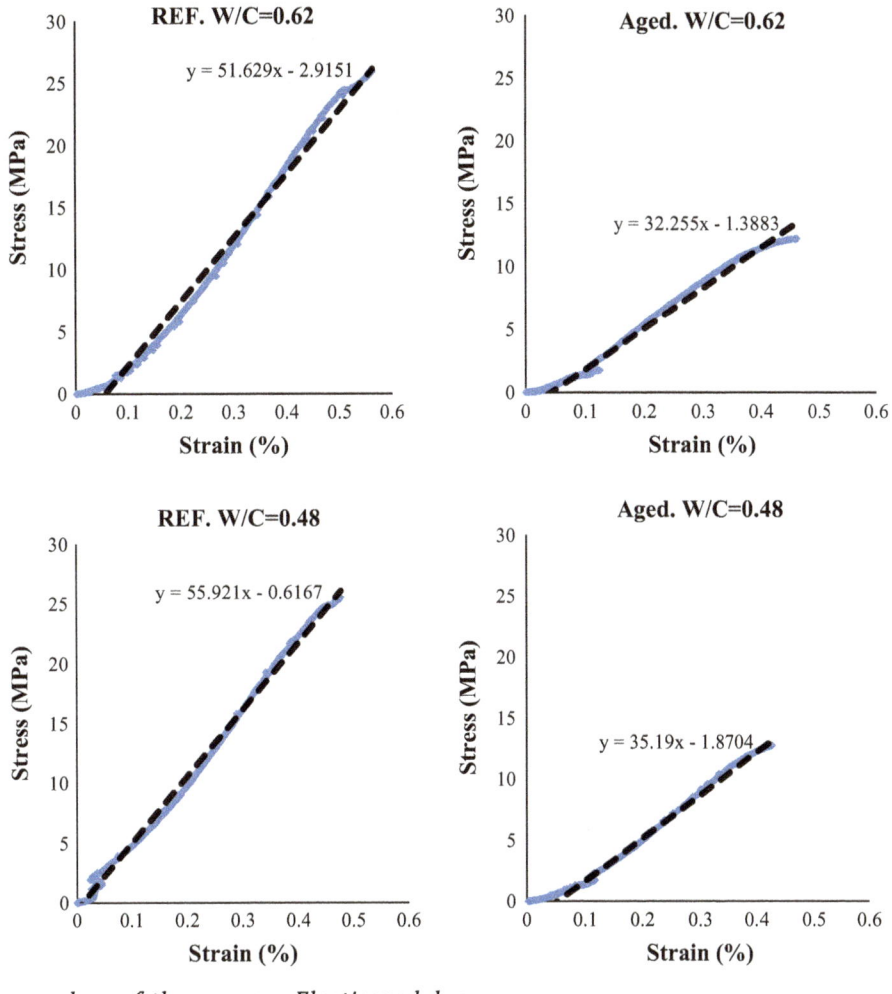

slope of the curves = Elastic modulus

Fig. 10 Stress-strain curves for concrete specimens.

Table 4 Natural frequency and calculated E-modulus of the reference and aged specimens.

		Mass (g)		Length (m)		F (kHz)		E (GPa)	
		Ref	Aged	Ref	Aged	Ref	Aged	Ref	Aged
Concrete	W/C = 0.48	390	325	0.078	0.078	27.4	21.6	46.5	24.1
	W/C = 0.62	350	330	0.076	0.075	29.3	22.1	46.5	24.6
Mortar	W/C = 0.5	323	326	0.072	0.076	25.7	17.6	30.4	15.18
Paste	W/C = 0.5	230	220	0.076	0.08	17.6	13.2	11.4	6.24

Acknowledgments

The authors greatly appreciate financial support from Swedish Nuclear Fuel and Waste Management Company (SKB). Assistance with mechanical tests, BET and MIP analysis by Lars Wahlström, Ann Wendel and Liu Wei respectively are hereby appreciated and acknowledged.

References

Adenot, F., & Buil, M. (1992). Modelling of the corrosion of the cement paste by deionized water. *Cement and Concrete Research, 22,* 489–496. doi:10.1016/0008-8846(92)90092-a.

ASTM C39/C39M-14a, Standard Test Method for Compressive Strength of Cylindrical Concrete Specimens.

ASTM C496/C496M-11, Standard Test Method for Splitting Tensile Strength of Cylindrical Concrete Specimens.

Babaahmadi, A., Tang, L., Abbas, Z., Zack, T., & Mårtensson, P. (2015). Development of an electro-chemical accelerated ageing method for leaching of calcium from cementitious materials. *Materials and Structures*. doi:10.1617/s11527-015-0531-8.

Berner, U. R. (1992). Evolution of pore water chemistry during degradation of cement in a radioactive waste repository environment. *Waste Management, 12*, 201–219. doi: 10.1016/0956-053x(92)90049-o.

Carde, C., Escadeillas, G., & François, R. (1997). Use of ammonium nitrate solution to simulate and accelerate the leaching of cement pastes due to deionized water. *Magazine of Concrete Research, 49*, 295–301.

Carde, C., & François, R. (1997). Effect of the leaching of calcium hydroxide from cement paste on mechanical and physical properties. *Cement and Concrete Research, 27*, 539–550. doi:10.1016/s0008-8846(97)00042-2.

Carde, C., François, R., & Torrenti, J.-M. (1996). Leaching of both calcium hydroxide and C-S-H from cement paste: Modeling the mechanical behavior. *Cement and Concrete Research, 26*, 1257–1268. doi:10.1016/0008-8846(96)00095-6.

Choi, Y. S., & Yang, E. I. (2013). Effect of calcium leaching on the pore structure, strength, and chloride penetration resistance in concrete specimens. *Nuclear Engineering and Design, 259*, 126–136. doi:10.1016/j.nucengdes.2013.02.049.

Emborg, M., Jonasson, J. E., & Knutsson, S. (2007) *Long-term stability due to freezing and thawing of concrete and bentonite at the disposal of low and intermediate level nuclear waste in SFR 1 (in Swedish)* vol. R-07-60. SKB Technical Report R-07-60, Swedish Nuclear and Waste Management Company

EN 933-1 Tests for geometrical properties of aggregates.

Faucon, P., Adenot, F., Jacquinot, J. F., Petit, J. C., Cabrillac, R., & Jorda, M. (1998). Long-term behaviour of cement pastes used for nuclear waste disposal: Review of physico-chemical mechanisms of water degradation. *Cement and Concrete Research, 28*, 847–857. doi:10.1016/s0008-8846(98)00053-2.

Faucon, P., Le Bescop, P., Adenot, F., Bonville, P., Jacquinot, J. F., Pineau, F., & Felix, B. (1996). Leaching of cement: Study of the surface layer. *Cement and Concrete Research, 26*, 1707–1715. doi:10.1016/S0008-8846(96)00157-3.

Gustafson, G., Hagström, M., & Abbas, Z. (2008). Beständighet av cementinjektering

Haga, K., Sutou, S., Hironaga, M., Tanaka, S., & Nagasaki, S. (2005). Effects of porosity on leaching of Ca from hardened ordinary Portland cement paste. *Cement and Concrete Research, 35*, 1764–1775. doi:10.1016/j.cemconres.2004.06.034.

Heukamp, F. H., Ulm, F. J., & Germaine, J. T. (2001). Mechanical properties of calcium-leached cement pastes: Triaxial stress states and the influence of the pore pressures. *Cement and Concrete Research, 31*, 767–774.

Hinsenveld, M. (1992). A shrinkage core model as a fundamental representation of leaching mechanism in cement stabilized waste. Doctoral thesis, University of Cincinnati, Cincinnati, OH.

Höglund, L.-O. (2001). Project SAFE: Modeling of long-term concrete degradation processes in the Swedish SFR repository vol R-01-08. SKB Report, Svensk Kärnbränslehantering AB.

Lagerblad, B. (2001). *Leaching performance of concrete based on studies of sample from old concrete constructions, TR-01-27.* Stockholm, Sweden: Swedish Nuclear Fuel and Waste Management.

Langton, C., & Kosson, D. (2009). Review of mechanistic understanding and modeling and uncertainty analysis methods for predicting cementitious barriers performance. doi:10.2172/974326

Mainguy, M., Tognazzi, C., Torrenti, J.-M., & Adenot, F. (2000). Modelling of leaching in pure cement paste and mortar. *Cement and Concrete Research, 30*, 83–90. doi:10.1016/S0008-8846(99)00208-2.

Maltais, Y., Samson, E., & Marchand, J. (2004). Predicting the durability of Portland cement systems in aggressive environments—Laboratory validation. *Cement and Concrete Research, 34*, 1579–1589. doi:10.1016/j.cemconres.2004.03.029.

Marinoni, N., Pavese, A., Voltolini, M., & Merlini, M. (2008). Long-term leaching test in concretes: An X-ray powder diffraction study. *Cement & Concrete Composites, 30*, 700–705. doi:10.1016/j.cemconcomp.2008.05.004.

Morga, M., & Marano, G. C. (2015). Chloride penetration in circular concrete columns. *International Journal of Concrete Structures and Materials, 9*, 173–183. doi:10.1007/s40069-014-0095-y.

Nguyen, V. H., Colina, H., Torrenti, J. M., Boulay, C., & Nedjar, B. (2007). Chemo-mechanical coupling behaviour of leached concrete: Part I: Experimental results. *Nuclear Engineering and Design, 237*, 2083–2089. doi:10.1016/j.nucengdes.2007.02.013.

NT BUILD 492. (1999). Concrete, mortar and cement-based repair materials: Chloride migration coefficient from non-steady-state migration experiments.

Park, S., & Choi, Y. (2012). Influence of curing-form material on the chloride penetration of off-shore concrete. *International Journal of Concrete Structures and Materials, 6*, 251–256. doi:10.1007/s40069-012-0026-8.

Peyronnard, O., Benzaazoua, M., Blanc, D., & Moszkowicz, P. (2009). Study of mineralogy and leaching behavior of stabilized/solidified sludge using differential acid neutralization analysis: Part I: Experimental study. *Cement and Concrete Research, 39*, 600–609. doi:10.1016/j.cemconres.cemconres.2009.03.016.

Pham, S., & Prince, W. (2014). Effects of carbonation on the microstructure of cement materials: influence of measuring methods and of types of cement. *International Journal of Concrete Structures and Materials, 8*, 327–333. doi:10.1007/s40069-014-0079-y.

Pritzl, M., Tabatabai, H., & Ghorbanpoor, A. (2014). Laboratory evaluation of select methods of corrosion prevention in reinforced concrete bridges. *International Journal of Concrete Structures and Materials, 8*, 201–212. doi:10.1007/s40069-014-0074-3.

Reardon, E. J. (1992). Problems and approaches to the prediction of the chemical composition in cement/water systems. *Waste Management, 12*, 221–239. doi:10.1016/0956-053x(92)90050-s.

Revertegat, E., Richet, C., & Gégout, P. (1992). Effect of pH on the durability of cement pastes. *Cement and Concrete Research, 22*, 259–272. doi:10.1016/0008-8846(92)90064-3.

Ryu, J.-S., Otsuki, N., & Minagawa, H. (2002). Long-term forecast of Ca leaching from mortar and associated degeneration. *Cement and Concrete Research, 32*, 1539–1544. doi:10.1016/s0008-8846(02)00830-x.

Saito, H., & Deguchi, A. (2000). Leaching tests on different mortars using accelerated electrochemical method. *Cement and Concrete Research, 30*, 1815–1825. doi:10.1016/S0008-8846(00)00377-X.

Saito, H., Nakane, S., Ikari, S., & Fujiwara, A. (1992). Preliminary experimental study on the deterioration of cementitious materials by an acceleration method. *Nuclear Engineering and Design, 138*, 151–155. doi:10.1016/0029-5493(92)90290-c.

Sellier, A., Buffo-Lacarrière, L., Gonnouni, M. E., & Bourbon, X. (2011). Behavior of HPC nuclear waste disposal structures in leaching environment. *Nuclear Engineering and Design, 241*, 402–414. doi:10.1016/j.nucengdes.2010.11.002.

Tang, L. (1996). Electrically accelerated methods for determining chloride diffusivity in concrete—Current development. *Magazine of Concrete Research, 48*, 173–179.

Trägårdh, J., & Lagerblad, B. (1998). *Leaching of 90-year old concrete in contact with stagnant water, TR 98-11*. Stockholm, Sweden: Swedish Nuclear fuel and Waste.

Torrent, R. (2007). Non-destructive evaluation of the penetrability and thickness of the concrete cover—State-of-the-Art Report of RILEM Technical Committee 189-NEC vol rep040.

Ulm, F.-J., Lemarchand, E., & Heukamp, F. H. (2003). Elements of chemomechanics of calcium leaching of cement-based materials at different scales. *Engineering Fracture Mechanics, 70*, 871–889. doi:10.1016/s0013-7944(02)00155-8.

Wittmann, F. H. (1997). Corrosion of cement-based materials under the influence of an electric field. *Materials Science Forum, 247*, 107–126. doi:10.4028/www.scientific.net/MSF.247.107.

Zhang, Q., & Ye, G. (2012). *Dehydration kinetics of Portland cement paste at high temperature J Therm Anal Calorim, 110*, 153–158. doi:10.1007/s10973-012-2303-9.

3

Automated Surface Wave Measurements for Evaluating the Depth of Surface-Breaking Cracks in Concrete

Seong-Hoon Kee[1],*, and Boohyun Nam[2]

Abstract: The primary objective of this study is to investigate the feasibility of an innovative surface-mount sensor, made of a piezoelectric disc (PZT sensor), as a consistent source for surface wave velocity and transmission measurements in concrete structures. To this end, one concrete slab with lateral dimensions of 1500 by 1500 mm and a thickness of 200 mm was prepared in the laboratory. The concrete slab had a notch-type, surface-breaking crack at its center, with depths increasing from 0 to 100 mm at stepwise intervals of 10 mm. A PZT sensor was attached to the concrete surface and used to generate incident surface waves for surface wave measurements. Two accelerometers were used to measure the surface waves. Signals generated by the PZT sensors show a broad bandwidth with a center frequency around 40 kHz, and very good signal consistency in the frequency range from 0 to 100 kHz. Furthermore, repeatability of the surface wave velocity and transmission measurements is significantly improved compared to that obtained using manual impact sources. In addition, the PZT sensors are demonstrated to be effective for monitoring an actual surface-breaking crack in a concrete beam specimen subjected to various external loadings (compressive and flexural loading with stepwise increases). The findings in this study demonstrate that the surface mount sensor has great potential as a consistent source for surface wave velocity and transmission measurements for automated health monitoring of concrete structures.

Keywords: surface waves, surface wave transmission, surface-breaking crack, concrete, non-destructive evaluation.

1. Introduction and Motivation

Surface wave measurements have been widely used to develop non-destructive evaluation (NDE) techniques for concrete structures in civil engineering due to their useful features (Graff 1991). Surface waves are mechanical waves that propagate along the surface of concrete with most of their energy confined near the surface, which enables one-sided access of concrete structures. The particle vibration amplitude of surface waves exponentially decreases with distance from the free surface boundary and with a frequency-dependent penetration depth, which is particularly useful to identify and characterize surface-breaking or subsurface defects in concrete structures (Achenbach 2000, 2002). In infinite media, surface waves are non-dispersive, that is the wave velocity does not change with frequency. In practice, this assumption is valid when the thickness H of the

solid body of interest is sufficiently larger than the wavelength λ of the surface wave (i.e., $H > 2\lambda$). In thin plates or layered systems, the velocity of surface wave changes with frequency. Surface wave velocity measurements have been demonstrated to be effective for characterizing mechanical properties of concrete in many civil engineering applications (ACI committee 228 1998).

In surface wave measurements, an impact source is used to generate incident surface waves, and two receivers are used to measure the surface waves propagating in concrete (see Fig. 1). Selecting an appropriate impact source is of great importance for successfully measuring surface waves in concrete structures. The normal impact of a ball on the concrete surface has been widely used as a source for surface wave-based NDE techniques. According to the Hertzian impact theory (McLaskey and Glaser 2010), the impulse force (or force pulse) is approximated by a "half sine" pulse, and the duration of the pulse depends on the geometry and material properties of a ball and a massive material that the ball hits; however, the most influential factor is the radius of the ball. Therefore, the correct ball size can be selected so that the frequency bandwidth generated by the impact source appropriately covers a frequency range of interest for the surface wave measurements. In practice, the impact source is usually operated by hand; however, there are several limitations to this method. First, it is difficult to generate consistent waves with an impact source controlled by a human hand. Furthermore, it is impossible to conduct surface wave

[1]Department of Architectural Engineering, Dong-A University, Busan 604-714, Korea.
*Corresponding Author; E-mail: shkee@dau.ac.kr
[2]Department of Civil, Environmental and Construction Engineering, University of Central Florida, Orlando, FL 32816, USA.

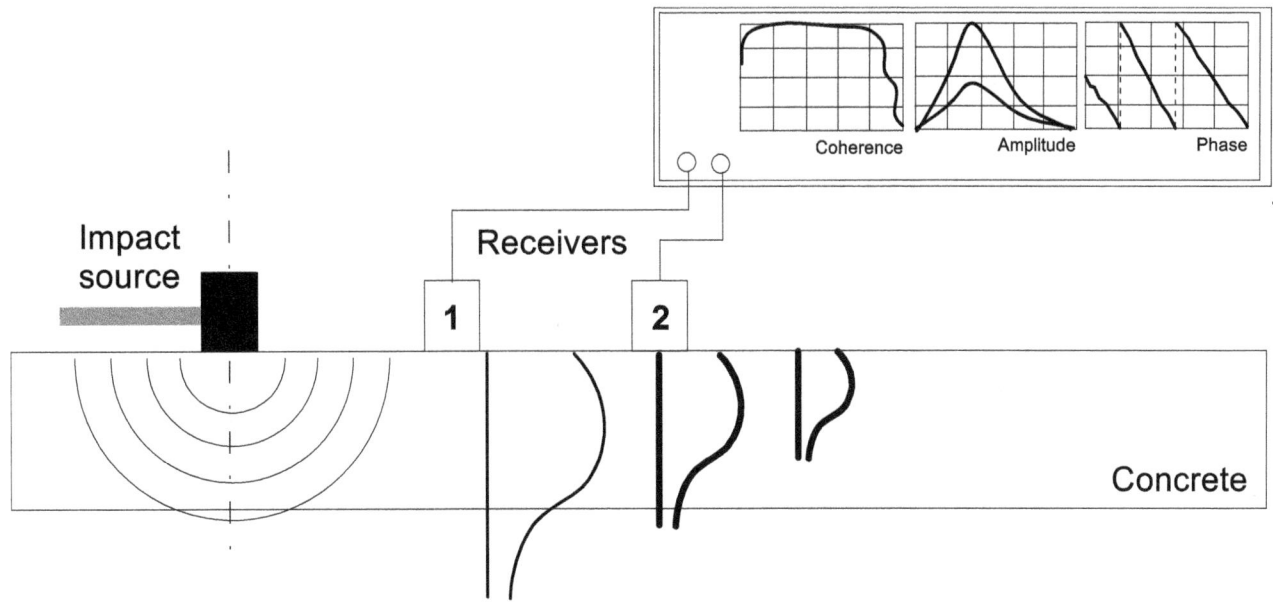

Fig. 1 Source and receiver configuration for surface wave measurements.

measurements in hard-to-access regions of concrete structures, due to safety reasons or spatial limitations. Regardless of the feasibility of these methods, it would be time and labor intensive to be used in large civil structures.

A possible solution to the aforementioned problems can be obtained by using "surface-mounted piezoelectric transducers (PZT)". PZT has been successfully applied to the structural health monitoring of concrete structures by using stress wave-based methods (Liao et al. 2011; Okafor et al. 1996; Song et al. 2006). The wave propagation properties were studied to detect and evaluate the cracks and damages inside concrete structures. Wang et al. (2001) studied the debonding behavior between steel rebar and concrete by using PZT (lead zirconate titanate) patches fixed to generate and receive elastic waves in concrete, and obtained the modulus of elasticity by utilizing the wave propagation characteristics. Song et al. (2007, 2008) developed smart aggregates to perform structural health monitoring for concrete structures. The mortar-typed aggregate was embedded in concrete structures during casting and successfully used for monitoring damage in concrete structures by measuring an energy-based damage index. Dong et al. (2011) developed a cement-based piezoelectric ceramic composite and effectively applied it as a sensor for health monitoring of concrete structures. Hou et al. (2012, 2013) developed a marble-based smart aggregate for seismic compressive and shear stresses. Recently, Kee and Zhu (2013) developed PZT embedded sensors using a PZT disc that can be used as ultrasonic transmitting and receiving transducers for ultrasonic pulse velocity tests. In this study, the author developed a surface-mount sensor using a PZT disc for generating incident surface waves for surface wave velocity and transmission measurements in concrete structures.

The primary objective of this study is to investigate the feasibility of an innovative surface-mount sensor made of a PZT disc (hereafter refer to as "surface-mount sensor") as a

consistent source for surface wave velocity and transmission measurements in concrete structures. Two surface-mount sensors were attached to a concrete slab with dimensions of 1500 by 1500 by 200 mm (width X length X thickness). The slab possessed a surface-breaking crack located in the middle of the slab, which extended to varying depths. Two surface-mount sensors on either side of the surface-breaking crack worked as actuators driven by an ultrasonic pulse-and-receiver, and two accelerometers worked as a receiver. A series of surface wave measurements was performed to investigate the performance of the surface-mount sensors as a consistent source. An additional aspect to the investigation was to test the ability of surface-mount sensors to perform automated monitoring of an actual surface-breaking crack in a concrete beam subjected to various external loadings (compressive and flexural loading with stepwise increasing) (ElSafty and Abdel-Mohti 2013; Soltani et al. 2013).

2. Background

2.1 Ultrasonic Surface Waves (USW) Method

The surface wave velocity methods involve determining the relationship between the wavelength and velocity of surface vibrations at varying vibration frequencies. The resulting relation is called a dispersion curve, which is generally determined by the spectral analysis of surface waves (SASW) (Nazarian and Desai 1993; Nazarian and Stokoe 1986). For plate-like concrete structures (deck, slab, wall, etc.), the ultrasonic surface wave (USW) technique has been demonstrated to be effective for evaluating material damages caused by many sources: ASR, DEF, freeze-and-thaw, and corrosion of reinforcing steel (Gucunski et al. 2013). The USW test consists of recording the response of the concrete, at two receiver locations, to an impact on the surface of the concrete in structures (See Fig. 1). The surface wave velocity can be obtained by measuring the phase difference $\Delta\Phi$ between two different sensors (sensor 1 and sensor 2) ($C = 2\pi fd/\Delta\Phi$; where f is frequency, and d is the distance between two sensors). The frequency range of interest in the USW technique is a high-frequency range compared to the thickness of the tested object, in which surface waves are non-dispersive. In cases of relatively homogeneous materials, the velocity of the surface waves does not vary significantly with frequency. Therefore, the surface wave velocity can be precisely related to the elastic modulus of concrete, using the measured or assumed mass density and Poisson's ratio of the material. A complex process called inversion is not necessary in the USW technique, leading to a substantially reduced time required for data interpretation and post-processing. In the case of a sound and homogenous concrete plate, the velocity of the surface waves will show little variability, while significant variation in the phase velocity will be an indication of the presence of a defect or other anomaly.

2.2 Surface Wave Transmission (SWT) Method

The surface wave transmission (SWT) method has been demonstrated to be effective for evaluating surface-breaking or sub-surface defects in concrete structures. The SWT method uses the frequency-dependent penetration depth of surface waves. When incident surface waves (R_i) propagate across a surface-breaking crack, the low-frequency components of the incident surface waves will transmit to the forward scattering field with attenuation (R_{tr}), while the high-frequency components will reflect back (R_r). Consequently, the transmission coefficient of surface waves Tr across a surface-breaking crack, which is defined as the ratio of spectral amplitudes of R_{tr} to R_i, depends on the frequency of surface waves and dimensions of the defect in the concrete. For example, an analytical solution relating Tr and the normalized crack depth (h/λ, h is the depth of a surface-breaking crack) was given by Achenbach and his colleagues (Achenbach et al. 1980; Angel and Achenbach 1984; Mendelsohn et al. 1980). It was demonstrated through numerical simulations and experimental studies that the SWT method is effective for evaluating the depth of surface-breaking cracks in concrete structures (Hevin et al. 1998; Kee 2011; Kee and Zhu 2011; Popovics et al. 2000; Shin et al. 2008; Song et al. 2003). Test configuration of the SWT method is the same as that of the USW method, consisting of two receivers and an impact source. However, the measured amplitude of the surface waves is sensitive to the coupling condition of the sensors and the magnitude of impact force, which may cause significant errors in predicted values. It has been demonstrated that the self-calibrating procedure is effective for eliminating undesirable effects due to sensors and sources in the SWT method (Popovics et al. 2000). Recently, Kee and Zhu (2011, 2010) proposed the air-coupled sensing method, which significantly improves signal consistency and test speed in transmission measurements of surface waves in concrete.

2.3 Preparation of Piezoelectric Sensor

A piezoelectric sensor is an active element that converts electrical energy to mechanical energy, and mechanical energy to electrical energy (i.e., the piezoelectricity). The piezoelectricity causes a sensor to produce electric charges when subjected to stress (receiving action) and conversely, to generate mechanical vibrations when an electrical voltage is applied (actuator action). In this study, a piezoelectric disc is used as an actuator for generating incident surface waves in concrete.

A piezoelectric disc is one of the most widely used piezoelectric elements, as shown in Fig. 2. The thickness of a piezoelectric disc is much smaller than its lateral dimensions. The sensors used in this study are commercial piezoelectric warning devices, which generate sound when a voltage is applied continuously. One side of the piezoelectric element is attached to a metal plate (brass) for reinforcing the thin piezoelectric disc. A piezoelectric disc is polarized in the thickness direction. In the actuating mode, the disc expands/contracts in the thickness direction, i.e., along the axis of polarization, when a voltage is applied to the two surfaces of the ceramic disc, as shown in Fig. 2b. At the same time, the disc contracts/expands in the transverse direction. In the receiving mode, mechanical vibrations generate electric charges in the piezoelectric disc.

Two piezoelectric sensors were attached to the concrete surface to generate incident surface waves in concrete structures. The piezoelectric disc has a thickness of 0.2 mm (7.87 mils) and a diameter of 22 mm (866.14 mils). Two wires were then soldered to the electrodes on the piezoelectric disc. Since concrete surfaces are often under conductive environments, electrical shielding and waterproofing were needed. The waterproof procedure follows the description given by Jung (Jung 2005). First, five layers of polyurethane coating (M-coat A by VISHAY®) were applied to the surface of the piezoelectric discs. Each coating had to be fully air-dry before applying a subsequent one.

3. Experimental Program

3.1 Preparation of Specimens

Two concrete specimens were prepared in the laboratory to investigate the performance of the surface-mount sensor as an impact source in surface wave measurements. One concrete specimen (specimen 1) prepared in the laboratory had lateral dimensions of 1500 mm^2 and a thickness of 200 mm (see Fig. 3). A notch-type crack was made at the mid-section of the specimen by inserting a 0.5 mm thick Zinc sheet before pouring the concrete. The notch-type crack was designed to have depths increasing stepwise from 10 to 100 mm, at intervals of 10 mm (Fig. 3b). The Zinc sheet was coated by a thin plastic film to avoid chemical bonding between the Zinc sheet and concrete during the cement hydration process, and to facilitate the extraction of the sheet from the concrete. The Zinc sheet was removed about 6 h after casting, slightly earlier than the final setting of the

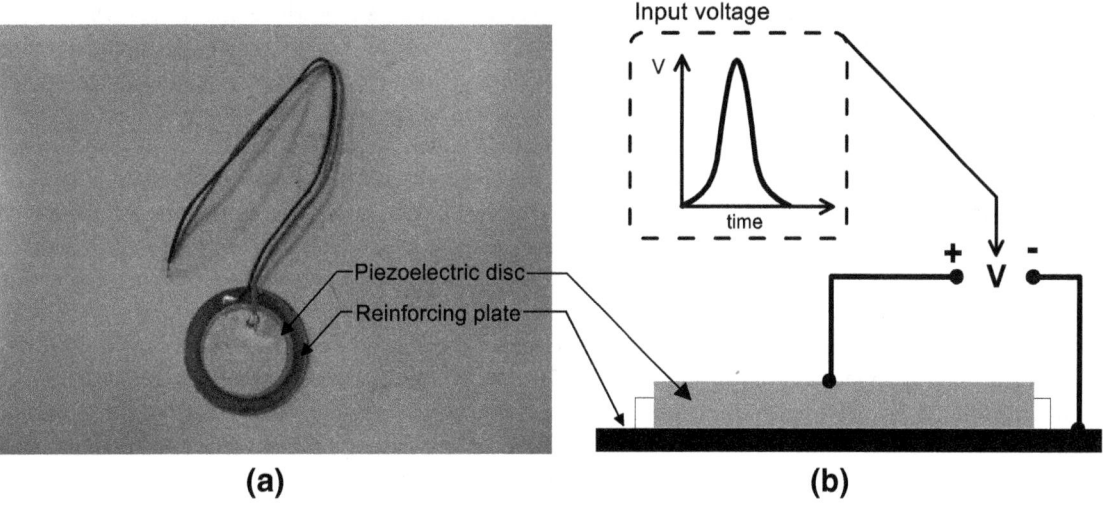

Fig. 2 Structure of piezoelectric disc sensor: **a** a photo of the piezoelectric disc, and **b** a piezoelectric disc connected to a voltage.

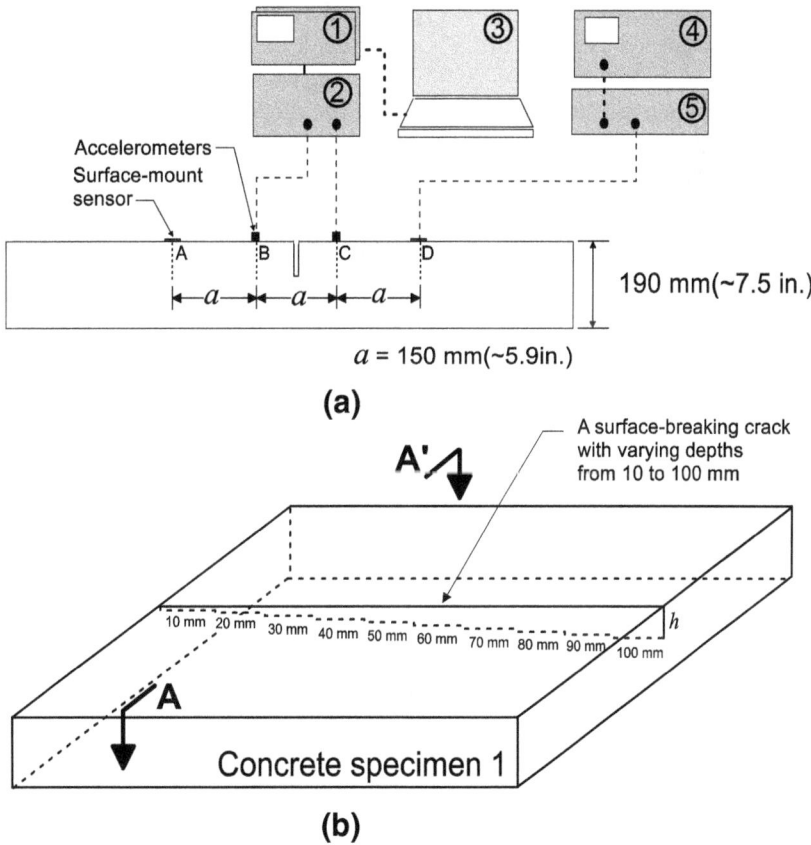

Fig. 3 Sectional view of a concrete slab (specimen 1), and data acquisition and signal generation systems for surface wave measurement using surface-mount sensors (**a**), and **b** isometric view of the concrete specimen 1.

concrete, based on observations from a series of preliminary tests before fabricating the actual specimen. Finally, the width of the crack in the hardened concrete specimen, measured with a crack width gauge after 7 days, was approximately 0.5 mm. The proposed surface-mount sensors were attached on either side of the notch crack (see Fig. 3a). Concrete material used for specimen 1 was normal weight, ready-mixed concrete made of type I/II cement, river sand, and coarse aggregate with a maximum size of 19 mm. The design compressive strength of concrete was 20 MPa. Compressive strength measured according to ASTM C39

(ASTM C39 2014) at the time of testing ranges from 22.4 to 24.3 MPs, with a mean value of 23.58 MPa. P wave velocities measured with a pair of 54 kHz ultrasonic transducers in the through-transmission mode were in the range of 4331 and 4386 m/s.

In addition, a concrete beam (specimen 2) with dimensions of 400 by 1500 mm and a thickness of 190 mm was prepared in the laboratory, as shown in Fig. 5. Concrete used for the specimen 2 was normal weight, ready-mixed concrete made from Type I/II cement, river sand, and coarse aggregate with a maximum size of 19 mm. The design compressive strength of the concrete was 20 MPa. Three cylinder specimens were used to measure concrete compressive strength according to ASTM C39 (ASTM C39 2014), resulting in a measured compressive strength (at the time of testing) ranging from 22.3 to 25.58 MPa, with a mean value of 22.84 MPa. P-wave velocities measured with a pair of 54 kHz ultrasonic transducers were in the range of 4328 and 4375 m/s.

Two layers of longitudinal reinforcing bars (13.3 mm diameter) were used for the top and bottom layers, respectively. A real surface-breaking crack was designed to appear in the middle of the concrete specimen by three-point bending (see Fig. 4a). Before applying external loadings, two proposed surface-mount piezoelectric sensors were attached to the concrete on either side of the expected crack location. To ensure generation of a single flexural crack in the middle of the concrete specimen, the reinforcing bars were unbonded to the concrete by wrapping a thin, 400 mm-long plastic film around the middle section of the reinforcing bars. After cracking, it is reasonable to assume that concrete in the crack section cannot provide any tensile strength, and only the top reinforcing bars participate in the load-resistance mechanism. Assuming a constant strain distribution in the unbonded steel reinforcing bars, shear stresses in the unbonded concrete region disappear after cracking, which prevents initiation of additional shear cracks or other flexural cracks in the middle of the concrete specimen. Consequently,

a single vertical surface-breaking crack will occur in the middle section of the specimen. In addition, transverse reinforcing bars (No. 3) were placed to avoid abrupt shear failure and to ensure flexural failure of the beam.

3.2 Surface Wave Measurements

The test setup for surface wave measurements using surface-mount sensors consists of a function generator, a power amplifier, a pair of surface-mount sensors and accelerometers (as sources and receivers, respectively), a digital oscilloscope, a computer, and LabVIEW program (see Figs. 4a, 5a). The two surface-mount sensors were located at A and D, and two accelerometers (PCB 352C65) were located at B and C on concrete specimens 1 and 2 (see Figs. 3a, 4a). A function generator (EXTX/2A60) was used to drive the surface-mount sensors. Stress waves are highly attenuated as they propagate through concrete. Therefore, a power amplifier (Trek PZD250) was used to amplify the signal generated by the function generator. Gaussian functions with a duration T of 100 μs were used as an input signal for the function generator. The acquired signals were digitized by a high-speed digital oscilloscope (NI-PXI 5101) at a sampling frequency of 1 MHz and total signal length of 0.001 s. The digitized data were then transferred to a laptop computer for data storage and post-processing. To eliminate the effects of experimental variations regarding sensor and source, the

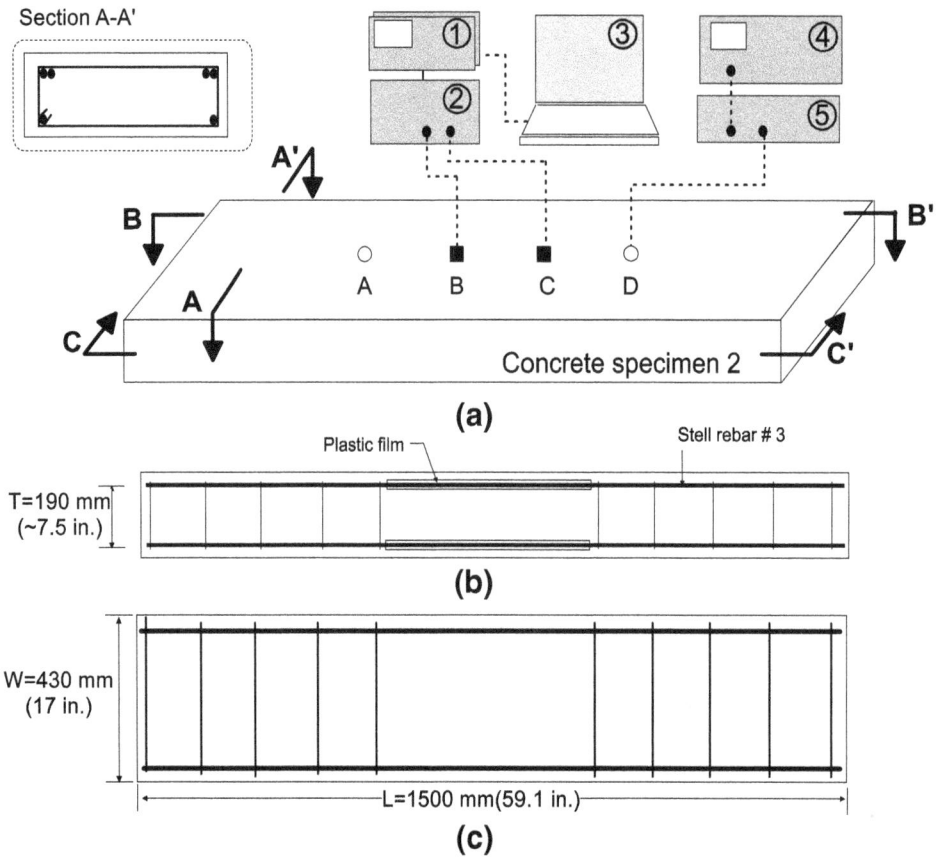

Fig. 4 Isometric view of a concrete beam (specimen 2), and data acquisition and signal generation systems for surface wave measurement using surface-mount sensors (**a**), sectional view of the specimen 2 (*B-B'*), and (**c**) sectional view of the specimen 2 (*C–C'*).

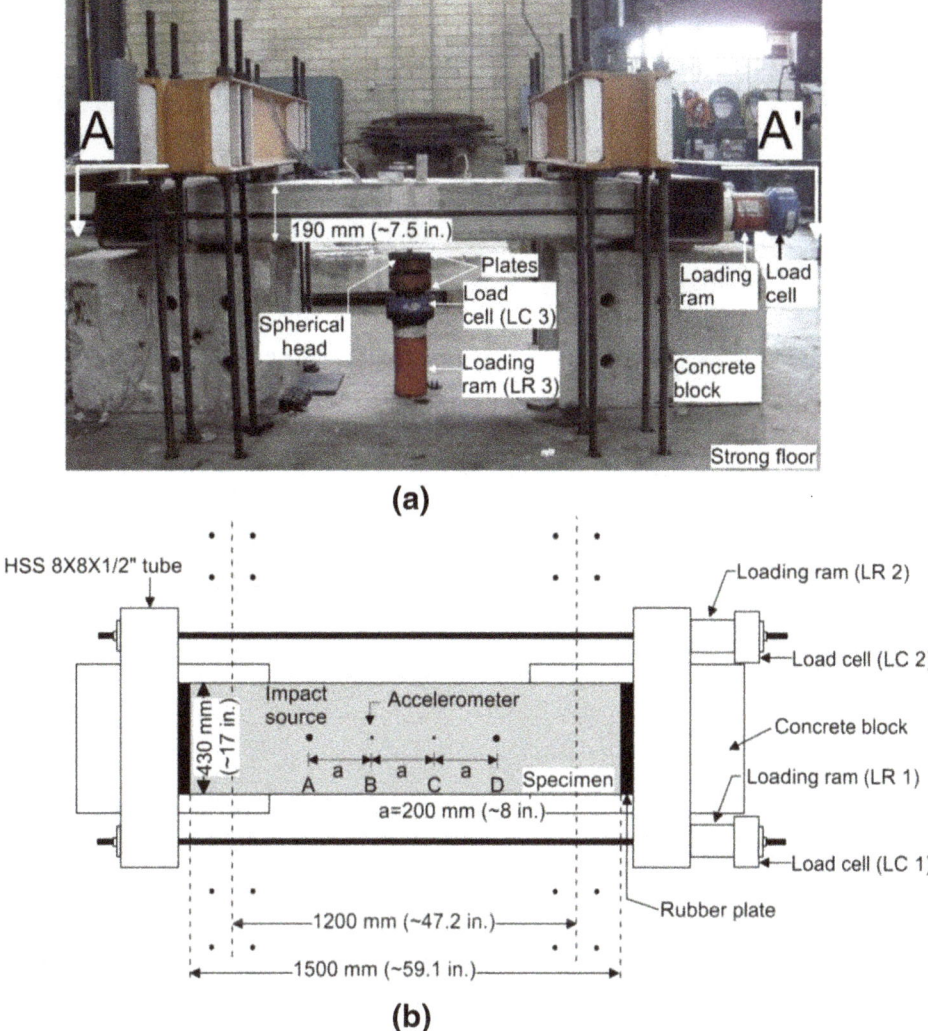

Fig. 5 Test setup of a concrete beam (specimen 2) for applying compressive forces and generating a single surface-breaking crack: **a** a photo of a test setup, and **b** plan view from the section *A-A'* shown in (**a**).

self-calibrating procedure (Popovics et al. 2000) was used to measure surface wave transmission and velocity. Two accelerometers were placed at locations B and C of the test specimens 1 and 2 to measure surface waves. First, the stress waves generated by the impact source at A propagated towards the sensor at B, and then towards the far sensor at C, denoted as $\mathbf{V}_{AB}$, and $\mathbf{V}_{AC}$, respectively. For example, the typical time signals $\mathbf{V}_{AB}$ and $\mathbf{V}_{AC}$ measured across a surface-breaking crack with a 20 mm depth in the specimen 1 are shown in Fig. 6a. Surface wave components were extracted from the full waveform in time domain by applying a hanning window (bold lines in Fig. 6a), and converted to frequency domain signals, denoted as $\mathbf{S}_{AB}$, and $\mathbf{S}_{AC}$ respectively. Similarly, stress waves generated by the surface-mount sensor at D were measured by accelerometers at C and B, denoted as $\mathbf{S}_{DC}$ and $\mathbf{S}_{DB}$. The frequency domain signals of $\mathbf{S}_{AB}$, $\mathbf{S}_{AC}$, $\mathbf{S}_{DC}$ and $\mathbf{S}_{DB}$ were used to calculate the transmission functions of surface waves propagating through the cracked region BC in concrete specimens ($\mathbf{S}_{AC}/\mathbf{S}_{AB}$ and $\mathbf{S}_{DB}/\mathbf{S}_{DC}$). The modulus (amplitude) and phase angle of the transmission functions are shown in Fig. 6b, c, respectively. In the cracked region, the phase angle of $\mathbf{S}_{AC}/\mathbf{S}_{AB}$ (or $\mathbf{S}_{DB}/$

$\mathbf{S}_{DC}$) almost linearly increases in a frequency range between 0 and about 45 kHz, in which the slope of the phase spectra is comparable to the theoretical value of the solid concrete with the surface wave velocity of 2200 m/s. In contrast, the modulus of $\mathbf{S}_{AC}/\mathbf{S}_{AB}$ (or $\mathbf{S}_{DB}/\mathbf{S}_{DC}$) decreases with increasing frequency, which is clearly differentiated from the theoretical curve of the solid concrete. Therefore, it can be seen that the modulus components of the transmission function are more informative of the presence and characteristics of surface-breaking cracks in concrete then the phase components.

In this study, the surface wave transmission ratio between B and C was calculated by averaging signals in the frequency domain obtained from opposite sides according to the self-calibrating procedure (Popovics et al. 2000) as follows,

$$\mathbf{Tr}_{BC} = \sqrt{\frac{\mathbf{S}_{AC}\mathbf{S}_{DB}}{\mathbf{S}_{AB}\mathbf{S}_{DC}}} \tag{1}$$

In this study, five repeated signal data sets were collected at the same test location to investigate the repeatability of signals generated by the surface-mount sensors. The transmission coefficient measured from cracked regions was further normalized by the reference, producing the

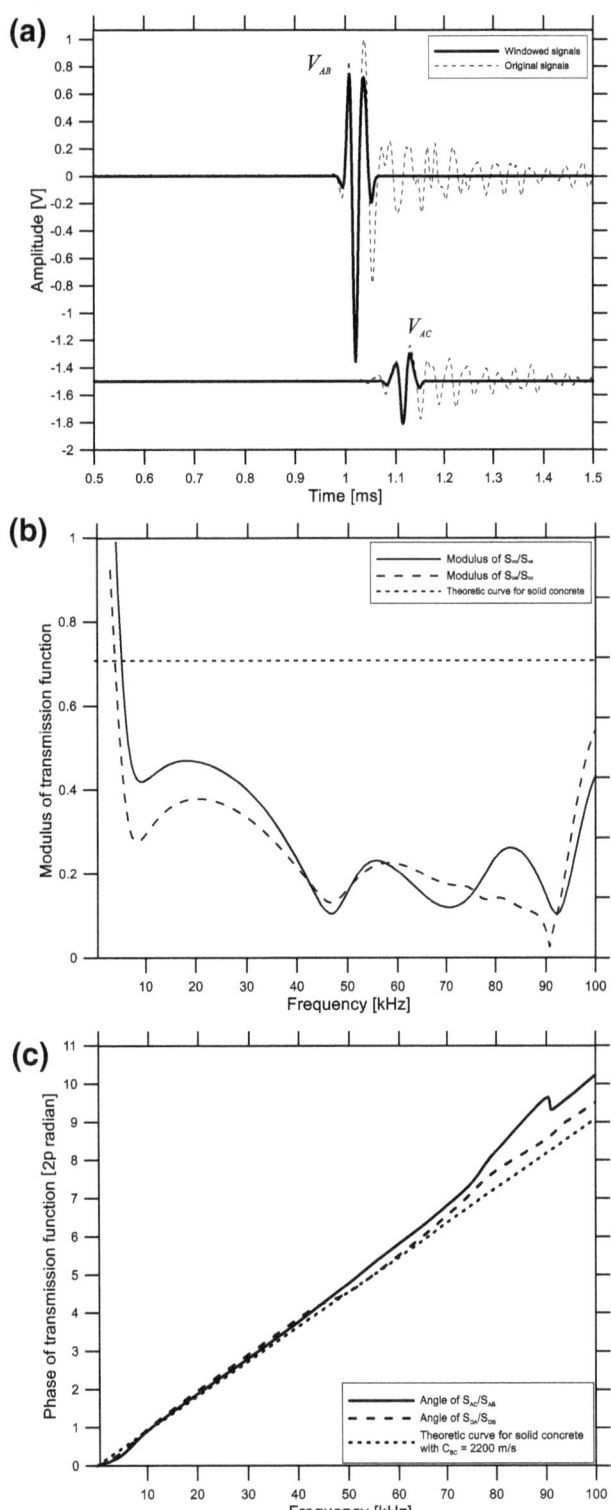

Fig. 6 Typical signals generated by the surface-mount piezo-electric transducers, measured using accelerometers across a surface breaking crack with a depth of 20 mm in the specimen 1: **a** time signals, **b**, **c** the modulus and phase angle of the transmission functions (S_{AC}/S_{AB} or S_{DB}/S_{DC}).

normalized transmission coefficient $\mathbf{Tr}_n$. It has been demonstrated that effects due to geometric attenuation and material damping can be effectively reduced by using the normalization process.

The phase velocity of the surface wave was calculated in the frequency domain by using the spectral analysis of surface waves (SASW). First, the phase difference between $\mathbf{S_{AB}}$ and $\mathbf{S_{AC}}$ by a source at A (ϕ_{BC}) and between $\mathbf{S_{DC}}$ and $\mathbf{S_{DB}}$ by a source at D (ϕ_{CB}) was calculated. Then the average phase velocity $\mathbf{C_{BC}}$ was calculated using the average phase difference as follows,

$$\mathbf{C_{BC}} = 2\pi f \frac{BC}{(\phi_{BC} + \phi_{CB})/2} \tag{2}$$

where BC is the distance between two sensors on the locations B and C in Figs. 3, 4, 5 (BC = 200 mm in this study).

4. Results and Discussion

4.1 Repeatability of Signals Generated by the Surface-Mount Sensors

In this study, the repeatability of measured signals generated by the piezoelectric sensors was evaluated by the coherence function as follows,

$$\gamma_j(f) = \frac{\left|\sum_{i=1}^{5} \mathbf{P}_{1j}^i(f)\right|}{\sqrt{\sum_{i=1}^{5} \mathbf{P}_{11}^i(f) \times \sum_{i=1}^{5} \mathbf{P}_{jj}^i(f)}} \tag{3}$$

where $\mathbf{P}_{1j}^i(f)$ is the cross-power spectrum of the reference signal and the signal measured at the jth test step; $\mathbf{P}_{11}^i(f)$ and $\mathbf{P}_{jj}^i(f)$ represent the auto-power spectrum of these signals; i is the index of the five repeated time domain signals; and f is frequency. The coherence function represents the degree of correlation between two signals as a function of frequency. The resulting γ ranges from 0 to 1.0, in which a value close to 1.0 indicates a good signal coherence. There are several factors that makes the coherence function less than one, including (i) measurements with incoherent noise, (ii) inconsistent coupling of source-and-receiver transducer and (iii) additional inputs in materials (i.e., additional scattering by any changes in internal defects).

Figure 7a shows the signal coherence γ of the measured signals generated by the surface-mount sensors on a crack-free region in specimen 1. For comparison purposes, the coherence curve of signals generated by a manual impact source (a steel ball of 13 mm diameter) is shown as a dash line in Fig. 6. It is demonstrated that the piezoelectric surface-mount sensors produce a very repeatable signal in a wideband frequency range of 0–100 kHz where $\gamma \geq 0.999$, a criterion for a useful frequency range in this study. In contrast, the measured signals generated by a manual impact source shows good signal consistency in the frequency range of 10–30 kHz: the useful frequency range depends on the diameter of the steel ball used for impact sources.

It is also observed that the presence of a surface-breaking crack decreases the magnitude of γ in high-frequency ranges: the useful frequency range becomes narrower as the depth of a crack increases from 0 to 50 mm (see Fig. 7b). However, it is not necessarily attributed to the coupling

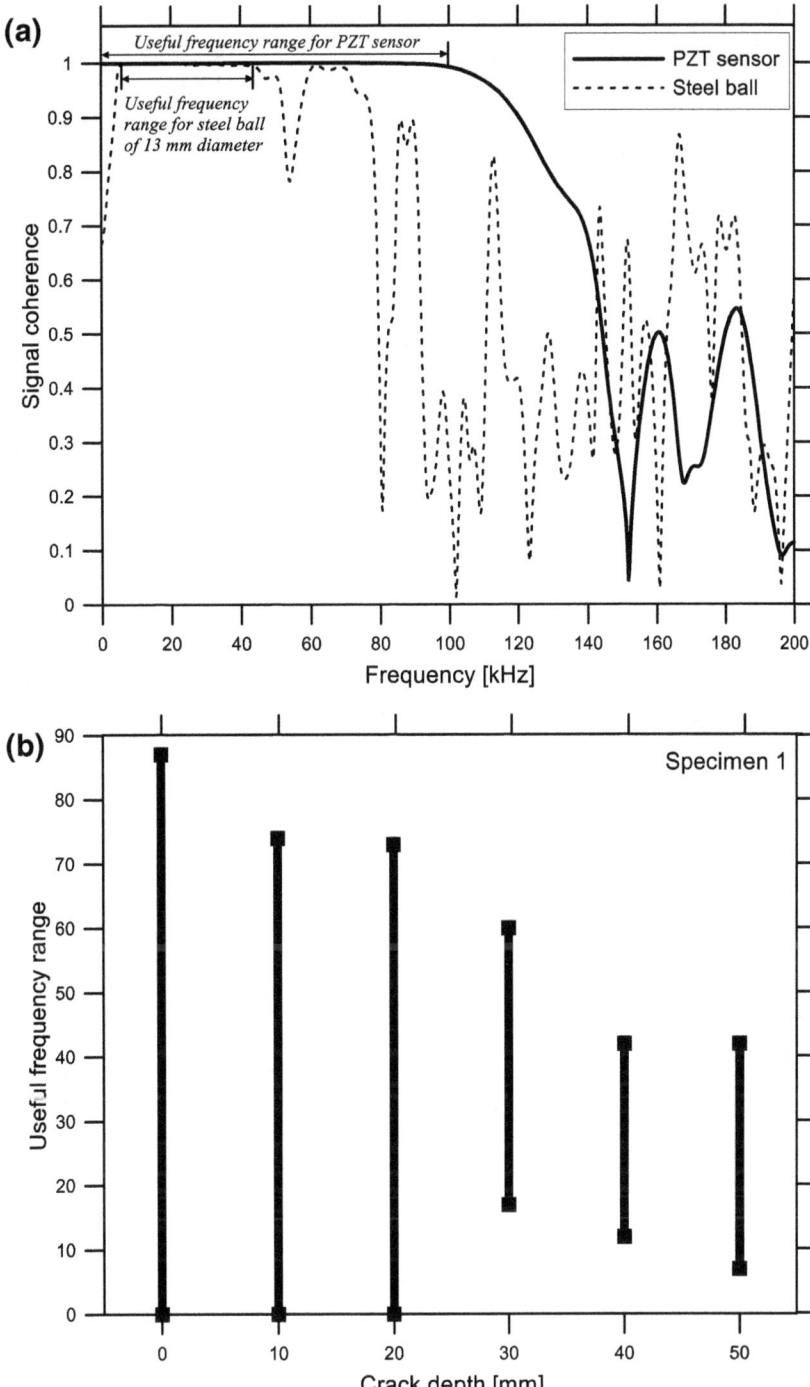

Fig. 7 Comparison of signal coherence generated by the surface-mount transducers and manual impact sources (**a**), and variations of the useful frequency range in a signal coherence function with increasing depth of the surface-breaking crack in concrete (**b**).

performance of the surface-mount sensor because the surface-breaking crack does not affect the surface-mount source and accelerometers. Instead, this phenomenon can be explained by the fact that the high-frequency components of surface waves are reflected back to the backward scattering field: consequently, the presence of a surface-breaking crack can significantly decrease the energy of transmitted surface waves in a higher frequency range.

4.2 Consistency of the Measured Surface Wave Parameters (Velocity and Transmission)

The consistency of measured results is of great interest when exploring the performance of surface-mount sensors. Ten repeated surface wave measurements were conducted at each test step using the same test setup and data acquisition system, in which surface-mount sensors located at A and D were alternatively used for generating incident waves in concrete. At each test step, the velocity and transmission coefficient of surface waves were determined by using Eqs. 1 and 2, respectively. In this study, the coefficient of variation (COV, the standard deviation divided by the mean value of a set of samples) was used as a means of consistency for the resulting surface wave velocity and transmission coefficients.

Figure 8a shows COVs of the two sets of surface wave transmission coefficients (Tr_{PZT} and Tr_{SB}) measured in the crack-free region of specimen 1. The COVs of Tr_{PZT} and Tr_{SB} remain low and flat, with values of 1.24 ± 0.39 % ($\mu \pm \sigma$) and 2.29 ± 0.65 % ($\mu \pm \sigma$) in a useful frequency range for each method (i.e., 0–100 kHz for Tr_{PZT} and 10 kHz to 30 kHz for Tr_{SB}), where μ and σ are the mean value and the standard deviation of COV, respectively. Similarly, the COVs of the phase velocity of surface waves ($C_{R,PZT}$ and $C_{R,SB}$) are shown in Fig. 8b. The COVs of $C_{R,PZT}$ and $C_{R,SB}$ are 0.04 ± 0.023 % ($\mu \pm \sigma$) and 0.3 ± 0.15 %, respectively, in the useful frequency range of each method. Therefore, Fig. 8 illustrates that the surface-mount sensors produce equivalent or improved consistency of surface wave transmission and velocity measurements compared to the measurements using a manual impact source.

Furthermore, it is observed that the presence of a surface-breaking crack may affect the COV of both the Tr_{PZT} and C_{RPZT} (Fig. 9). For specimen 1, the COVs of Tr_{PZT} and C_{RPZT} gradually increase up to 5 and 4 %, compared to reference values measured on the crack-free region where crack depth increases from 0 to 50 mm. For test specimen 2, the COVs of Tr_{PZT} and $C_{R,PZT}$ at 20 kHz in the crack-free surface are 1.2 and 0.01 %, respectively. However, onset of a surface-breaking crack in the middle of the concrete (a depth of about 150 mm) significantly increases the COVs of Tr_{PZT} and $C_{R,PZT}$ to about 10 and 3 %, respectively. It

appears that higher variability in cracked concrete is due to a combination of two reasons: (i) energy loss of transmitted surface waves due to a surface-breaking crack, and (ii) scattering of the surface wave at the tip of the crack. However, as will be demonstrated in the next section, the COV levels of Tr_{PZT} and $C_{R,PZT}$ are still acceptable compared to the sensitivity of each parameter to the presence and severity of a surface-breaking crack Fig. 10.

4.3 Sensitivity of the Measured Surface Wave Parameters to the Depth of a Surface-Breaking Crack

Figure 10 shows the relationship between the normalized transmission coefficient of surface waves $Tr_{PZT,n}$ and the normalized crack depth, h/λ. The results in Fig. 10 were obtained from the surface wave measurements across five surface-breaking cracks with depths of 10, 20, 30, 40, and 50 mm, respectively. Shown in Fig. 10 are only twelve transmission coefficient values for each crack depth in a useful frequency range (i.e., from 20 kHz to 80 kHz with intervals of 5 kHz). Therefore, 60 dots are shown as solid circles in Fig. 10. The approximate expression that describes the relationship between $Tr_{PZT,n}$ and h/λ is established by a non-linear regression of the experimental data as follows, and shown as a dash line in Fig. 10.

$$tr_n = 0.1515 \, (h/\lambda)^{-0.7152} \qquad 0.1 \le h/\lambda \le 1.4 \qquad (4)$$

In comparison, a theoretical model obtained from a series of numerical simulations (FEM) is shown as a solid line in the same figure.

In a surface wave transmission test, many transmission values can be obtained within a frequency range; thus, multiple redundant estimates of crack depth may be calculated from a single measurement. In this study, the depth of a

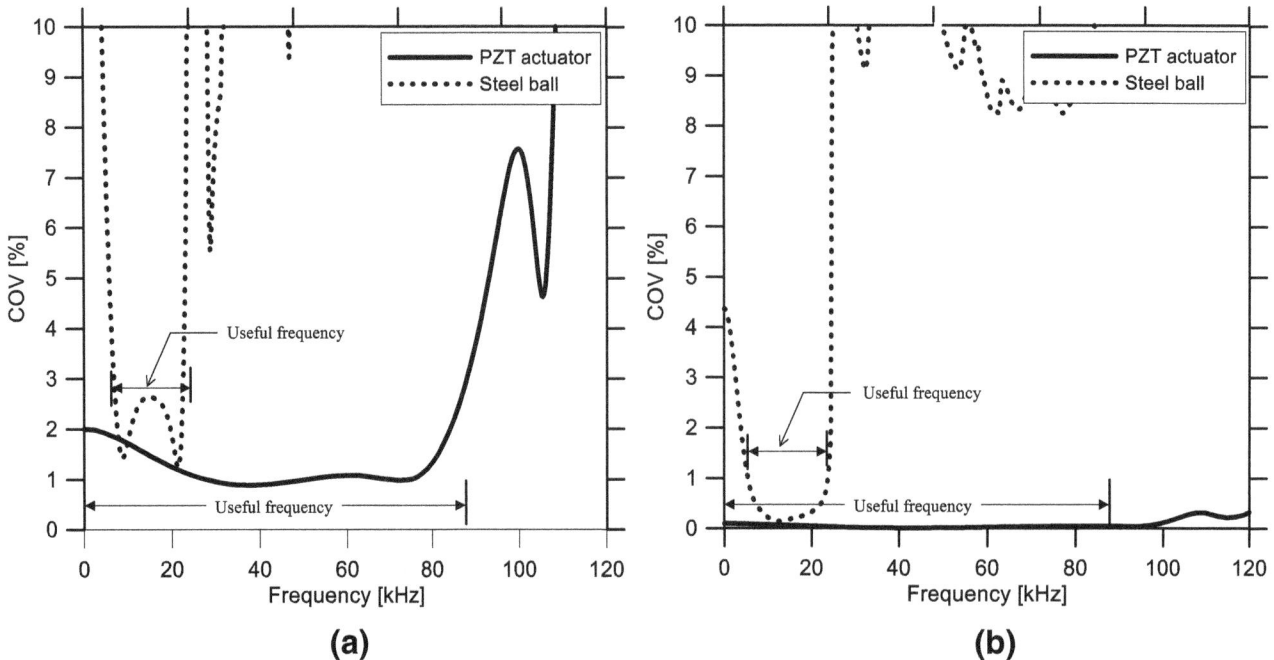

Fig. 8 The coefficient of variation of the surface wave transmission and velocity.

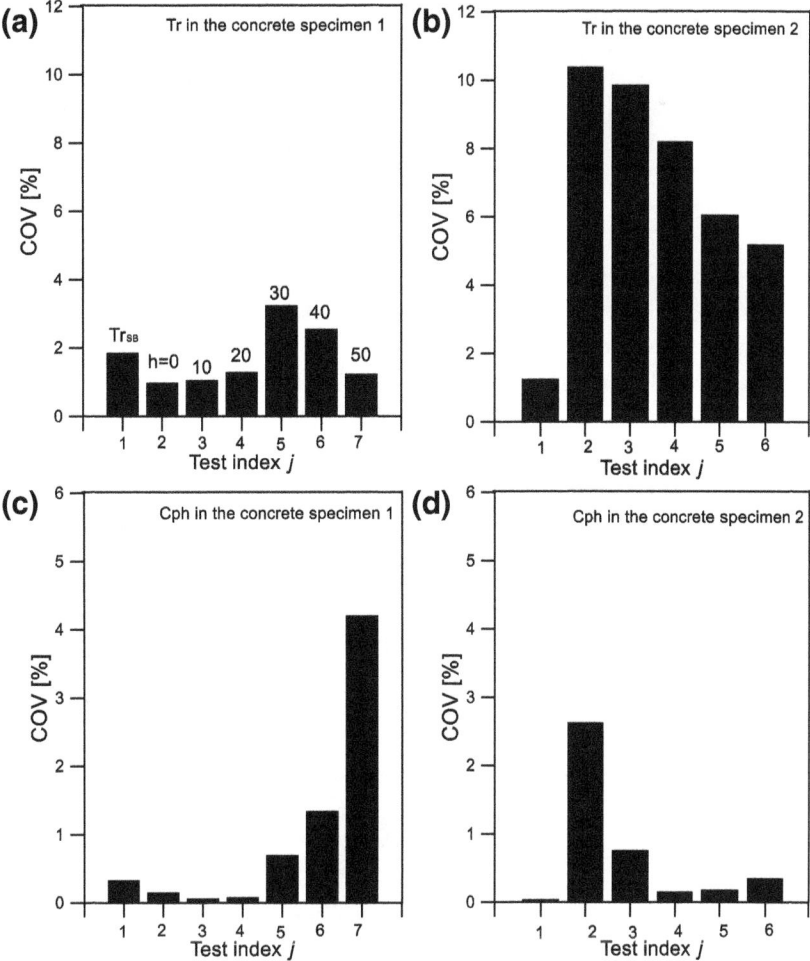

Fig. 9 The coefficient of variation (COV) of the surface wave transmission Tr_{PZT} and velocity C_{RPZT}: **a**, **b** COV of Tr_{PZT} measured in concrete specimens 1 and 2, respectively, **b** COV of C_{RPZT} measured in concrete specimens 1 and 2, respectively.

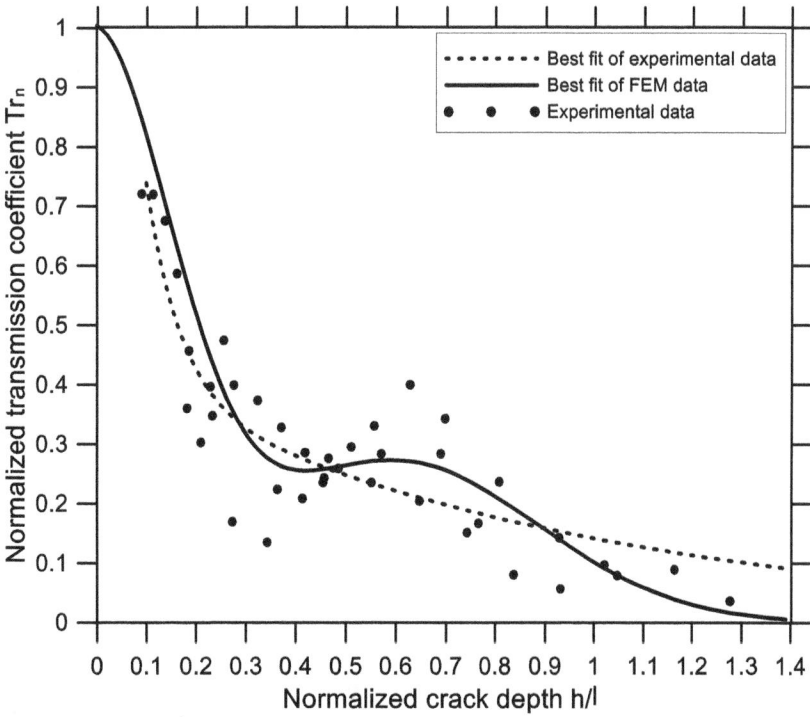

Fig. 10 Normalized transmission coefficient $Tr_{PZT,n}$ versus normalized crack depth h/λ.

surface-breaking crack was determined by using the least square method. The optimum depth result was determined to minimize the sum of square residuals of the transmission function (SSR),

$$SSR = \sum_{i=1}^{N} \left(\frac{tr_n(f_i, h/\lambda_i) - tr'_n(f_i)}{tr_n(f_i, h/\lambda_i)} \right)^2 \qquad (5)$$

where tr_n is the transmission ratio in the proposed calibration curve in Eq. (4), Tr_n, is the measured transmission ratio calculated using Eq. (1), i is an index of input values, and f_i and λ_i are the frequency and wavelength with the index i. As a result, the crack depths estimated for the notch-type cracks (using the surface wave transmission measurement h_{tr}) were 8, 18, 26, 42 and 48 mm, respectively. The estimated values are about 80–90 % compared to the as-built crack depths. Therefore, the proposed model can be effective for evaluating the depth of a surface-breaking crack in concrete.

4.4 Effect of Crack Depth on Surface Wave Velocity Measurement

Figure 11 shows the variation in surface wave velocity at 20 kHz with increasing crack depths obtained from numerical simulation and experiment in this study. It was observed that C_R obtained from numerical simulation remains stable around the reference velocity of 2250 m/s until crack depth is less than about 120 mm. For the experimental data, the surface wave velocity at 20 kHz was 2208 m/s in the crack free-region. Increasing the crack depth up to 50 mm does not affect the surface wave velocity at 20 kHz. This observation is consistent with observations by previous researchers (Masserey and Mazza 2007) that the surface wave velocity is only sensitive to a crack deeper than about 80 % of the wavelength of surface waves. Compared to the surface wave transmission coefficient, the phase velocity is less sensitive to the presence of cracks. However, it is reasonable to say that some degradation of the surface wave velocity (about

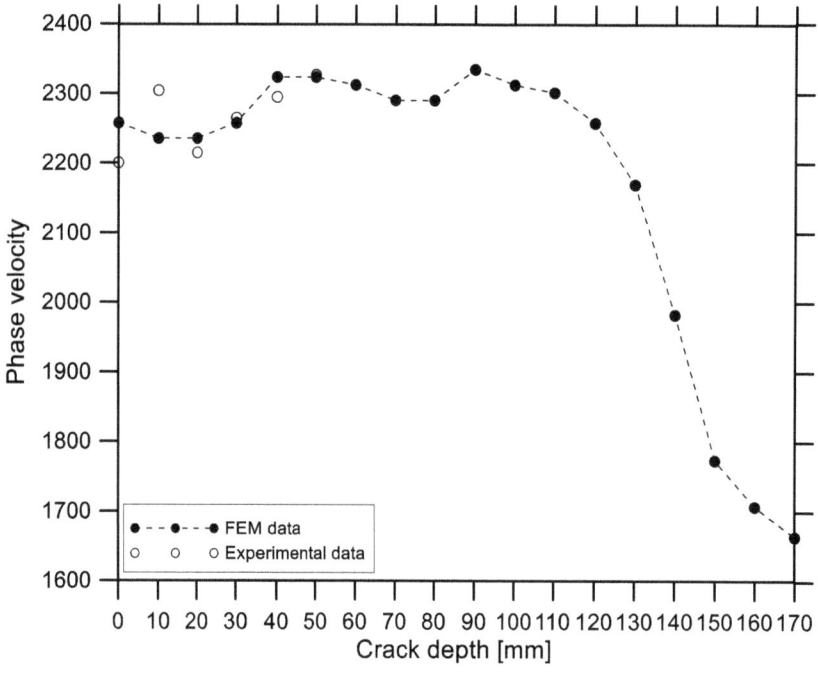

Fig. 11 Phase velocity of surface waves C_R versus crack depth h.

Table 1 Loading history in test stages 1, 2, and 3 for specimen 2

Test stage 1		Test stage 1		Test stage 1	
Test steps i	Loading P_1 [kips (kN)]	Test steps i	Loading P_2 [kips (kN)]	Test steps i	Loading P_3 [kips (kN)]
0	0	7	1 (4.48)	12	0
1	4 (17.79)	8	2 (8.89)	13	4 (17.79)
2	8 (35.58)	9	3 (13.34)	14	8 (35.58)
3	12 (53.37)	10	3.5 (15.56)	15	12 (53.37)
4	16 (71.17)	11[a]	–	16	16 (71.17)
5	20 (88.96)			17	20 (88.96)
6	24 (106.75)			18	24 (106.75)

[a] Onset of cracking.

10 %) across a surface-breaking crack is evidence of a deep crack, comparable to the wavelength of surface waves.

4.5 Application to Monitoring a Concrete Beam under Various Loadings

Described in this section is the application of the surface-mount sensor to concrete subjected to various loadings, which are common in actual concrete structures. Surface wave transmission and velocity were monitored in a stepwise manner in the various test steps described in Table 1. Figure 12a, b are plots illustrating the variations in the normalized transmission coefficients and phase velocity at a frequency of 20 kHz, respectively, in different test steps.

Before cracking, there was only a slight change in Tr_n and Cr_R with increasing compressive loadings up to 106.75 kN (24 kips). Furthermore, with increasing tensile stress up to 3 MPa (420 ksi), which corresponds to a cracking moment of the concrete specimen, both Tr_n remain almost constant (± 2 % of the reference Tr_n) until the onset of the first surface-breaking crack on the top surface of specimen 2. The crack depth measured on the side surfaces of the concrete specimen was about 120 mm, which shows close agreement with the depth measured from a core sample taken at the surface after testing (see Fig. 13). In addition, C_R tends to decrease slowly with increasing compressive and bending loadings, and exhibit small variations (see Fig. 12b). Once a crack appears, Tr_n and C_R suddenly decrease to about 10 and 85–90 % of the reference values, respectively. It appears that Tr_n is much more sensitive to the presence of a surface-breaking crack in concrete than C_R. In summary, monitoring Tr_n and C_R can be effective for identifying the onset of a surface-breaking crack in concrete, and the approximate

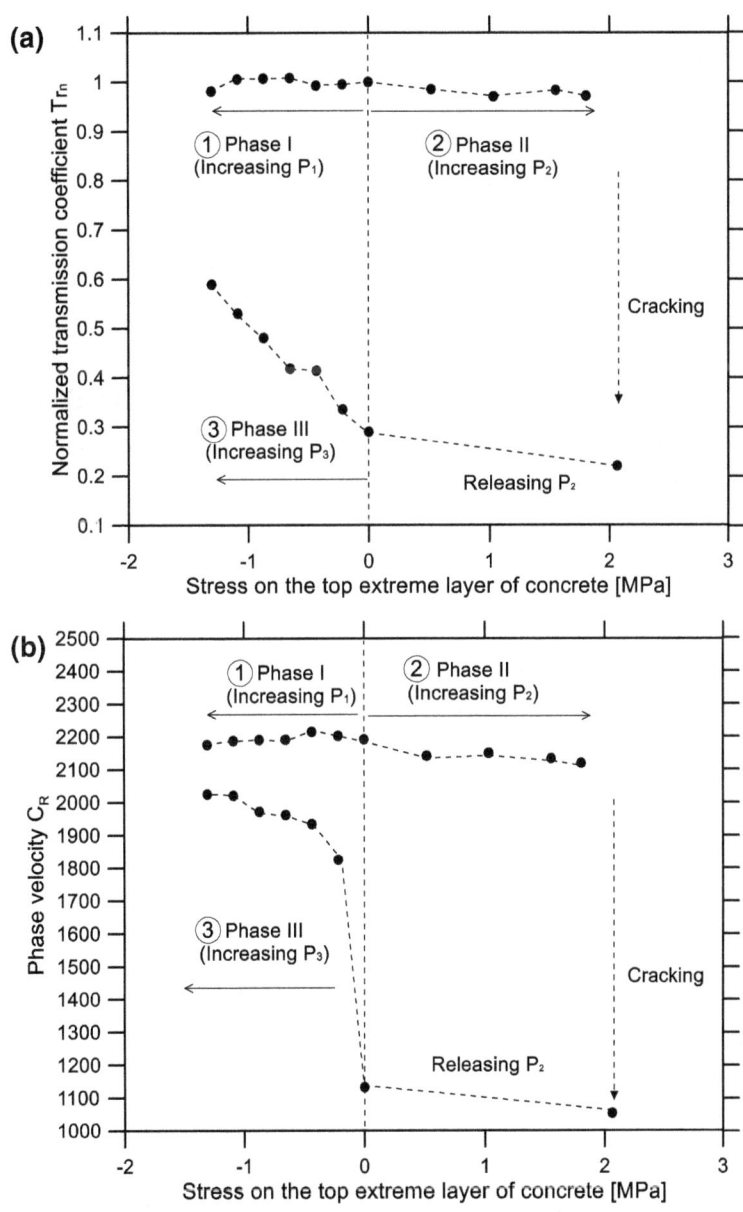

Fig. 12 Variation of transmission coefficient and phase velocity of surface waves at 20 kHz with stress on top extreme layer of concrete: **a** normalized transmission coefficient versus stress, and **b** phase velocity versus stress.

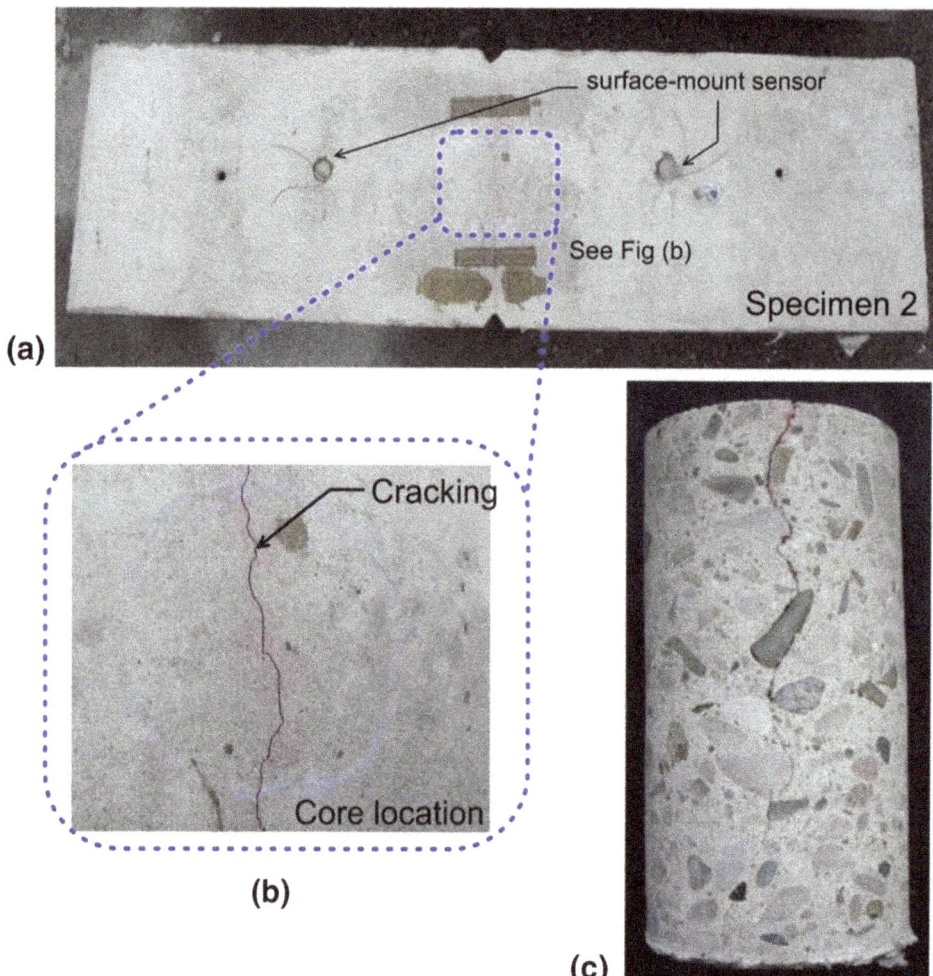

Fig. 13 Core samples extracted after completing tests from the test specimen 2: **a** concrete specimen 2 after testing, **b** location of core extraction, and **c** core sample.

depth can be estimated by using the SWT method. In addition, this process can be automated by using surface-mount sensors.

However, the effect of external loadings may pose difficulties in the interpretation of Tr_n and C_R. As shown in Fig. 12, Tr_n and C_R increase with an increase in the application of external compression P_3. At the last loading step of P_3, Tr_n and C_R were recovered to 90 and 95 %, respectively, of the values before cracking. Some portions of the incident surface waves (i.e., crack interfacial waves) are transmitted through the interface of an actual crack, which commonly has a partially closed interface. Increasing the compressive force gradually closes the concrete crack, increasing the interfacial stiffness of the crack. It was observed that both Tr_n and C_R are enhanced, owing to the crack interfacial waves, which may lead to substantial errors in predicting the depth of a surface-breaking crack in actual concrete structures.

One interesting finding is related to the potential for combining the results from the SWT and SASW as a more reliable crack depth estimation approach for testing actual structures. As observed in the theoretical results, the surface wave velocity is only sensitive to a crack deeper than about

80 % of the wavelength of surface waves. In addition, it was observed that the surface wave velocity is less sensitive to the interfacial stiffness of a surface-breaking crack than the surface wave transmission. Therefore, it is reasonable to say that some degradation of the surface wave velocity (about 10 %) across a surface-breaking crack is evidence of a deep crack, compared to the wavelength of surface waves.

5. Conclusions

In this paper, experimental results are presented to investigate the applicability of piezoelectric sensors as a consistent impact source for automated surface wave measurements in concrete structures. The conclusions are summarized as follows,

(1) The piezoelectric surface-mount sensors produce excellent signal coherence $\gamma \geq 0.999$ in a wideband frequency range from 0 to 120 kHz. In contrast, signals generated by manual impacts have good signal consistency in a narrower frequency range of 10–30 kHz, and are dependent on the diameter of the steel ball. It was observed that the surface-mount

sensors have consistent coupling under various stress states and damage levels of concrete.

(2) Experimental variability in the transmission and velocity measurements measured using the piezoelectric surface-mount sensors are equivalent or less than that from manual impacts. The COVs of the Tr_{PZT100} and $C_{R,PZT100}$ are $1.24 \pm 0.39\%$ ($\mu \pm \sigma$), and $0.04 \pm 0.023\%$, respectively, in a useful frequency range of 0–120 kHz. In contrast, the COVs of the Tr_{SB} and $C_{R,SB}$ are $2.29 \pm 0.65\%$ ($\mu \pm \sigma$), and $0.3 \pm 0.15\%$, respectively, in a useful frequency range of 10 kHz to 30 kHz.

(3) The proposed model for the surface wave transmission coefficient is demonstrated to be effective for evaluating the depth of a surface-breaking crack in concrete. However, special care is needed to apply the surface wave transmission method to a partially closed crack in actual concrete structure because of interference of the crack interfacial waves with transmitted surface waves.

(4) Results from experiments and numerical simulations show that the surface wave velocity is only sensitive to cracks deeper than about 80 % of the wavelength of surface waves. However, it is reasonable to say that some degradation of the surface wave velocity (about 10 %) across a surface-breaking crack is evidence of a deep crack comparable to the wavelength of surface waves. Therefore, a fusion of the results from the SWT and SASW can be used as a more reliable crack depth estimation approach for testing actual structures.

Acknowledgments

This research was supported by a grant (15DRP-B066470-03) from Infrastructure and transportation technology promotion research Program funded by Ministry of Land, Infrastructure and Transport of Korean government.

References

Achenbach, J. D. (2000). Quantitative nondestructive evaluation. *International Journal of Solids and Structures, 37*(1–2), 13–27.

Achenbach, J. D. (2002). Modeling for quantitative non-destructive evaluation. *Ultrasonics, 40*(1–8), 1–10.

Achenbach, J. D., Keer, L. M., & Mendelsohn, D. A. (1980). Elastodynamic analysis of an edge crack. *Journal of Applied Mechanics, 47*(3), 551–556.

ACI Committee 228 (1998). Nondestructive test methods for evaluation of concrete in structures. *Report ACI 228.2R-98*, American Concrete Institute, Farmington Hills, MI.

Angel, Y. C., & Achenbach, J. D. (1984). Reflection and transmission of obliquely incident Rayleigh waves by a surface-breaking crack. *The Journal of the Acoustical Society of America, 75*(2), 313–319.

ASTM C39. (2014). *Standard test method for compressive strength of cylindrical concrete specimens*. West Conshohocken: ASTM International.

Dong, B., Xing, F., & Li, Z. (2011). Cement-based piezoelectric ceramic composite and its seosor applications in civil engineeriing. *ACI Materials Journal, 108*(5), 543–549.

ElSafty, A., & Abdel-Mohti, A. (2013). Investigation of likelihood of cracking in reinforced concrete bridge decks. *International Journal of Concrete Structures and Materials, 7*(1), 79–93.

Graff, K. (1991). *Wave motion in elastic solid*. New York: Dover Publications.

Gucunski, N., Imani, A., Romero, F., Nazarian. S., Yuan, D., Wiggenhauser, H., et al. (2013). Nondestructive testig to identify concrete bridge deck deterioration. *SHRP 2 Report S2-R06A-RR-1*.

Hevin, G., Abraham, O., Petersen, H. A., & Campillo, M. (1998). Characterization of surface cracks with Rayleigh waves: A numerical model. *NDT and E International, 31*(4), 289–298.

Hou, S., Zhang, H. B., & Ou, J. P. (2012). A PZT-based smart aggregate for compressive seismic stress monitoring. *Smart Materials and Structures, 21*, 105035.

Hou, S., Zhang, H. B., & Ou, J. P. (2013). A PZT-based smart aggregate for seismic shear stress monitoring. *Smart Materials and Structures, 22*, 065012.

Jung, M. J. (2005). *Shear wave velocity measurements of normally consolidated kaolinite using bender elements*. Master of Science in Engineering, The University of Texas at Austin, Austin.

Kee, S.-H. (2011). *Evaluation of crack-depth in concrete using non-contact surface wave transmission measurement*. Doctor of Philosophy, The University of Texas at Austin, Austin, TX.

Kee, S.-H., & Zhu, J. (2010). Using air-coupled sensors to determine the depth of a surface-breaking crack in concrete. *The Journal of the Acoustical Society of America, 127*(3), 1279–1287.

Kee, S.-H., & Zhu, J. (2011). Effects of sensor locations on air-coupled surface wave transmission measurements. *Ultrasonics, Ferroelectrics and Frequency Control, IEEE Transactions on, 58*(2), 427–436.

Kee, S.-H., & Zhu, J. (2013). Using piezoelectric sensors for ultrasonic pulse velocity measurements in concrete. *Smart Materials and Structures, 22*(11), 115016.

Liao, W. I., Wang, J. X., Song, G., Gu, H., Olmi, C., Mo, Y. L., et al. (2011). Structural health monitoring of concrete columns subjected to seismic excitations using piezoceramic-based sensors. *Smart Materials and Structures, 20*(12), 125015.

Masserey, B., & Mazza, E. (2007). Ultrasonic sizing of short surface cracks. *Ultrasonics, 46*(3), 195–204.

McLaskey, G. C., & Glaser, S. D. (2010). Hertzian impact: Experimental study of the force pulse and resulting stress waves. *Journal of the Acoustical Society of America, 128*(3), 1087–1096.

Mendelsohn, D. A., Achenbach, J. D., & Keer, L. M. (1980). Scattering of elastic waves by a surface-breaking crack. *Wave Motion, 2*(3), 277–292.

Nazarian, S., & Desai, M. R. (1993). Automated surface wave method: Filed testing. *Journal of Geotechnical Engineering, ASCE, 119*(7), 1094–1111.

Nazarian, S., & Stokoe, K. H., II (1986). In-situ determination of elastic moduli of pavement systems by spectral-analysis-of-surface-wave method (practical aspects). *Research Report 368-1F*, University of Texas at Austin, Center for Transportation Research.

Okafor, A. C., Chandrashekhara, K., & Jiang, Y. P. (1996). Delamination prediction in composite beams with built-in piezoelectric devices using modal analysis and neural network. *Smart Materials and Structures, 5*(3), 338–347.

Popovics, J. S., Song, W.-J., Ghandehari, M., Subramaniam, K. V., Achenbach, J. D., & Shah, S. P. (2000). Application of surface wave transmission measurements for crack depth determination in concrete. *ACI Materials Journal, 97*(2), 127–135.

Shin, S. W., Zhu, J., Min, J., & Popovics, J. S. (2008). Crack depth estimation in concrete using energy transmission of surface waves. *ACI Materials Journal, 105*(5), 510–516.

Soltani, A., Harries, K. A., & Shahrooz, B. M. (2013). Crack opening behavior of concrete reinforced with high strength reinforcing steel. *International Journal of Concrete Structures and Materials, 7*(4), 253–264.

Song, G. B., Gu, H. C., & Mo, Y. L. (2008). Smart aggregates: multi-functional sensors for concrete structures—a tutorial and a review. *Smart Materials and Structures, 17*(3), 033001.

Song, G., Gu, H., Mo, Y. L., Hsu, T. T. C., & Dhonde, H. (2007). Concrete structural health monitoring using embedded piezoceramic transducers.". *Smart Materials and Structures, 16*(4), 959–968.

Song, G., Mo, Y. L., Otero, K., & Gu, H. (2006). Health monitoring and rehabilitation of a concrete structure using intelligent materials. *Smart Materrials & Structures, 15*(2), 309–314.

Song, W.-J., Popovics, J. S., Aldrin, J. C., & Shah, S. P. (2003). Measurement of surface wave transmission coefficient across surface-breaking cracks and notches in concrete. *The Journal of the Acoustical Society of America, 113*(2), 717–725.

Wang, C. S., Wu, F., & Chang, F. K. (2001). Structural health monitoring from fiber-reinforced composites to steel reinforced concrete. *Smart Materials and Structures, 10*(3), 548–552.

4

Collapse Vulnerability and Fragility Analysis of Substandard RC Bridges Rehabilitated with Different Repair Jackets Under Post-mainshock Cascading Events

Mostafa Fakharifar[1], Genda Chen[2],*, Ahmad Dalvand[3], and Anoosh Shamsabadi[4]

Abstract: Past earthquakes have signaled the increased collapse vulnerability of mainshock-damaged bridge piers and urgent need of repair interventions prior to subsequent cascading hazard events, such as aftershocks, triggered by the mainshock (MS). The overarching goal of this study is to quantify the collapse vulnerability of mainshock-damaged substandard RC bridge piers rehabilitated with different repair jackets (FRP, conventional thick steel and hybrid jacket) under aftershock (AS) attacks of various intensities. The efficacy of repair jackets on post-MS resilience of repaired bridges is quantified for a prototype two-span single-column bridge bent with lap-splice deficiency at column-footing interface. Extensive number of incremental dynamic time history analyses on numerical finite element bridge models with deteriorating properties under back-to-back MS-AS sequences were utilized to evaluate the efficacy of different repair jackets on the post-repair behavior of RC bridges subjected to AS attacks. Results indicate the dramatic impact of repair jacket application on post-MS resilience of damaged bridge piers—up to 45.5 % increase of structural collapse capacity—subjected to aftershocks of multiple intensities. Besides, the efficacy of repair jackets is found to be proportionate to the intensity of AS attacks. Moreover, the steel jacket exhibited to be the most vulnerable repair intervention compared to CFRP, irrespective of the seismic sequence (severe MS-severe or moderate AS) or earthquake type (near-fault or far-fault).

Keywords: earthquake damage, rapid repair, RC column, performance-based seismic engineering.

1. Introduction

Past earthquakes have demonstrated catastrophic structural failure of mainshock-damaged structures subjected to cascading events, defined as events likely to be triggered by the main earthquake, such as aftershocks, explosions and tsunamis (Li et al. 2014; Ribeiro et al. 2014). This study is focused on multiple earthquakes consisting of the mainshock (MS) and aftershocks. Multiple earthquakes occur after the first ground shaking, where all the accumulated strain in the fault system is not relieved at the onset of the first rupture of the faulted area. Thus, sequential ruptures take place until the fault system is completely stabilized (Abdelnaby and Elnashai 2014; Li et al. 2014; Di Sarno 2013). The sequential fault ruptures after the first earthquake, causes multiple earthquakes which are often referred to as foreshock (FS), MS and aftershock (AS) events. The March, 2011 Tohuku, Japan subduction earthquake with a moment magnitude (M_w) of 9 was succeeded by hundreds of aftershocks including at least thirty aftershocks greater than M_w 6 (USGS 2012). While the aftershocks are usually smaller in magnitude compared to the MS, however, aftershocks may have higher intensity, with a longer effective duration; different spectral shape, energy content; and specific energy density compared to the MS (Nazari et al. 2013; Li et al. 2014). The large subduction earthquakes, such as the Maule-Chile earthquake (Feb. 2010, M_w 8.8), Sumatra–Andaman earthquake (Dec. 2004, M_w 9.2) and Tokachi-Oki earthquake (Sep. 2003, M_w 8.1), feature very long duration earthquakes consisting of series of foreshocks, mainshock and aftershocks (USGS 2012; Yang 2009). Even for the modern constructions, the shallow crustal earthquakes (including single earthquake event) are only included for the design and the subduction earthquakes are not considered (White and Ventura 2004; Yang 2009).

MS-damaged structures are vulnerable to aftershocks even when only minor damage is present from the MS (Li et al. 2012, 2014; Di Sarno 2013; Abdelnaby and Elnashai 2014).

[1] Department of Civil, Architectural and Environmental Engineering, Missouri University of Science and Technology, Rolla, MO 65409, USA.

[2] Missouri University of Science and Technology, Rolla, MO 65409, USA.

*Corresponding Author; E-mail: gchen@mst.edu

[3] Department of Engineering, Lorestan University, Khorramabad, Iran.

[4] Office of Earthquake Engineering, California Department of Transportation, Sacramento, CA 95816, USA.

The May 12, 2008 Wenchuan earthquake with a M_w of 7.9 was followed by numerous aftershocks, where the August 5, 2008 aftershock (M_w larger than 6) (USGS 2012) contributed to collapse of many of the structures sustaining damage from the MS event, and causing more loss of life (about 90,000 dead or missing) (Li et al. 2012; Yang 2009). While the delay between a MS and the largest AS could range from minutes to months and it is difficult to predict, however magnitudes of AS are relatively easy to predict (Scholz 2002; Li et al. 2012). Bridge columns are one of the most vulnerable structural elements during seismic events (Priestley et al. 1996; Buckle et al. 2006; Fakharifar et al. 2013, 2014a; Lin et al. 2013). Subsequently, rapid repair of the damaged bridge columns, specifically substandard columns (insufficient confinement in the plastic hinge and longitudinal reinforcement lap-spliced at column-footing interface) after the first seismic event is vital for the post-MS integrity and collapse prevention of the bridge under subsequent aftershocks. Different rehabilitation methods for RC bridge columns, utilizing thin steel jackets (typical 0.5–1.2 mm thick) (Ying et al. 2006), carbon fiber reinforced polymer (CFRP) wraps, conventional thick steel jackets (typical 10–50 mm) (Priestley et al. 1996), reinforced concrete jackets, prestressing strands (PS) and shape memory alloy (SMA) wraps have been investigated in different studies (Ying et al. 2006; Priestley et al. 1996; Buckle et al. 2006; Vosooghi and Saiidi 2012; Fakharifar et al. 2014a, c, 2015a; Andrawes et al. 2009). Shear deficient columns are excluded from this study, as the shear deficient columns require rehabilitation prior to any seismic event.

The post-MS load carrying capacity of earthquake damaged structures depends on the extent of damage present (Aschheim and Black 1999; Terzic and Stojadinovic 2013; Priestley et al. 1996; Buckle et al. 2006; Vosooghi and Saiidi 2012; Fakharifar et al. 2014a; He et al. 2013; Grelle and Sneed 2013). While numerous studies have investigated the significance and efficacy of retrofit jackets application on RC bridge columns (Priestley et al. 1996; Buckle et al. 2006; Vosooghi and Saiidi 2012; Fakharifar et al. 2014a; He et al. 2013; ElGawady et al. 2009; Saatcioglu and Yalcin 2003 amongst many others), this study is a preliminary attempt to investigate the efficacy of different repair jackets application on MS-damaged RC bridge piers subjected to AS attacks of multiple intensities. Study by Huang and Andrawes (2014) indicated very promising results that an improvement as high as 117 % in bridge overall performance subjected to real strong mainshock-aftershock sequences can be achieved through applying SMA confinement to piers.

It is an unrealistic assumption that the structure would be repaired to its intact/undamaged state prior to the next seismic event, which is commonly assumed in seismic loss estimation (Nazari et al. 2013). Hence, this study aims at quantifying the impact of repair jackets on the post-MS collapse capacity of repaired bridges with substandard reinforcement detailing under AS attacks. While considering the effects of multiple (successive) earthquakes or sometimes referred to as long-duration earthquakes on the nonlinear response of structures has been the topic of numerous research studies (Chang et al. 2012; Di Sarno 2013; Li et al.

2014), however, to the best of the authors' knowledge, there is no comprehensive study on different repair jackets application on MS-damaged RC bridge structures subjected to AS attacks of multiple intensities of different earthquake types. Few studies have attempted to account for the prior damage effect, particularly when the structures are being repaired after the first seismic attack. The first method to include the existing damage condition prior to repair is to modify the material properties of the repaired region—for example reducing the structure's initial stiffness—to account for the effect of previous damage and repair (Aschheim and Black 1999). The second approach to incorporate the degraded strength and stiffness of members due to existing damage is through the use of time-dependent repair elements that can be activated (birth) [or deactivated (death)] within the user-defined time intervals of interest during static and dynamic time-history analysis (Lee et al. 2009, 2011).

Many studies have indicated the seismic vulnerability of MS-damaged bridge columns, if subjected to potential aftershocks, as substantial damage could accumulate during the MS event (Vosooghi and Saiidi 2012; Di Sarno 2013; He et al. 2013; Li et al. 2014). A MS-AS (or often termed FS-AS) sequence is illustrated in Fig. 1a. The Arias intensity (Arias 1970) plot for the MS only and MS-AS sequence are depicted in Fig. 1b, which shows that the energy accumulates over a longer time for the MS-AS sequence (longer duration ground motion) as compared to the MS only (shorter ground motion). The energy flux plot in Fig. 1c, as a measure of seismic energy content, shows the second increase due to AS effect compared to one ground motion. Safety and integrity of a bridge structure subjected to such seismic event heavily depends on the capacity of bridge columns to dissipate the seismic input energy through inelastic hysteretic cycles. Figure 2a illustrates an RC single column bridge bent subjected to a MS-AS sequence. Figure 2b depicts the bridge at incipient collapse after the MS attack and the repair jacket application in the plastic hinge region of the damaged column prior to the AS attack. Figure 2c illustrates the impact of existing damage from the MS event on reducing the lateral load carrying capacity of the damaged structure subjected to AS. It is clear that if repair jackets (e.g. CFRP wraps or steel plate jackets) are implemented, the repaired bridge column could sustain considerably higher drift ratios compared to the unrepaired bridge column. On the contrary, the unrepaired bridge bent experienced rapid strength deterioration with degrading post-peak strength. It could be noticed that there is a direct correlation between damage state from the MS event and the bridge resilience to the AS event, and appropriately designed/implemented repair jackets could enhance the bridge post-repair seismic performance.

2. Research Significance

The overarching goal of this study is to evaluate the damage from MS-AS sequences of multiple intensities on substandard bridge piers (local behavior) and quantify the efficacy of different repair jackets on the post-MS collapse

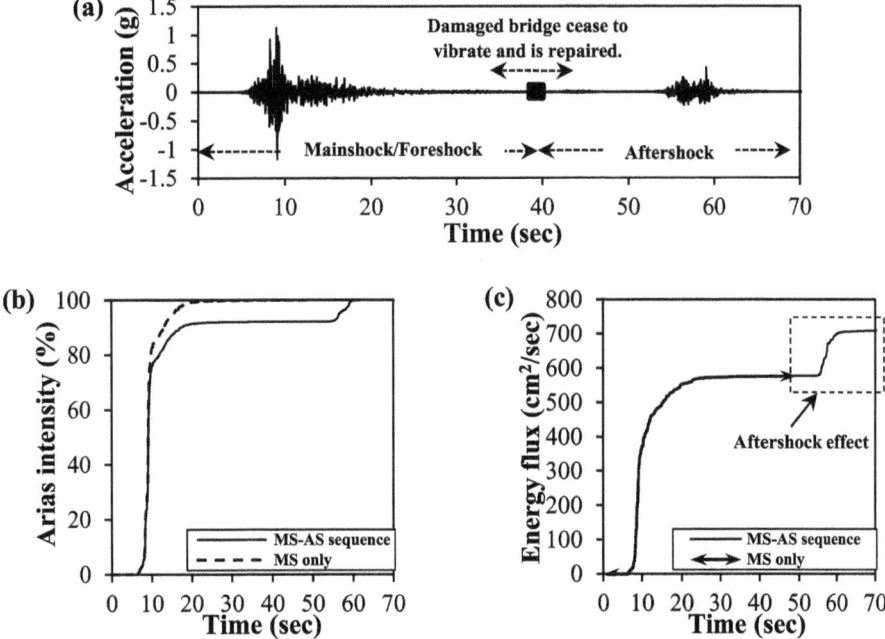

Fig. 1 **a** A typical mainshock-aftershock pair acceleration time-history sequence, **b** arias intensity for mainshock (MS) only and mainshock-aftershock (MS-AS) sequence, and **c** energy flux for MS and MS-AS sequence.

capacity of rehabilitated bridges (system behavior) under aftershocks. To achieve this goal, a two-span, substandard single-column bent RC bridge, typical of older construction practice, is utilized as a reference bridge to illustrate the impact of repair jacket application on damaged bridges subjected to AS attacks. The state of damage and structural collapse capacity under the MS-AS excitations is quantified through incremental dynamic analysis procedure. Different repair jackets, namely CFRP, conventional thick steel, and a proposed hybrid jacket are considered as repair interventions. Different repair jackets were considered to evaluate potential disparities on the post-repair behavior of bridges upon introducing different confining jackets, which results in different levels of confining pressure and confinement type (steel or FRP-based confined models). Results from different seismic sequences of various intensities are presented in detail, and effects of different earthquake types on structural collapse capacity and fragility of the repaired and unrepaired bridge are highlighted.

3. Prototype Highway Bridge Structure

One type of the most common class of highway bridges was selected for this investigation. The bridge was a two-span bridge seated on two abutments and one single column-bent (see Fig. 3). The two spans were 36.5 m long each, with a single cell box girder with a width of 8.22 m. The column longitudinal reinforcement was lap-spliced ($20d_b$, where d_b is the longitudinal reinforcement diameter) at column footing interface. The column is 6.70 tall with a 1.20 diameter. The longitudinal reinforcement of the column cross section consisted of 24 No. 8 ($d_b = 25.4$ mm) lap-spliced reinforcement bars, and the transverse reinforcement

consisted of No. 4 ($d_b = 12$ mm) spiral deformed bars spaced every 102 mm (Fig. 3b). The column is seismically deficient in accordance to modern seismic codes, as the reinforcement is spliced at the potential plastic hinge region. The column was investigated to ensure shear failure is not the governing mode of failure, as shear failure is a catastrophic failure, and retrofitting measures must be implemented prior to any seismic event (e.g. MS). The column aspect ratio is well above 2.5 (i.e., flexural dominant column, exhibiting flexural failure due to formation of a plastic hinge). The material properties constitute the reinforcing steel with a design yield strength of 300 MPa, while the design compressive strength of the concrete was 28 MPa.

4. Confining Repair Jackets for Bridge Pier Rehabilitation

4.1 Hybrid Repair Jacket

Recently, the authors have proposed and experimentally validated a hybrid jacket for the repair of severely damaged RC bridge piers (Fakharifar et al. 2015b). The hybrid jacket comprised of passive and active confining pressure. The hybrid jacket application on the plastic hinge region of a severely damaged RC bridge column is illustrated in Fig. 4. The hybrid jacket is composed of a thin cold-formed steel sheet wrapped around the column with prestressing strands placed over the column. While the prestressing strands can prevent buckling of the confining steel sheet, the thin steel sheet can in turn prevent the prestressing strands from penetrating into cracked cover concrete (i.e., preventing loss of confining pressure in the active strands). The thin cold formed sheet metal is directly wrapped and welded around

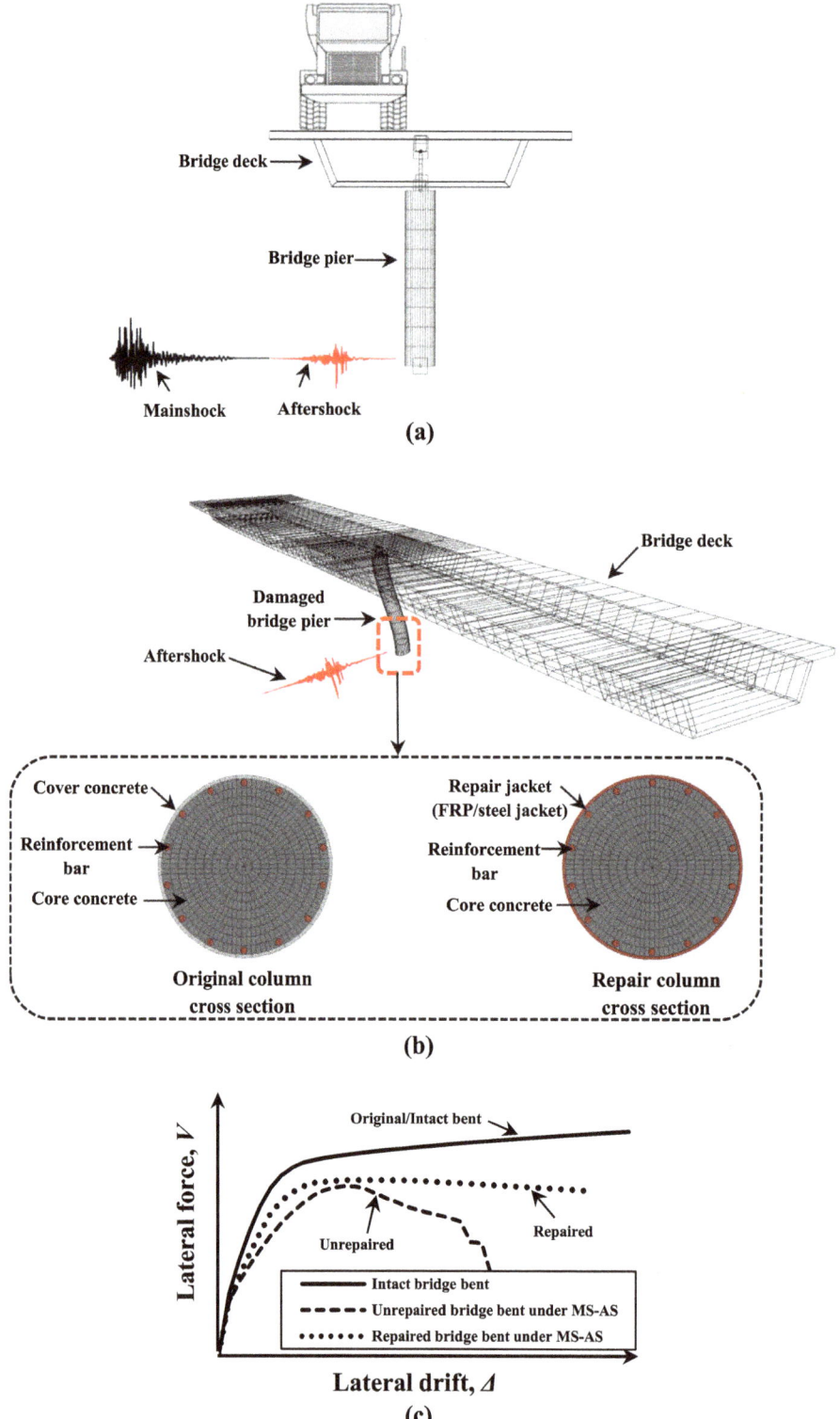

Fig. 2 Repair jacket application on mainshock-damaged single column RC bridge pier, **a** bridge pier subjected to mainshock-aftershock sequence; **b** bridge pier at incipient collapse after the mainshock with/without repair jacket in the plastic hinge region, and **c** representative lateral force–displacement relationship of a bridge bent in the original, unrepaired and repaired condition.

the existing column. The repair is completed by placing and prestressing the strands around the column. The hybrid jacket exhibits advantageous features over traditional repair jackets such as feasibility of rapid application after a MS event (no heavy equipment required); no additional seismic weight introduced due to small thickness of sheet metal and negligible weight of cables (similar to FRP jackets); and no

requirement for column surface preparation and adhesive material application (required for FRP jackets). The most prominent structural feature of the hybrid jacket is resisting the shear crack opening in both the vertical and horizontal directions (see Fig. 5), while the prestressing strands (similarly any unidirectional jacket such as FRP or SMA wires) can resist the shear crack opening in the transverse direction

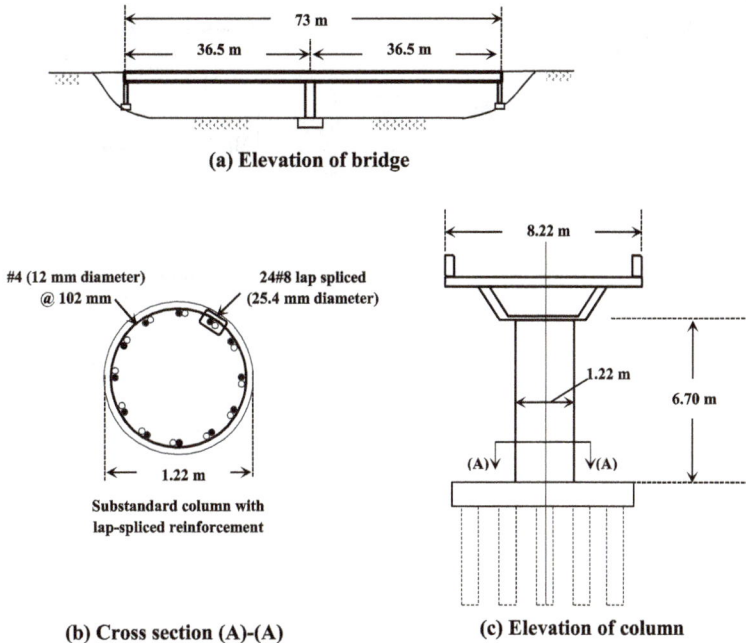

(a) Elevation of bridge

(b) Cross section (A)-(A)

(c) Elevation of column

Fig. 3 Details of the studied prototype bridge.

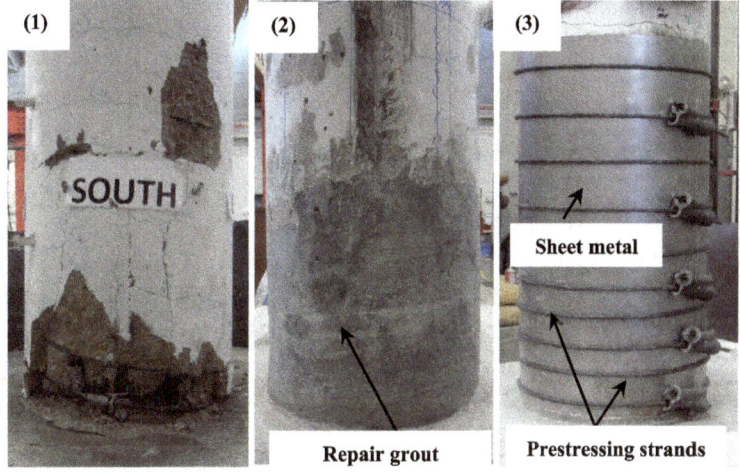

Fig. 4 Details of the hybrid repair jacket: *1* damaged column, *2* original column cross section restored, *3* thin sheet metal wrapped around the column with prestressing strands placed over column.

only (Jirawattanasomkul 2013). Thus, due to the enhanced aggregate interlock provided, concrete shear strength in the hybrid jacketed column—similarly conventional steel jacketed column—is larger than that in column with prestressing strands (FRP or SMA) only.

4.2 Repair Techniques

Three different repair techniques were considered, including a CFRP repair jacket, a conventional thick steel repair jacket, and hybrid repair jacket (discussed in Sect. 4.1). The CFRP jacket was chosen as the first repair technique due to the excellent high strength-to-weight ratio which is becoming increasingly attractive for structural repair/retrofit projects (Priestley et al. 1996; Buckle et al. 2006; He et al. 2013; Grelle and Sneed 2013). The steel jacket was the first and the most widely used retrofit technique (Priestley et al. 1996; Buckle et al. 2006; ElGawady

2009; Fakharifar et al. 2013;), due to its availability, price, and ductile behavior when compared with FRP materials with brittle failure (Fakharifar et al. 2015c). However, in addition to the labor intense work required for steel plate or RC jacketing, one major concern, is the increase of the original cross section (i.e., increased stiffness and reduced natural vibration period (Haroun and Elsanadeby 2005) results in alteration of the system dynamic response). The additional flexural strength of a member with a thick repair jacket would increase the plastic shear demand on the retrofitted/repaired column. Thus, the conventional steel or RC concrete jackets need to be detailed so as not to trigger other failure modes such as cap beam damage or cause damage to other capacity protected elements (Priestley et al. 1996; Buckle et al. 2006; Fakharifar et al. 2014b). Therefore, active prestressing strands were considered as another repair technique. Active cables have the advantage of providing the

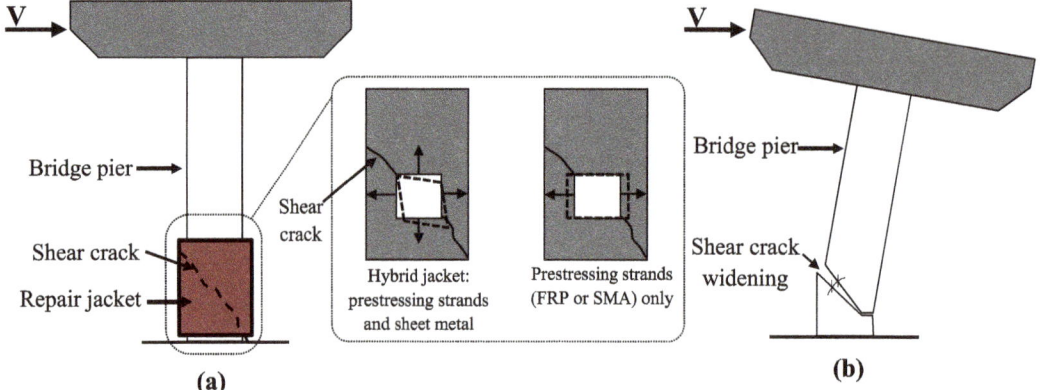

Fig. 5 **a** stress state in the repair jacket along the shear crack for confining pressure from hybrid jacket and prestressing strands, **b** increase of shear deformations due to widening of shear cracks in an RC column with existing damage subjected to aftershock attacks.

active confining pressure without increasing the column cross section (Saatcioglu and Yalcin 2003). However, tests proved that for retrofitted columns with prestressing strands, penetration of the prestressed cables into the cracked cover concrete caused loss of confining force in the active cables (Buckle et al. 2006). To incorporate the economical price of steel jackets with the benefit of active confining pressure, the hybrid confining jacket was proposed and considered as the third seismic repair intervention in this study.

4.3 Seismic Repair Design and Repair Performance Objectives

The bridge column repair design aimed at restoring the shear strength, providing the required confining pressure in the plastic hinge and inhibiting the lap splice failure. Repair jacket design procedure is briefly described as follow.

The height of plastic hinge to be repaired, L_{pz}, was calculated according to California Department of Transportation Seismic Provisions (CALTRANS 2004).

$$L_{pz} = \frac{3}{8} AR \times D \geq 1.5D \tag{1}$$

where AR is the column aspect ratio, and D is the column diameter.

The shear strength of the column was based on those of individual components and checked against the factored shear (Vosooghi and Saiidi 2012). That is,

$$\frac{V^o}{\phi} < V_c + V_s + V_j \tag{2}$$

where V^o is the base shear, $\phi = 0.85$, and V_c, V_s and V_j is the shear resisted by concrete, existing transverse reinforcement, and repair jacket, respectively. Due to unknown extent of damage from a MS event, the shear resistance of the concrete and spirals were neglected, and the repair jacket was designed to resist the shear demand.

The shear resistance for the different studied repair techniques was determined as follows (Priestley et al. 1996; Buckle et al. 2006; Vosooghi and Saiidi 2012).

$$V_{cj} = 0.5\pi t_j f_{cj} D \cot \theta \tag{3}$$

$$V_{sj} = 0.5\pi t_j f_{yj} D \cot \theta \tag{4}$$

$$V_{sp} = 0.5\pi A_{ps} f_{ps} D s^{-1} \cot \theta \tag{5}$$

where V_{cj}, V_{sj} and V_{sp} is the shear enhancement from CFRP jacket, steel jacket and prestressing strands, respectively; t_j is the jacket thickness (CFRP or steel); D is the column diameter; f_{cj} is the hoop stress in the CFRP jacket; f_{yj} is the steel jacket yield stress; A_{ps} is the cross sectional area of prestressing strand; f_{ps} is the level of prestressing stress in the strand; s is the vertical spacing between prestressing strands; and θ is the angle of the critical inclined shear-flexure crack to the column axis. For the hybrid repair jacket shear resistance was calculated as the contribution from both the steel jacket and the prestressing strands (i.e., $V_{sj} + V_{sp}$).

The required confining stress (f_l) to prevent the lap-splice failure (Priestley et al. 1996) for the substandard bridge was determined from Eq. (6).

$$f_l = \frac{A_b f_s}{\mu p l_s} \tag{6a}$$

$$p = \frac{\pi D'}{2n} + 2(d_b + c) \leq 2\sqrt{2}(d_b + c) \tag{6b}$$

where A_b is the cross sectional area of reinforcement bar, $f_s = 1.7$ times reinforcement yield stress ($=1.7f_y$), μ is the coefficient of friction (assumed as 1.4), l_s is the lap splice length, p is the crack surface perimeter, n is the number of longitudinal lapped bars, d_b is the reinforcement bar diameter, D' is the core diameter (outside to outside dimension of the circular transverse reinforcement), and c is the concrete cover thickness.

The required jacket thickness and spacing between prestressing strands to provide the adequate confining pressure in the plastic hinge region of damaged columns were calculated as follows (Priestley et al. 1996; Buckle et al. 2006).

$$t_{cj} \geq \frac{215D}{E_{cj}} \tag{7}$$

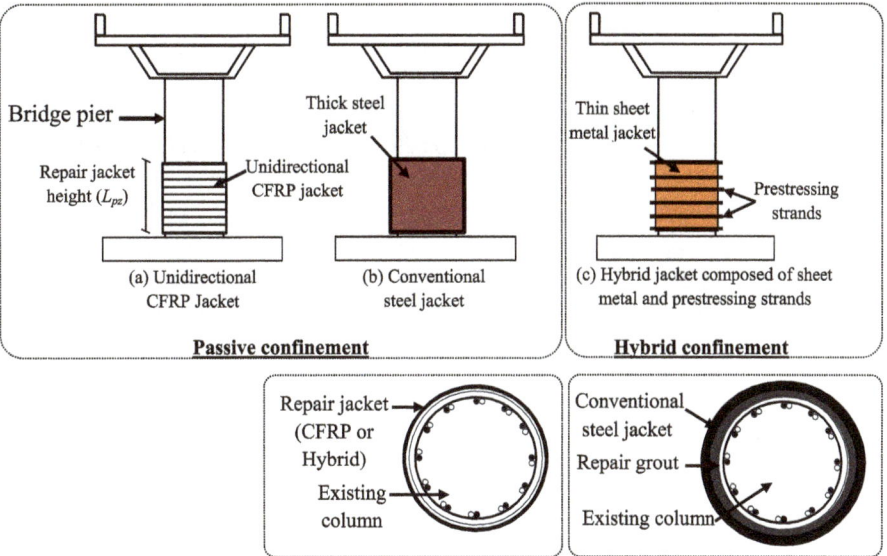

Fig. 6 Schematic of different repair techniques: **a** CFRP jacket, **b** conventional steel jacket, and **c** hybrid jacket.

$$t_{sj} \geq \frac{f_l D}{400} \tag{8}$$

$$s \leq \frac{A_{ps}\left(f_{ps} + 200\right)}{D} \tag{9}$$

where t_{cj} is the CFRP jacket thickness; t_{sj} is the steel jacket thickness; E_{cj} is the CFRP modulus of elasticity; and the remaining parameters were defined previously. The maximum spacing between the prestressing strands should not exceed six times the diameter of the longitudinal reinforcing bar (Buckle et al. 2006).

Figure 6 illustrates the schematic of the studied repair jackets. The jacket height for the studied repair techniques was calculated as $L_{pz} = 1220$ mm.

The CFRP jacket was assumed to have a tensile strength of 3800 MPa, modulus of elasticity of 227 GPa, and a rupture strain of 1.67 %. The selected composite CFRP repair system consisted of unidirectional high strength carbon fabric with fibers perpendicular to the column longitudinal axis (Fig. 6a). The required jacket thickness for the studied bridge column was calculated as 5.81 mm (using nominal fiber thickness).

A 7.5-mm thick A36 steel with minimum yield strength of 400 MPa was required for the thick steel repair technique. For practicality, a 10-mm thick steel jacket with yield strength of 400 MPa and modulus of elasticity of 200 GPa, with a 25 mm gap left between the column and the jacket to be filled with repair grout (Fig. 6b), was required.

For the hybrid jacket, the thin sheet metal with yield strength of 680 MPa; tensile strength of 771 MPa; and elastic modulus of 207 GPa was considered. Considering field-application and formability of the sheet metal around the existing column without specialized equipment in the field, a 1220 mm wide sheet with a thickness of 1.27 mm was calculated. The nominal 12 mm diameter seven-wire strands with the ultimate tensile strength of 1937 MPa and modulus of elasticity of 200 GPa were considered for repair

design. The spacing between strands and the required prestressing pressure in the cables were calculated as 1045 MPa and 100 mm, respectively (Fig. 6c).

5. Nonlinear Numerical Finite Element Modeling

An extensively calibrated—at the component and system level—nonlinear finite element model of the bridge is developed and subjected to sequences of MS-AS. The numerical FE models in this study were developed and analyzed using SeismoStruct nonlinear analysis program (Seismosoft 2013a). The program is capable of predicting the large displacement behavior of space frame structures under static and dynamic loading while taking into account both geometric nonlinearities (P-Δ effect) and material inelasticity (Ferracuti et al. 2009). The fiber based modelling approach using the 3D inelastic beam-column element discretized by uniaxial fiber elements was utilized to develop the numerical models. The applicable nonlinear fiber discretized beam/column elements calibrating parameters where computed according to the analytical column modeling strategies proposed by Berry and Eberhard (2007) and Haselton et al. (2008). Accuracy of the numerical models is significantly important for accurate and realistic assessment of the studied bridges during the MS-AS event. Many studies (Terzic and Stojadinovic 2013; Fakharifar et al. 2014c; Shamsabadi et al. 2009) have demonstrated the accuracy of adequately calibrated fiber elements in predicting the lateral force–displacement hysteresis relationship of different structural systems under static and dynamic loading conditions. Hilbert-Hughes-Taylor integration scheme (Scott and Fenves 2006) was utilized to acquire the global flexibility matrix for the flexibility based (FB) elements; considering the potential strain softening or localized deformations phenomena.

The Menegotto–Pinto (1973) steel model with Filippou et al. (1983) hardening rules and Fragiadakis et al. (2008)

additional memory rule for higher numerical instability/accuracy under transient seismic loading was used for the reinforcing steel material. To incorporate the accumulated damage from the MS in the longitudinal reinforcing steel (for simulated damaged column), the steel properties were modified (Vosooghi and Saiidi 2010, 2012) to consider the Bauschinger effect (Kent and Park 1973), which resulted in lowering the reversed yield stress and the reversed stiffness. For concrete the nonlinear variable confinement model of Madas and Elnashai (1992) with Martinez-Rueda (Martinez-Rueda and Elnashai 1997) cyclic rules was used. The variable confinement algorithm enables the explicit computation of relation between the concrete lateral dilation and straining of the transverse reinforcement computed at every analysis step. The Ferracuti and Savoia (2005) model was utilized as the FRP confined concrete model. This model follows the constitutive relationship and cyclic rules proposed by Mander et al. (1988) and Yankelevsky and Reinhardt (1989) under compression and tension, respectively. This model implements the Spoelstra and Monti (1999) model for the confining effect of FRP wrapping. The FRP material was modelled as linear elastic material up to rupture with no compressive strength.

5.1 Nonlinear Numerical Model Details

The fiber discretized cross section of a numerical RC column and representative stress–strain behavior associated to fibers are shown in Fig. 7. For confined core fibers either steel-confined (for thick steel and hybrid jackets), or FRP-confined would be considered. Column elements were modelled through inelastic frame elements. The effect of lap-splice reinforcement was desired to be somewhat represented in the numerical models hysteresis behavior. For RC columns with spliced reinforcement, a multilinear smooth polygonal hysteresis loop, developed in the work of Sivaselvan and Reinhorn (1999, 2001) through a zero length rotational spring (available in Seismostruct) could be used to

model the plastic hinge to include the hysteresis pinching associated with reinforcement slippage. However, the state of rebar stress at splice failure was needed to be determined. FEMA 356 (2000) provisions on lap-spliced bars was used to determine the rebar stress at failure, however the predicted response was not accurate especially in the post-peak degrading behavior. Study by Cho and Pincheira (2006) indicated that FEMA 356 underestimates the lapped rebar stress. Using bond-slip models are a proven method for inelastic analysis of RC columns with lap-splice failure (Xiao and Ma 1997; Harajli et al. 2002; Cho and Pincheira 2006). However, bond-slip model could not be directly incorporated in the numerical model, since a zero length stress–displacement section was not available. Besides, the effect of the lateral confining pressure from different repair jackets was also important. To achieve this, the stress–strain relationship of the longitudinal steel reinforcement for lap spliced reinforcement was reduced to simulate the expected backbone of lateral force–displacement relationship. It's noteworthy to mention that the Xiao model (Xiao and Ma 1997) could be used to establish the average bond stress–lateral displacement relationship for columns with lap-spliced reinforcement in the as-built and with the consideration of the lateral confining pressure conditions in retrofitted ones, and then reduce the steel properties accordingly. When ignoring the lap-splice, monotonically increasing lateral force–displacement relationship with ascending post-peak behavior was observed. However, reduced material properties captured the bar-pullout and the associated softening behavior evident from lap-spliced columns hysteresis behavior. The method used was approximate, however resulted in fairly accurate predictions of the numerical models, as shown later in the paper. Note that fiber discretized sections with uniaxial constitutive laws were used in the finite element model and the fiber element is incapable of accurately aggregating shear deformations. However, the studied columns in the current investigation constitute

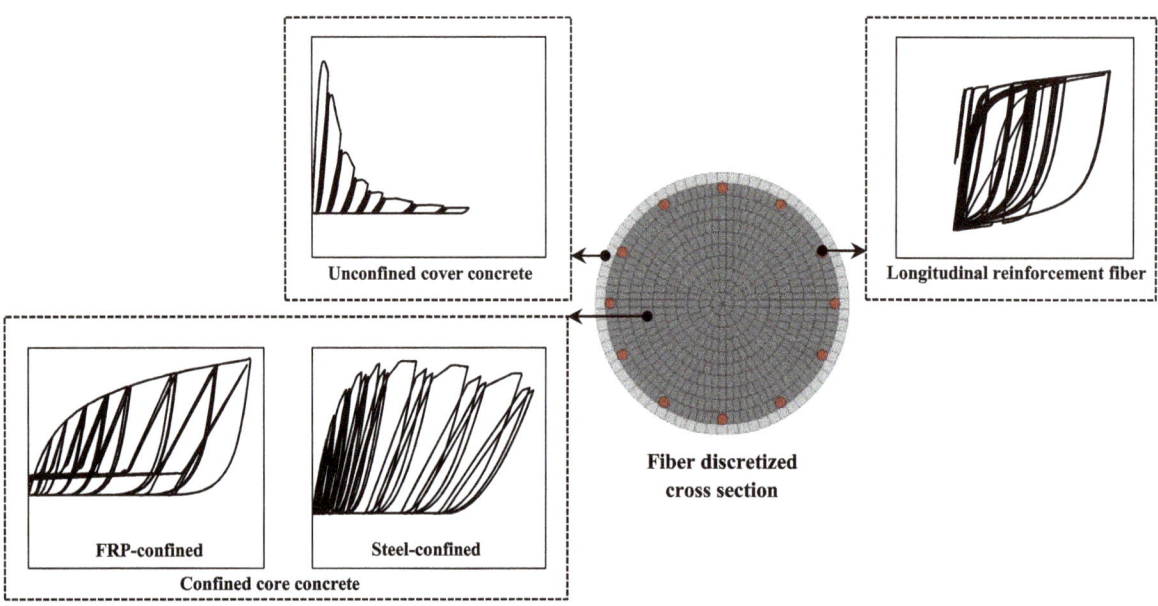

Fig. 7 Representative stress–strain responses of material fibers in an RC column cross section.

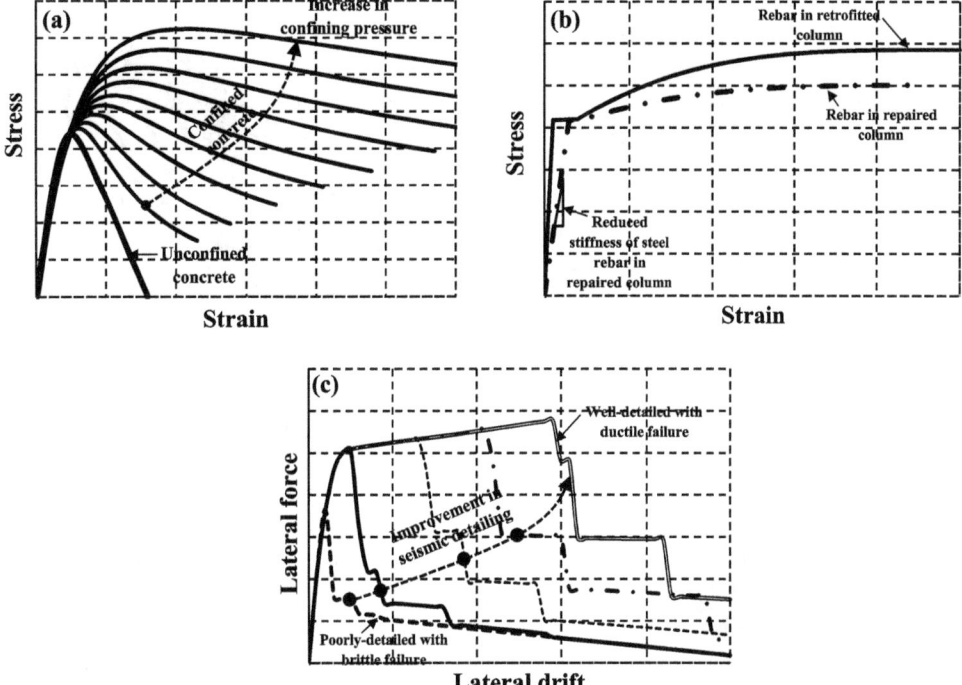

Fig. 8 Representative constitutive models and lateral force–drift relationship: **a** unconfined and confined concrete model, **b** stress state in steel reinforcement in retrofitted and repaired column, and **c** lateral force–drift responses according to seismic reinforcement detailing.

primarily flexural dominant (column aspect ratio well above 2.5) hysteresis response.

The increase of lateral confining pressure on the constitutive confined concrete model in this study is illustrated in Fig. 8a. A typical stress–strain relationship for the virgin steel reinforcement versus damaged reinforcement is illustrated in Fig. 8b. For the retrofit case, the jacket is applied on the intact column cross section prior to any major event, while for the repair scenario the repair jacket is implemented after column experiences some extent of damage. As it can be seen, in the case of a repaired column the initial stiffness of steel reinforcement is reduced, as well as the post-yield strain hardening. Representative lateral force-drift relationship for an identical RC column depending on the reinforcement detailing, is also illustrated in Fig. 8c. It can be seen brittle failure could be avoided through inclusion of seismic reinforcement, such as external jackets.

5.2 Verification of Numerical Models with Experimental Test Results

Extensive verification of the numerical models for their capability of capturing and simulating the global and local forces and deformations were implemented first. The authors have also verified their analytical models for FRP jacketed concrete columns under both monotonic and dynamic timehistory loading elsewhere (Fakharifar et al. 2014c). To evaluate the accuracy of the developed numerical models for evaluating the behavior of bridges subjected to MS-AS sequences, experimental models were selected from full scale RC column/bridge structures subjected to multiple series of ground motions.

Experimental tests on RC columns tested under incrementally increasing reversed cyclic loading and constant axial load with lap spliced reinforcement ($l_s = 20d_b$) as part of another study by the authors was selected to evaluate the accuracy of the developed numerical models in predicting the force–displacement relationship of columns with lap-spliced reinforcement. Figure 9a depicts backbone of the predicted force–displacement relationship from the numerical model versus experimental test results. It is clear that, ignoring the slip effect would not capture post-capping degradation in the numerical model. However, the reduced steel allows capturing the post-peak strength degradation.

An RC cantilever column with continuous reinforcement (Column BG-2) tested under constant axial load and unidirectional quasi-static incrementally increasing lateral displacement up to failure was selected (Saatcioglu and Grira 1999). The selected column was chosen for verification since unlike regular RC columns with stirrups or spirals, welded reinforcement grids were used as confinement reinforcement. Therefore the equivalent confining pressure from the grids was considered in the model. Figure 9b indicates the high accuracy of the numerical model in predicting the hysteretic behavior of the experimental column specimen in terms of strength, stiffness and displacement capacity. Besides, it was concluded that alternative forms of confinement can also be considered in the numerical analysis, as long as the appropriate lateral confinement pressure is provided in the model.

A full-scale RC bridge pier subjected to series of six incremental ground motions developed by PEER and NEES (PEER 2010) as part of a blind prediction contest was

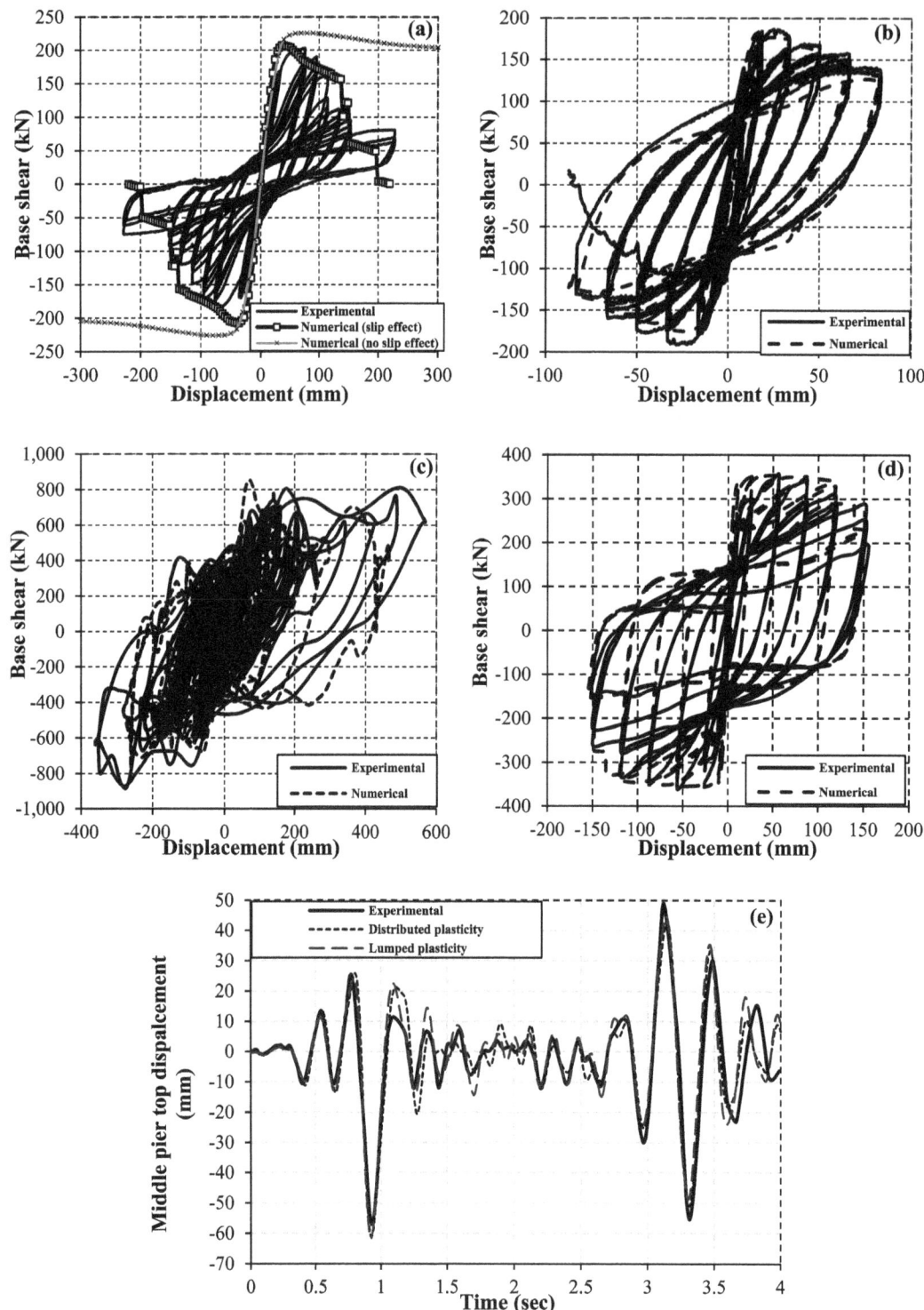

Fig. 9 Comparison of numerical models versus different experimental test specimens: **a** RC column with lap-spliced reinforcement under cyclic load reversals, **b** RC column under cyclic load reversals (Saatcioglu and Grira 1999), **c** RC column under shake table test (PEER 2010), **d** concrete-filled steel tube column under cyclic load reversals (Marson and Bruneau 2004), **e** full scale bridge shake table test (Pinto et al. 1996), **f** A2, **g** A4, **h** A8; and **i** A10 column specimens (Kunnath et al. 1997).

selected. The input ground motions commenced with the low intensity excitations succeeded by high intensity motions to bring the bridge pier to the incipient collapse state. The numerical model was verified adopting the proposed recommendations by Bianchi et al. (2011). Figure 9c demonstrates the capability of the numerical model in reproducing the dynamic hysteresis behavior of the studied

bridge pier. The selected bridge pier specimen exhibits significantly high level of nonlinearity (both material inelasticity and geometric nonlinearity). Damage was observed in the form of concrete spalling, concrete cracking, longitudinal bar buckling (thirteen of the eighteen bars were buckled), and longitudinal bar fracture in the plastic hinge region formed at column-footing interface (Schoettler et al. 2012).

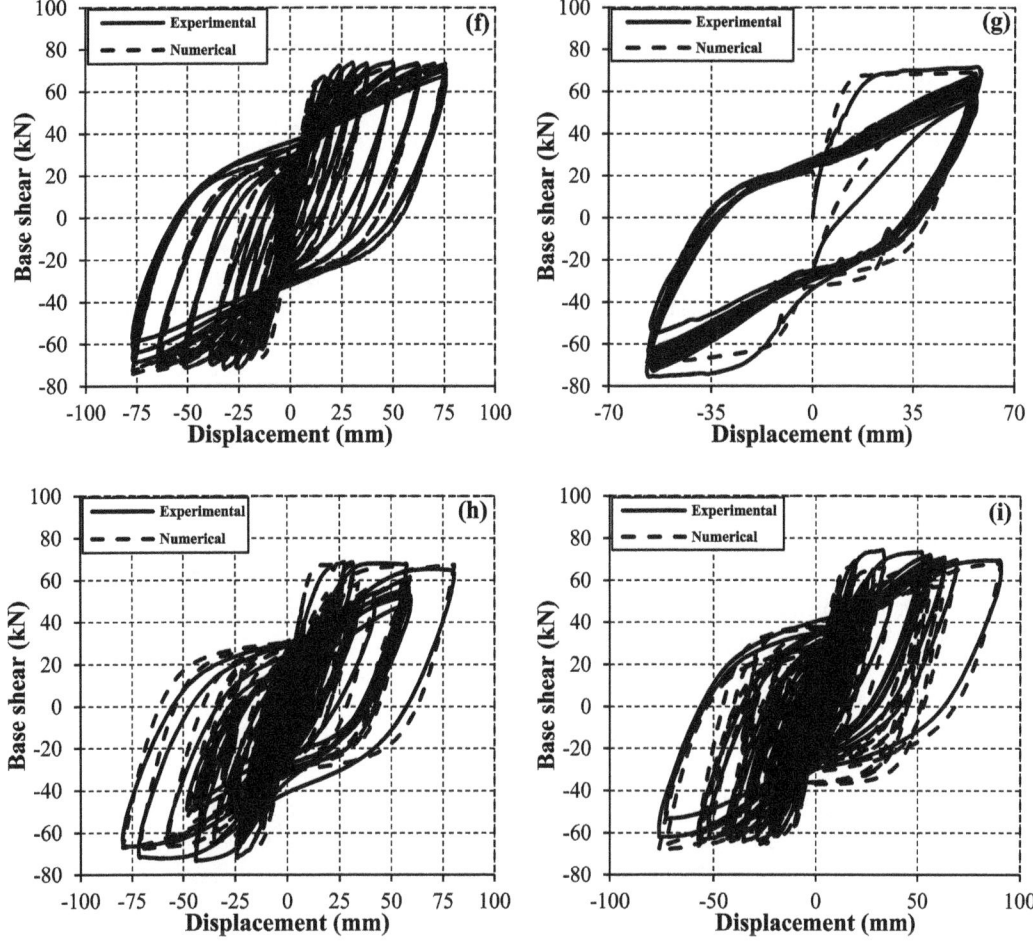

Fig. 9 continued

The numerical model could successfully capture the seismic response of the pier specimen subjected to six seismic sequences of various intensities.

To evaluate the adequacy of the developed numerical models in the case of steel-jacketed RC columns, a full scale concrete-filled steel tube bridge pier under incrementally increasing reversed cyclic loading and constant axial load (Marson and Bruneau 2004) was selected. Figure 9d depicts the predicted force–displacement relationship of the numerical model with reasonable accuracy as compared to the experimental test results.

To evaluate the accuracy of the numerical models in predicting the system response, the large scale bridge specimen tested in pseudo-dynamic type test was selected (Pinto et al. 1996). The bridge specimen was subjected to two successive input motions on shake table tests, including the design earthquake, followed by the 1.2 times the design level earthquake. The bridge deck was modelled using elastic frame elements with mass lumped along the centerline (Seismosoft 2013b). Both distributed plasticity (inelastic distributed element) and lumped plasticity (inelastic plastic hinge element) elements were adopted to model the bridge columns. The predicted displacement time history of the middle pier from the distributed plasticity and lumped plasticity models are compared versus experimental results in Fig. 9e. The results indicate the adequacy of both types of

element formulations in predicting the displacement time history and the maximum displacement with reasonable accuracy.

Finally, to further evaluate the accuracy of the developed numerical models under cumulative seismic damage (expected for MS-AS), series of RC bridge column specimens tested under various loading histories were chosen (A2, A4, A8 and A10 column specimens) (Kunnath et al. 1997). Each of the selected bridge column specimens were tested under a constant axial load and a unique lateral load path, specifically, specimen A2: standard cyclic load with three full symmetric cycles at increasing ductility up to failure; specimen A4: constant amplitude reversed cyclic displacement of $\pm3\Delta y$ (Δy = yield displacement level); specimen A8: random earthquake displacement histories, representing four scenario earthquake sequence (minor, minor, damaging and severe earthquakes), and A10: representing five scenario earthquake sequence (major, minor, moderate, minor and severe earthquakes) (Kunnath et al. 1997). Comparison of calculated force–displacement relationship of the studied column specimens against experimental results are illustrated in Fig. 9f–i for A2, A4, A8 and A10 column specimens, respectively. The obtained results indicate the high accuracy of the numerical models to simulate the obtained experimental results under different loading histories—including multiple earthquakes—in terms of strength, stiffness and ductility. The developed numerical models exhibited to provide sufficient accuracy, and these modeling assumptions were adopted for performance-based evaluation of the bridge in the current study.

6. Incremental Dynamic Analysis

The nonlinear static procedure, commonly referred to as static pushover analysis (SPA) has become a standard method to estimate seismic deformation demands in structures as well as their local and global capacities (Villaverde 2007). However, the SPA lacks a rigorous theoretical foundation, based on the incorrect assumption "that the nonlinear response of a structure can be related to the response of an equivalent single-degree-of-freedom system" (Villaverde 2007). The inherent inability of SPA analysis to correctly identify the seismic demands, specifically for structures that deform far into their inelastic range of behavior (expected for structures subjected to MS-AS sequence) has been clearly signified in many studies (Ferracuti et al. 2009; Villaverde 2007; Chopra and Goel 2002; Fakharifar et al. 2015c; amongst others). Therefore, incremental dynamic analysis (IDA), developed by Vamvatsikos and Cornell (2002), was incorporated to accurately evaluate the demand and capacity of the studied repaired bridges under different set of MS-AS sequences. IDA is capable of accurately estimating the failure/collapse capacity of structures subjected to seismic excitations of different intensities (Villaverde 2007; Vamvatsikos and Cornell 2002). An IDA requires multiple series of nonlinear dynamic time history analysis (NTHA) of the structure subjected to incrementally increasing ground motion intensity. The IDA analysis starts from a very low (non-negative) scaling factor multiplied by the intensity measure (IM) of each unscaled ground motion time history. The scaling factor for each run is incrementally increased until the structural median collapse (or until a defined failure limit state) is reached. Each data point on the IDA curve corresponds to a single NTHA for the structure subjected to one ground motion time history scaled to one intensity level. This procedure is repeated to obtain all the data points— from elastic to inelastic and finally to global dynamic instability (or numerical non-convergency)—for the full range of IDA curves.

The use of PGA as the IM for the structures with dominant first mode of vibration may result in biased structural response predictions (Vamvatsikos and Cornell 2002). For structures with a moderate period of vibration with maximum drift ratio as the damage measure (DM) and the 5 % damped first mode spectral acceleration ($Sa (T1,5 \%)$) as the IM subjected to a bin of relatively moderate to large magnitudes, M, results obtained from scaled and un-scaled records are similar. The use of $Sa (T1)$ as the IM is often recommended (Vamvatsikos and Cornell 2002; Tehrani and Mitchell 2013), and was adopted in this study.

Engineering demand parameter (EDP) is the output of the IDA analysis that elicits the state of structural damage in terms of its response under different earthquake records. For brevity, the IDA results are presented only in terms of maximum drift ratio as the demand parameter.

6.1 Damage/Limit State

Four damage states, namely minor, moderate, extensive and collapse as defined by HAZUS-MH (2003) were adopted herein, in which they were correlated with the quantitative engineering demand parameters. The drift-based limit states developed by Dutta and Mander (1998) were adopted to define the bridge columns drift limits. The drift limits considered for the bridge columns were 0.005, 0.007, 0.015, 0.025 and 0.05 for no damage, minor, moderate, extensive and collapse damage states, respectively. However, the constant drift limits better serve as a global (system) response quantity. For a more accurate evaluation, the strain limit sates (local response quantity) developed by Priestley et al. (2007) were incorporated as well. Using moment curvature and pushover analysis the strain limits were related to drift limit states. The maximum compressive strain in the confined core concrete and the maximum tensile strain in the longitudinal reinforcing steel were considered as the strain-based limit states (Priestley et al. 2007).

The bar buckling and bar fracture damage states were calculated according to the equations developed by Berry and Eberhard (2007). The equation proposed by Mackie and Stojadinovic (2007) for calculating the drift at column failure (i.e., lateral drift level beyond the peak strength that column reaches zero strength) was adopted to define the complete failure limit state. Incorporating the effect of load history on reinforcement bar buckling limit state—as shown in past studies (e.g., Rodriguez et al. 1999)—under the first seismic event may significantly alter the obtained results for the subsequent seismic event. Nevertheless, effects of seismic load-path history on reinforcement bar buckling were not considered.

6.2 Mainshock-Aftershock Seismic Ground Motion Records

The ground motions were selected from the Pacific Earthquake Engineering Research Center (PEER) Strong Motion Database, specifically, from the study by Baker et al. (2011) providing several standardized sets of ground motions for the PEER's Transportation Research Program. The selected ground motions are neither site-specific nor structure specific (i.e., unbiased) (Baker et al. 2011). The earthquake ground motions records in this study include both type of far-fault (also referred to as far-field) and near-fault records.

Two bins of earthquake records were considered to generate the MS-AS sequences, namely, ensemble 1 and ensemble 2. Ensemble 1 constitutes far-fault records where ensemble 2 included records with near-fault effects. Each ensemble constitutes records of large and moderate magnitude. The acceleration response spectra of ensemble 1 and 2 record sets (unscaled) are illustrated in Fig. 10.

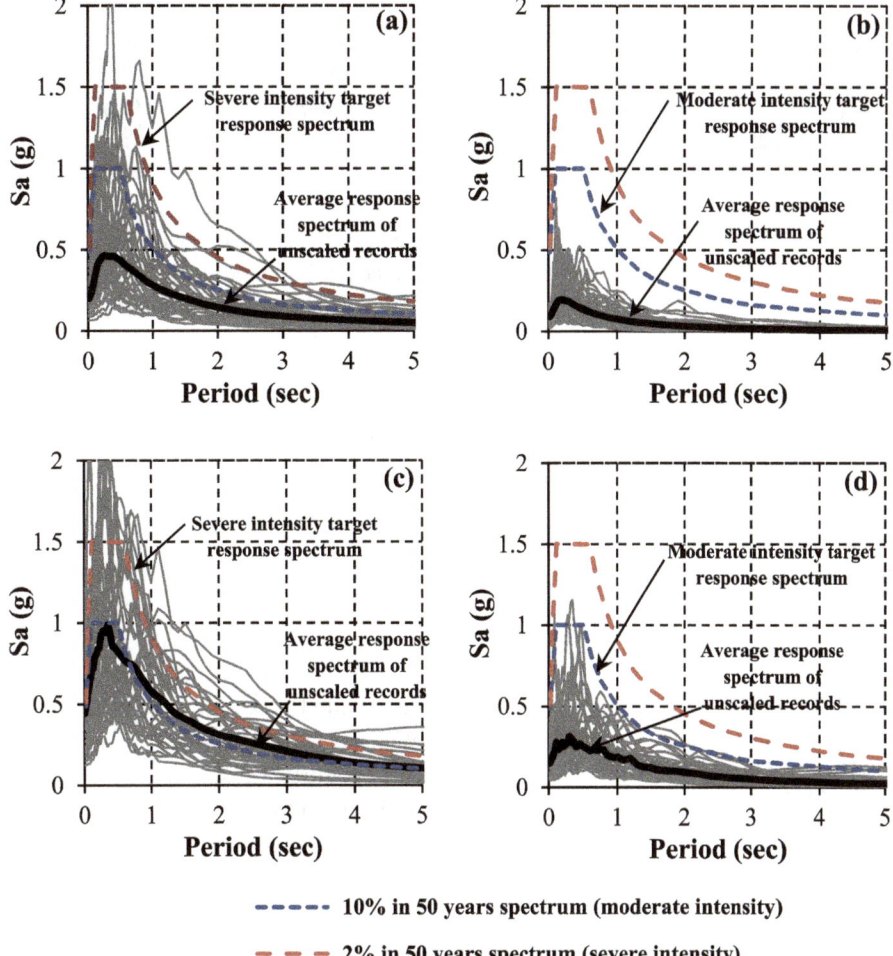

Fig. 10 Response spectra of selected unscaled ground motions and different hazard levels target response spectra: **a** ensemble 1 large (severe) magnitude, **b** ensemble 1 moderate magnitude; and **c** ensemble 2 large (severe) magnitude, **d** ensemble 2 moderate magnitude.

Two seismic sequences from ensemble 1 and two seismic sequences from ensemble 2 were considered, namely (1) severe MS-moderate AS, and (2) severe MS-severe AS. Eighty MS-AS seismic sequences from the ensemble 1 and eighty MS-AS seismic sequences from the ensemble 2 record sets were generated. Subsequently, for each ensemble, records were selected from the corresponding intensity record set. For instance for severe MS-moderate AS from ensemble 1, large magnitude (response spectra in Fig. 10a) and moderate magnitude (response spectra in Fig. 10b) records were paired. Similarly, for severe MS-severe AS from ensemble 1, large magnitude records followed by the same set of records were paired. Similar seismic sequences were generated from ensemble 2 bin of records. No paired seismic sequence from ensemble 1 to 2 and opposite were considered (i.e., no near fault MS-far fault AS or opposite). Hence, total of 160 paired accelerations (40 severe MS-moderate AS and 40 severe MS-severe AS from ensemble 1 and 2) were generated.

To simulate the free vibrating response of the bridge at the end of the MS, 20 s of zero acceleration was introduced between the MS and AS acceleration time histories, to assure that the bridge was at rest prior to the AS event.

For each intensity level, the selected ground motions were scaled to match the Los Angeles-California site response spectrum for two different hazard levels, namely 10 % probability of exceedance in 50 years (moderate earthquake), and 2 % probability of exceedance in 50 years (severe earthquake) (Fig. 10). The target response spectra were utilized to scale the original records at each intensity level. The selected unscaled ground motions from the severe (Fig. 10a) and moderate (Fig. 10b) intensity record sets from ensemble 1 were scaled considering the 2 and 10 % in 50 years target response spectra, respectively, to calculate the scaled severe and moderate magnitude records for ensemble 1. The same procedure was adopted to scale the original records from the ensemble 2 set of records. The characteristics of ensemble 1 and 2 record sets are detailed below.

Ensemble 1 (far-fault): for this set the large magnitude suite of records consists of 40 broad-band ground motions that their response spectra match the median and log standard deviations predicted for a large magnitude ($M_w = 7$) strike-slip earthquake at a distance of 10 km. The moderate magnitude motions constitutes 40 broad-band ground motions with their response spectra matching the median and log standard deviations predicted for a moderate magnitude

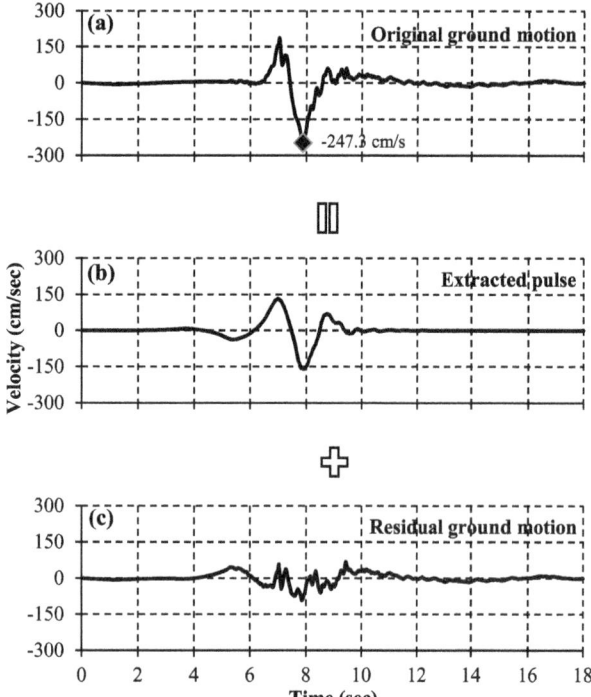

Fig. 11 A near-fault ground motion with large velocity pulse from ensemble 2 record set: **a** original ground motion, **b** extracted pulse, **c** and residual motion.

($M_w = 6$) strike-slip earthquake at a distance of 25 km. The selected set of ground motions as the far fault records represent a very broad range of spectral amplitudes up to two orders of magnitude. These set of records cover a broad range of intensities at sites located near active crustal earthquake sources (Baker et al. 2011). The acceleration response spectra of the severe and moderate intensity records from ensemble 1 ground motions are presented in Fig. 10a, b. Detailed information on the adopted records can be found elsewhere (Baker et al. 2011).

Ensemble 2 (near-fault): For this set the large magnitude ground motions consists of 40 records containing strong velocity pulses of varying periods. The velocity pulses are typically observed in the near fault ruptures due to directivity effects. This set of records are somewhat intense that constitute motions having a strike parallel peak ground velocity up to 250 cm/s. Figure 11 illustrates one original ground motion with strong velocity pulse along with the extracted pulse and non-pulse (residual) part of the original motion. The moderate magnitude ground motions consist of 40 records each obtained from scaling the large magnitude records (Bearman 2012). The acceleration response spectra of the ensemble 2 ground motions are presented in Fig. 10c, d. These set of near fault motions are similar to near-fault motions included in ATC-63 (2009), except these motions have a variety of pulse periods (1–12.9 s). This is important, since the pulse period, relative to the period(s) of vibration of a structure is not negligible. Besides, these set of records are somewhat intense that constitute motions having a strike parallel peak ground velocity of 250 cm/s (see Fig. 11a).

6.3 IDA Analysis Results Under Severe MS-Severe AS and Discussion

For brevity results from an extensive series of IDA analyses are presented only in terms of structural collapse capacity and maximum drift. For each series of analyses four cases of the studied bridge were considered: unrepaired bridge, column repaired with CFRP jacket, column repaired with conventional steel jacket, and column repaired with hybrid jacket. Each individual IDA curve corresponds to one MS-AS pair of ground motion. The IDA results were utilized to construct the IDA curves under back-to-back time histories. Extensive data from IDA analyses were generated and the IDA curves only for the severe MS-severe AS are plotted in Fig. 12. The IDA results under the severe MS-moderate AS are summarized at the end of this section.

The IDA results for the unrepaired case and the CFRP, conventional steel and hybrid repair cases subjected to severe MS-severe AS of ensemble 1 bin of records are presented in Fig. 12a–d, respectively. Similarly, IDA results under severe MS-severe AS from ensemble 2 suite of records for the unrepaired, CFRP, conventional steel and hybrid repaired cases are presented in Fig. 12e–h, respectively. To evaluate the impact of damage from the MS-AS sequence compared to an MS event only, the median of IDA results for the bridge subjected to severe MS only (i.e., no AS event) is superimposed on Fig. 12a, e accordingly (red dotted line).

The obtained mean structural collapse capacity for the MS only compared to MS-AS (Fig. 12a, e) is reduced as from 0.83 to 0.33 g; and from 0.69 to 0.36 g, for the ensemble 1 and 2 bin of records, respectively (60.0 and 47.8 % reduction in mean structural collapse capacity, respectively). This observation clearly indicates the effect of existing damage under a severe MS on the post-MS collapse capacity when severe AS is considered. Similarly, the obtained mean maximum drift level that the bridge could sustain prior to collapse for the MS only compared to MS-AS (Fig. 12a, e) is reduced from 3.42 to 2.35 %; and from 3.33 to 2.58 %, for the ensemble 1 and 2 bin of records, respectively (31.3 and 22.5 % reduction in maximum attainable drift level prior to collapse, respectively). The obtained mean structural collapse capacity/maximum drift for the original bridge subjected only to one seismic event (i.e., no AS) reveals the unsatisfactory performance of the bridge column with lapsplice deficiency, as expected. The original bridge could not reach drift capacity recommended by different seismic codes of practice (e.g. CALTRANS 2004). It is also shown that the accumulated damage under severe MS had impact on the post-MS collapse strength/deformability of the MS-damaged bridge when subjected to severe AS.

Usually the term MS-AS implies a severe MS followed by a minor/moderate AS, and the presented results herein is an extreme scenario, where the first severe earthquake is succeeded by another severe earthquake. Hence, the obtained results may not be realistic and represent the lowest bound to the mean structural collapse capacity. It is unlikely a severe MS is followed by a similar severe AS.

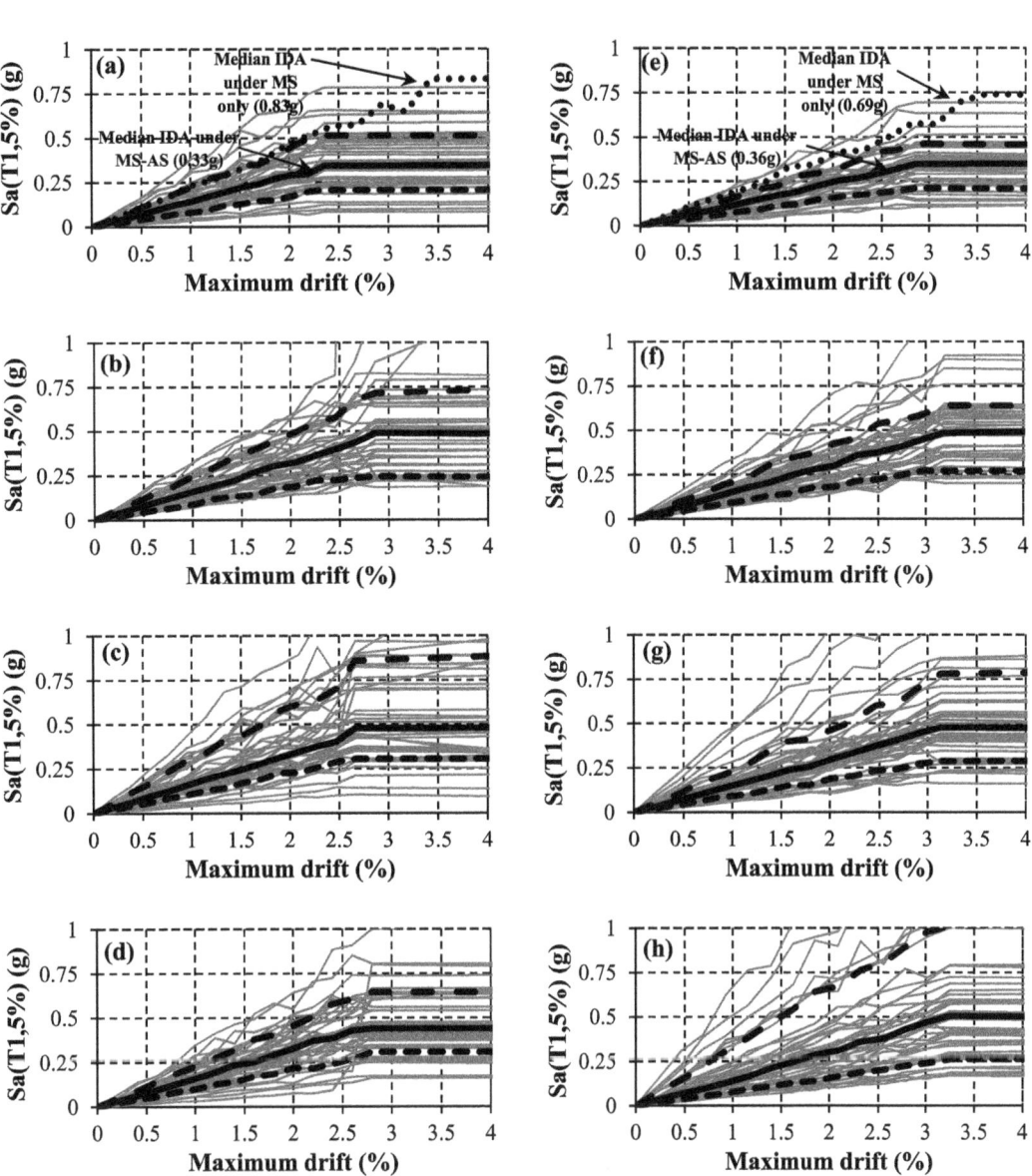

Fig. 12 IDA results (maximum drift) of the studied bridge under severe MS-severe AS sequence: ensemble 1 records **a** unrepaired, **b** CFRP, **c** conventional steel, and **d** hybrid; ensemble 2 records **e** unrepaired, **f** CFRP, **g** conventional steel, **h** hybrid jacket.

The mean structural collapse capacity for the CFRP, conventional steel, and hybrid repaired bridge columns under severe MS-severe AS from ensemble 1 record set (Fig. 12b–d) was 0.47, 0.48 and 0.43 g, respectively [i.e., 42.4, 45.5 and 30.3 % increase in the structural mean collapse capacity, respectively, compared to the unrepaired bridge (0.33 g)]. Similarly, the mean structural collapse capacity for the CFRP, conventional steel, and hybrid repaired bridge columns under the severe MS-severe AS from ensemble 2 record set (Fig. 12f, h) was 0.5, 0.46 and 0.5 g, respectively [i.e., 38.8, 27.7 and 38.8 % increase in the mean structural collapse capacity, respectively, compared to the unrepaired bridge (0.36 g)]. The obtained mean structural collapse capacities for the repaired bridge indicate the efficacy of confining repair jackets to be significant (up

to 45.5 %) for the post-MS collapse capacity of the earthquake-damage bridge.

Similar results on significance/necessity of repair jackets application were found for the severe MS-moderate AS sequences. However, the decrease in the mean structural collapse capacity between the MS only and severe MS-moderate AS was less significant compared to the severe MS-severe AS case. It is due to the fact that the major/primary damage is accumulated during the higher intensity seismic event, either during the FS or AS, and not depending on the sequence of events. For this reason, repair jackets were less effective for the severe MS-moderate AS compared to severe MS-severe AS sequence. The repair jackets were still required for the severe MS-moderate AS sequence as lapped reinforcement bars are seismically deficient in the plastic hinge region. The results under severe MS-moderate AS seismic sequences confirm that the primary damage occurred under the severe MS and not the moderate AS, but numerous inelastic excursions are critical for lap-spliced columns even when subjected to minor or moderate AS. For instance, checking some of the steel reinforcement fibers stress–strain histories under single seismic event indicated the steel linear response up to proximity of yielding. However, same fiber under two seismic events exhibited highly nonlinear (post-yield) stress–strain response.

Comparison of Fig. 12a, e indicate that the mean structural collapse capacity of the original bridge under MS-only from ensemble 2 (near-fault) is smaller compared to ensemble 1 (far-fault) (0.69 g for near-fault records versus 0.83 g for far-fault records). The forward directivity effect and permanent displacement effect resulting in long period and large velocity pulse of near-fault ground motions (e.g. Fig. 11) have been addressed in previous studies.

The abovementioned results clearly indicate that the CFRP, steel and hybrid repair jackets were essential for post-MS collapse resilience of earthquake damaged bridge piers subjected to potential aftershocks of moderate or severe intensity. However, the repair jackets were most effective for the severe MS-severe AS scenario, and conversely were least effective for severe MS-moderate AS. Overall, the obtained results prove the efficacy of the studied repair jackets on post-MS seismic resilience of substandard RC bridge piers under aftershocks.

7. Fragility Analysis of the Bridge

Eventually, a probabilistic seismic demand model (PSDM) was developed for the fragility curves for the studied bridge. Seismic fragility curves describe the likelihood of exceeding the capacity of the bridge structures under a given level of strong ground motion. Fragility curves were developed to assess the vulnerability of the as-built and repaired bridge under back-to-back MS-AS sequences and select the optimal repair technique for MS-damaged bridge piers subjected to aftershocks. The PSDM establishes a correlation between the EDPs and ground IMs. The PSDM model is regarded as the most rigorous and reliable analytical method to derive fragility functions (Shinozuka et al. 2000; Billah et al. 2013). The cloud approach (Nielson and DesRoches 2007) was used to develop the PSDM model utilizing the IDA results. Fragility curves for two intensities of MS-AS, namely severe MS-severe AS and severe MS-moderate AS are developed for four damage states, namely slight, moderate, extensive and collapse. Regression analyses were undertaken to acquire the mean and standard deviation for each damage state, by assuming the power law model as suggested by Cornell et al. (2002), which derives a logarithmic correlation between median EDP and the selected IM.

$$EDP = a(IM)^b \tag{10}$$

where a and b is the unknown coefficients estimated from the regression analysis of the response data points from IDA results. The intermediate responses from the IDA results were required, as the IDA analyses established the response from low to high levels of ground motion intensities. The relation by Baker and Cornell (2006) was adopted to estimate the dispersion of the demand, $\beta_{EDP|IM}$, conditioned upon the IM.

$$\beta_{EDP|IM} = \sqrt{\frac{\sum_{i=1}^{N}(\ln(EDP) - \ln(aIM^b))^2}{N-2}} \tag{11}$$

where N is the number of total analyses cases.

By having the PSDMs and the limit states at different damage states the fragilities were obtained from Eq. (12) (Nielson and DesRoches 2007) as follows.

$$P[LS|IM] = \Phi\left[\frac{\ln(IM) - \ln(IM_n)}{\beta_{comp}}\right] \tag{12}$$

where Φ is the standard normal cumulative distribution function; IM_n is the median value of the IM; and $\ln(IM_n)$ is the natural log of the median value of IM for the selected damage state (e.g. minor, moderate, extensive and collapse) as shown in Eq. 13 (Billah et al. 2013).

$$\ln(IM_n) = \frac{\ln(S_c) - \ln(a)}{b} \tag{13}$$

where a and b is the regression coefficients of the PSDMs; and β_{comp} is the dispersion component as shown in Eq. (14) (Nielson and DesRoches 2007).

$$\beta_{comp} = \frac{\sqrt{\beta_{EDP|IM} + \beta_c^2}}{b} \tag{14}$$

where S_c is the median and β_c is the dispersion value of the unrepaired/repaired bridge capacity, as the lognormal parameters for the damage states.

7.1 Fragility Analyses Results and Discussion

The fragility curves for the studied bridge in the repaired and unrepaired condition under severe MS-severe AS set of records from ensemble 1 (far-fault) and ensemble 2 (near-fault) for minor, moderate, extensive and collapse damage states are presented in Fig. 13a–h, respectively (Fig. 13a–d for ensemble 1 and Fig. 13e–h for ensemble 2). Similarly, the results under severe MS-moderate AS sequences from ensemble 1 and 2 for minor, moderate, extensive and collapse damage states are presented in Fig. 14a–h, respectively.

The obtained fragilities indicate that the fragility curves for the unrepaired scenario almost overlaps the repaired cases under the slight damage state, where the extent of damage is somewhat negligible (i.e., the damage state is cosmetic and repairable, and the onset of bar buckling did not occur). All the fragility curves irrespective of the repair jacket material, intensity of seismic sequences (i.e., severe MS-severe AS or severe MS-moderate AS) and earthquake

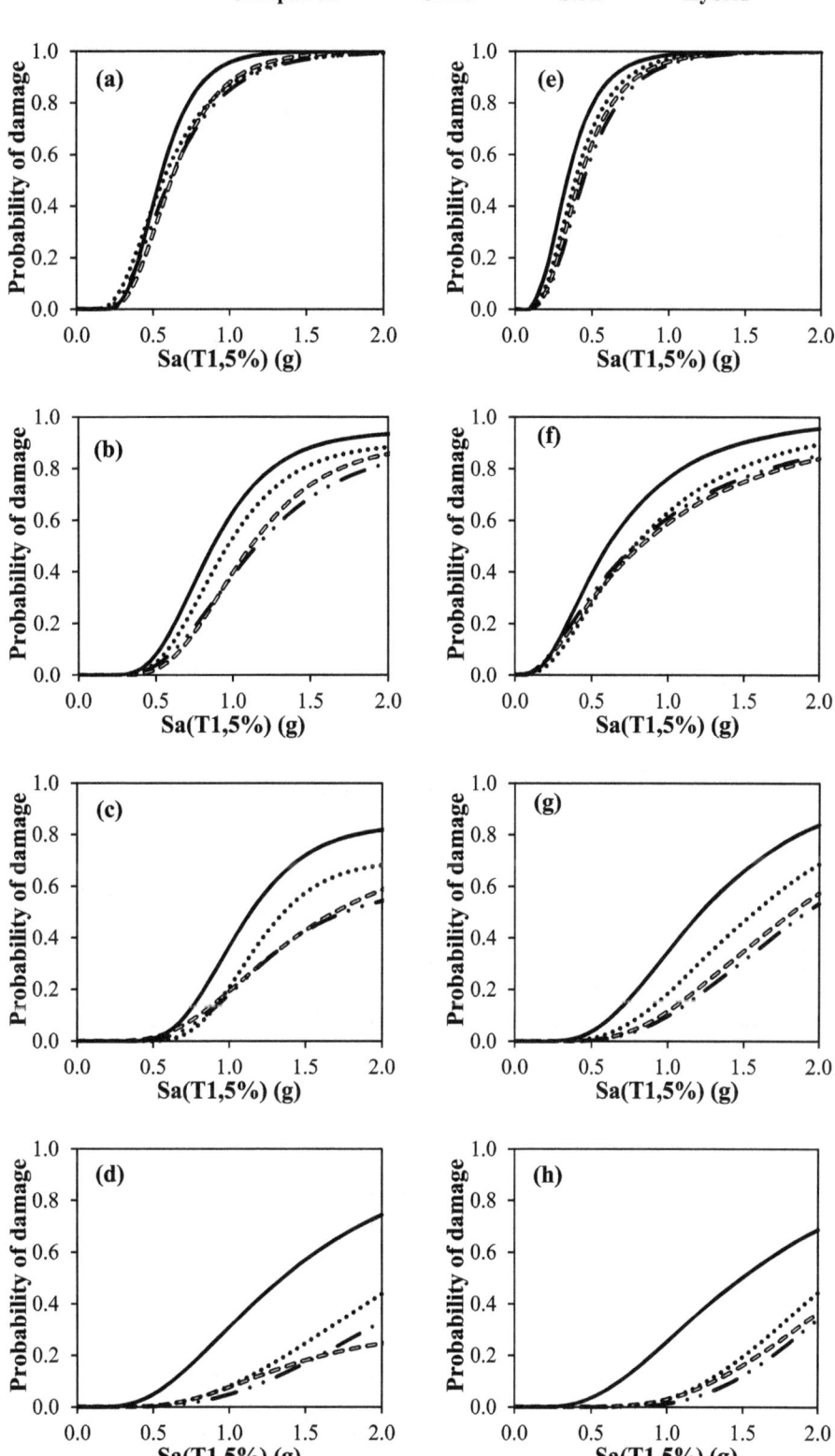

Fig. 13 Fragility curves for the unrepaired and repaired bridge under severe MS-severe AS sequence: ensemble 1 records **a** minor, **b** moderate, **c** extensive, and **d** collapse damage state; ensemble 2 records **e** minor, **f** moderate, **g** extensive, and **h** collapse damage state.

type (i.e., far-fault and near-fault) indicate the increase of difference (shift) between fragilities of the unrepaired and repaired bridge as the damage state progresses. While Fig. 13a, e (similarly, Fig. 14a, e) clearly indicate that the

fragility curves for the unrepaired bridge almost overlaps the repaired bridge under the minor damage state, the unrepaired bridge's vulnerability is significantly larger than the repaired bridge for the extensive and collapse damage states

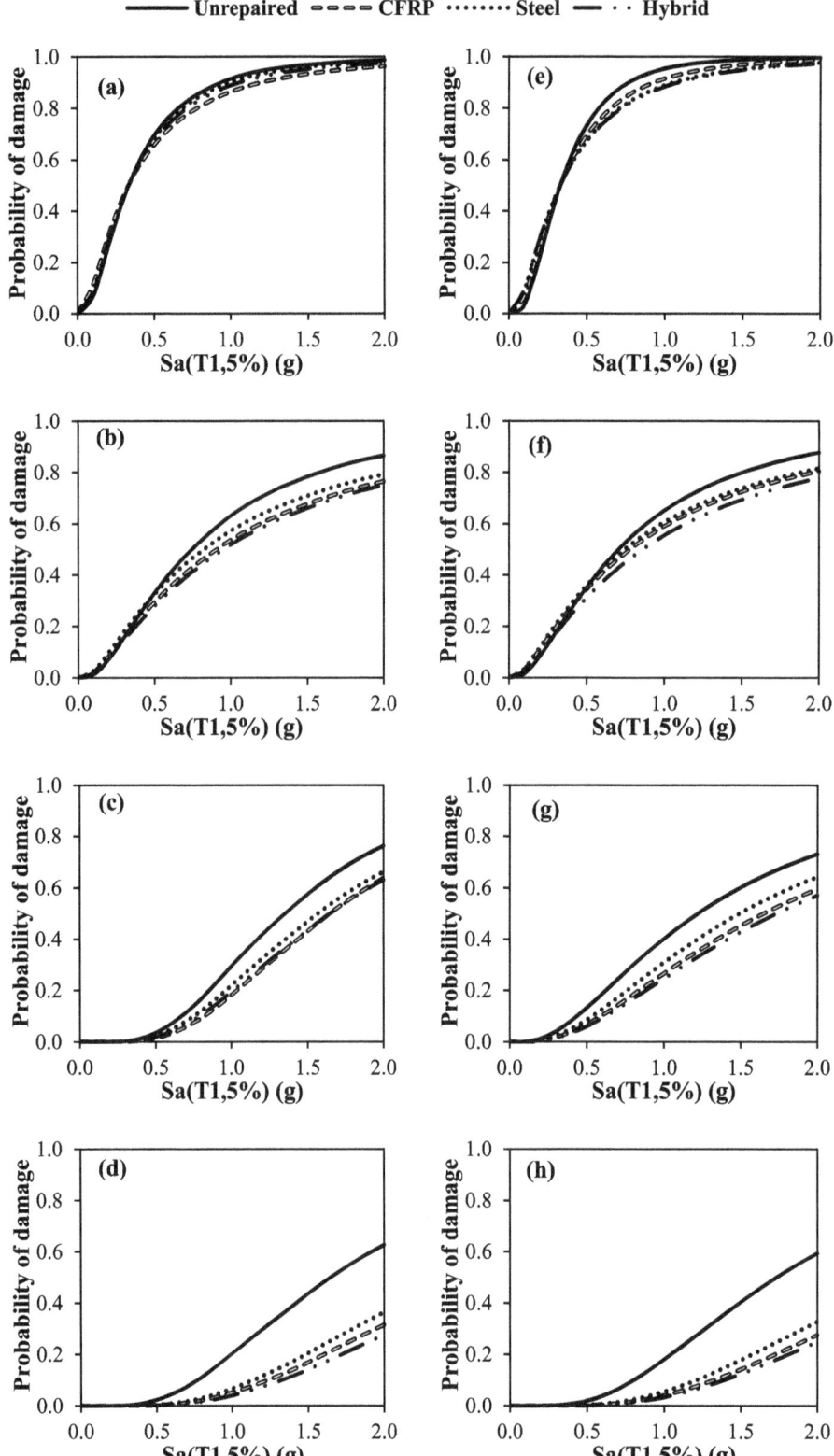

Fig. 14 Fragility curves for the unrepaired and repaired bridge under severe MS-moderate AS sequence: ensemble 1 records **a** minor, **b** moderate, **c** extensive, and **d** collapse damage state; ensemble 2 records **e** minor, **f** moderate, **g** extensive, and **h** collapse damage state.

(Figs. 13c, d, g, h, 14c, d, g, h). For instance, given the same probability of damage under the collapse damage state, the associated Sa of the unrepaired bridge is significantly less than that of the repaired bridge, irrespective of the repair jacket material. The fragilities accord to the obtained results from the IDA analyses results, where the repair jackets were more significant/critical for the severe MS-severe AS compared to the severe MS-moderate AS sequence. The fragilities clearly indicate the importance of repair jackets as the damage state increases. However, the influence of different intensity MS-AS sequences is clear through comparison of Figs. 13 versus 14. The obtained fragilities demonstrate the decrease of vulnerability for the unrepaired and repaired bridge under severe MS-moderate AS compared to severe MS-severe AS. It also suggests that the fragilities of unrepaired and repaired bridge are shifted more toward right under severe MS-moderate AS compared to severe MS-severe AS, which signals the lesser significance of repair jackets under moderate AS.

The fragilities also reveal that the steel jacketed bridge exhibits the highest vulnerability compared to CFRP and hybrid repair jackets at all damage states under both earthquake types. The increased vulnerability of the steel jacketed bridge over the other two repair interventions is due to the increased seismic action attracted by the thick steel jacketed bridge pier (inertia force of the dynamic system is increased due to thick jacket). For the conventional thick steel jacketing, the column cross section (deadweight) is increased, which in turn absorbs larger seismic actions (Priestley et al. 1996). On the other hand, the CFRP and hybrid jackets demonstrated comparable fragilities under all damage states and earthquake types. Although the hybrid jacket showed marginal decreased vulnerability compared to the CFRP jacket. Although the conventional steel jacket was shown to be the most vulnerable repair alternative, however, the steel jacket exhibited larger hysteresis loops compared to the CFRP jacket. Thus, fragility function for different repair interventions should be considered along with further investigation to better select the appropriate repair alternative. The reduced service stiffness of earthquake damaged RC bridge piers is an issue which steel jackets may be more advantageous over CFRP jackets, as they could provide higher service stiffness for the repaired RC pier (Youm et al. 2006).

Moreover, the fragility functions provide additional valuable information in regards to far-fault and near-fault records effects under different damage states. The obtained fragilities indicate that the vulnerability under near-fault records is larger than far-fault records for the minor and moderate damage states under severe MS-severe AS sequence (e.g. Fig. 13b vs. f). However, for collapse damage state, the fragilities under far-fault ground motions were slightly higher than those of the near-fault records (e.g. Fig. 13d vs. h). The obtained results indicate higher vulnerability for far-fault AS under collapse damage state for severe MS-severe AS. For minor and moderate damage states the near-fault record aftershocks are more prominent over far-fault aftershocks. For severe MS-moderate AS the near-fault records were more prominent. The obtained results suggest that maybe there is a relation between significance of earthquake type (far-fault or near-fault) and intensity of the seismic sequence considered (severe MS-severe AS or severe MS-moderate AS) at different damage states which requires further studies. Needless to mention, this observation on earthquake type effect at various damage states is limited only to the present study, and may not be generalized.

8. Conclusions

Results from a comprehensive study on the seismic collapse capacity and seismic fragility of mainshock-damaged highway RC bridges—repaired with repair jackets—that are representatives of older bridge construction practice are presented in this paper. Analytical models of the studied bridge were subjected to numerous series of mainshock-aftershock sequences of various intensities of different fault-types (i.e., far-fault and near-fault). The seismic collapse capacity of mainshock-damaged bridge was investigated through an extensive number of IDA analyses. Three different repair techniques (CFRP, conventional thick steel and hybrid jacket) were considered. The collapse capacity and drift capacity of repaired bridge pier under severe MS-severe AS and severe MS-moderate AS were quantified. Different repair jackets were compared to address the differences that may arise on the post-repair response of rehabilitated bridges subjected to potential aftershocks. This study is limited to bridge with regular geometry under two ensembles of record sets in one direction. Additional studies considering irregular bridge configuration, near-fault mainshocks and far-fault aftershocks and vice versa, in addition to subduction, inslab and interface record sets and the inclusion of multi-directional loading should be conducted to further understand contributions of other parameters on the seismic response of bridges. The impact of jacket application incorporated in more complex numerical models considering soil-structure interaction, bearings at expansion joints (pounding and unseating), and abutment backfill soil effects, with the inclusion of pile-footing-pier failure would be contributory in this area. According to the obtained results the following conclusions can be drawn from the present study:

1. The existing damage under a severe MS attack may significantly jeopardize the post-MS resilience of substandard bridge piers. Damage accumulation and destabilizing moments due to geometric nonlinearities (P-Δ effect) are significant for the post-MS drift capacity of the single column bridge piers with lap-spliced reinforcement at column-footing interface.

2. Severe MS-severe AS sequences could significantly reduce the mean structural collapse capacity/maximum drift capacity of substandard bridge piers compared to severe MS only attacks. MS-AS seismic sequences incur numerous inelastic excursions, irrespective of the AS intensity (severe or moderate) or fault type (near-fault or far-fault).

3. Repair jackets contribute on the post-MS collapse capacity of damaged bridges subjected to aftershocks. Repair jackets application on damaged bridge piers are less effective for the severe MS-moderate AS sequence compared to severe MS-severe AS scenario. However, repair jackets are still essential for severe MS-moderate AS sequence.

4. Under minor and moderate damage states, near-fault aftershocks were prominent over far-fault aftershocks for severe MS-severe AS sequence. Under extensive

and collapse damages states far-fault aftershocks appeared more significant compared to near-fault aftershocks for severe MS-severe AS sequence. However, near-fault aftershocks were prominent at all damage states under severe MS-moderate AS.

5. The unrepaired and repaired bridge exhibit approximately identical fragilities under the minor damage state. The fragilities for the unrepaired and repaired bridge largely deviate from each other under the extensive and collapse damage states.

6. The conventional thick steel jacket exhibits the highest fragility compared to CFRP repair jacket.

Acknowledgments

Some of the test data to validate the numerical model are provided by Professors R. Pinho and F. Bianchi. This help is highly appreciated. Financial support for this study was provided by the Department of Civil, Architectural, and Environmental Engineering at Missouri University of Science and Technology, and by the U.S. National Science Foundation under Award No. CMMI-1030399. The conclusions and opinions expressed in this paper are those of the authors only and do not necessarily reflect the official views or policies of the sponsors.

References

Abdelnaby, A., & Elnashai, A. (2014). Performance of degrading reinforced concrete frame systems under Tohoku and Christchurch earthquake sequences. *Journal of Earthquake Engineering 18*(7), 1009–1036. doi:10.1080/13632469.2014.923796.

Andrawes, B., Shin, M., & Wierschem, N. (2009). Active confinement of reinforced concrete bridge columns using shape memory alloys. *Journal of Bridge Engineering 15*(1), 81–89. doi:10.1061/ASCEBE.1943-5592.0000038.

Arias, A. (1970). A measure of earthquake intensity. In R. J. Hansen (Ed.), *Seismic design for nuclear power plants* (pp. 438–483). Cambridge, MA: MIT Press.

Aschheim, M., & Black, E. (1999). Effects of prior earthquake damage on response of simple stiffness-degrading structures. *Earthquake Spectra 15*(1), 1–24.

ATC-63. (2009). *Quantification of building seismic performance factors.* Redwood City, CA: FEMA P695.

Baker, J. W., & Cornell, C. A. (2006). *Vector-valued ground motion intensity measures for probabilistic seismic demand analysis.* Berkeley, CA: Pacific Earthquake Engineering Research Center, College of Engineering, University of California.

Baker, J. W., Lin, T., Shahi, S. K., & Jayaram, N. (2011). *New ground motion selection procedures and selection motions for the PEER transportation research program.* PEER report 2011/03, Berkeley, CA: Pacific Earthquake Engineering Research Center.

Bearman, C. F. (2012). *Post-earthquake assessment of reinforced concrete frames.* M.S. Thesis, Department of Civil and Environmental Engineering, University of Washington, Seattle, WA.

Berry, M. P., & Eberhard, M. O. (2007). *Performance modeling strategies for modern reinforced concrete bridge columns.* PEER report 2007/07. Berkeley, CA: Pacific Engineering Research Center, University of California.

Bianchi, F., Sousa, R., & Pinho, R. (2011). Blind prediction of a full-scale RC bridge column tested under dynamic conditions. In *Proceedings of the 3rd international conference on computational methods in structural dynamics and earthquake engineering (COMPDYN 2011) Corfu, Greece,* Paper no. 294.

Billah, A. M., Alam, M. S., & Bhuiyan, M. R. (2013). Fragility analysis of retrofitted multicolumn bridge bent subjected to near-fault and far-field ground motion. *Journal of Bridge Engineering.* doi:10.1061/(ASCE)BE.1943-5592.0000452.

Buckle, I., Friedland, I., Mander, J., Martin, G., Nutt, R., & Power, M. (2006). Seismic retrofitting manual for highway structures: Part 1-bridges (No. FHWA-HRT-06-032).

Calderone, A., Lehman, D. E., & Moehle, J. P. (2001). *Behavior of reinforced concrete bridge columns having varying aspect ratios and varying lengths of confinement.* Berkeley, CA: Pacific Earthquake Engineering Research Center.

California Department of Transportation. (2004). *Caltrans bridge design specification.* Sacramento, CA: California Department of Transportation.

Chang, L., Peng, F., Ouyang, Y., Elnashai, A. S., & Spencer, B. F, Jr. (2012). Bridge seismic retrofit program planning to maximize postearthquake transportation network capacity. *Journal of Infrastructure Systems 18*(2), 75–88.

Cho, J. Y., & Pincheira, J. A. (2006). Inelastic analysis of reinforced concrete columns with short lap splices subjected to reversed cyclic loads. *ACI Structural Journal 103*(2), 280–290.

Chopra, A. K., & Goel, R. K. (2002). A modal pushover analysis procedure for estimating seismic demands for buildings. *Earthquake Engineering and Structural Dynamics 31*(3), 561–582.

Cornell, C. A., Jalayer, F., Hamburger, R. O., & Foutch, D. A. (2002). Probabilistic basis for 2000 SAC federal emergency management agency steel moment frame guidelines. *Journal of Structural Engineering 128*(4), 526–533.

Di Sarno, L. (2013). Effects of multiple earthquakes on inelastic structural response. *Engineering Structures 56*, 673–681.

Dutta, A., & Mander, J. B. (1998). Seismic fragility analysis of highway bridges. In *Proceedings of the INCEDE-MCEER* center-to-center project workshop on earthquake engineering Frontiers in transportation systems (pp. 22–23).

ElGawady, M., Endeshaw, M., McLean, D., & Sack, R. (2009). Retrofitting of rectangular columns with deficient lap splices. *Journal of Composites for Construction 14*(1), 22–35.

Fakharifar, M., Chen, G., Arezoumandi, M., & ElGawady, M. (2015a). Hybrid jacketing for rapid repair of seismically damaged reinforced concrete columns. *Transportation Research Record: Journal of the Transportation Research Board.* doi:10.3141/2522-07.

Fakharifar, M., Chen, G., Lin, Z., & Woolsey, Z. (2014a). Behavior and strength of passively confined concrete filled tubes. In *The 10th U.S. National conference on earthquake engineering: July 21–25, 2014, Anchorage, AL.*

Fakharifar, M., Chen, G., Sneed, L., & Dalvand, A. (2015b). Seismic performance of post-mainshock FRP/steel repaired RC bridge columns subjected to aftershocks. *Composites Part B: Engineering.* doi:10.1016/j.compositesb.2014.12.010.

Fakharifar, M., Dalvand, A., Arezoumandi, M., Sharbatdar, M. K., Chen, G., & Kheyroddin, A. (2014b). Mechanical properties of high performance fiber reinforced cementitious composites. *Construction and Building Materials 71*, 510–520. doi:10.1016/j.conbuildmat.2014.08.068.

Fakharifar, M., Dalvand, A., Sharbatdar, M. K., Chen, G., & Sneed, L. (2015c). Innovative hybrid reinforcement constituting conventional longitudinal steel and FRP stirrups for improved seismic strength and ductility of RC structures. *Frontiers of Structural and Civil Engineering.* doi:10.1007/s11709-015-0295-9.

Fakharifar, M., Sharbatdar, M. K., & Lin, Z. (2013). Seismic performance and global ductility of reinforced concrete frames with CFRP laminates retrofitted joints. In *Structures congress 2013* (pp. 2080–2093). ASCE.

Fakharifar, M., Sharbatdar, M. K., Lin, Z., Dalvand, A., Sivandi-Pour, A., & Chen, G. (2014c). Seismic performance and global ductility of RC frames rehabilitated with retrofitted joints by CFRP laminates. *Earthquake Engineering and Engineering Vibration, 13*(1), 59–73.

FEMA 356. (2000). *Prestandard and commentary for the seismic rehabilitation of buildings.* Prepared by ASCE for Federal Emergency Management Agency, Washington, D.C.

Ferracuti, B., Pinho, R., Savoia, M., & Francia, R. (2009). Verification of displacement-based adaptive pushover through multi-ground motion incremental dynamic analyses. *Engineering Structures 31*(8), 1789–1799.

Ferracuti, B., & Savoia, M. (2005). Cyclic behaviour of FRP-wrapped columns under axial and flexural loadings. In *Proceedings of the international conference on fracture, Turin, Italy.*

Filippou, F. C., Popov, E. P., & Bertero, V. V. (1983). *Effects of bond deterioration on hysteretic behaviour of reinforced concrete joints.* Report EERC 83-19. Berkeley, CA: Earthquake Engineering Research Center, University of California.

Fragiadakis, M., Pinho, R., & Antoniou, S. (2008). Modelling inelastic buckling of reinforcing bars under earthquake loading. In M. Papadrakakis, D. C. Charmpis, N.

D. Lagaros, Y. Tsompanakis, & A. A. Balkema (Eds.), *Progress in computational dynamics and earthquake engineering.* Leiden, Netherlands: Taylor & Francis.

Grelle, S. V., & Sneed, L. H. (2013). Review of anchorage systems for externally bonded FRP laminates. *International Journal of Concrete Structures and Materials 7*(1), 17–33. doi:10.1007/s40069-013-0029-0.

Harajli, M., Hamad, B., & Karam, K. (2002). Bond-slip response of reinforcing bars embedded in plain and fiber concrete. *Journal of Materials in Civil Engineering 14*(6), 503–511.

Haroun, M. A., & Elsanadedy, H. M. (2005). Fiber-reinforced plastic jackets for ductility enhancement of reinforced concrete bridge columns with poor lap-splice detailing. *Journal of Bridge Engineering 10*(6), 749–757.

Haselton, C. B., Liel, A. B., Taylor Lange, S., & Deierlein, G. G. (2008). *Beam-column element model calibrated for predicting flexural response leading to global collapse of RC frame buildings.* PEER report 2007/03. Berkeley, CA: Pacific Engineering Research Center, University of California.

HAZUS-MH. (2003). *Multi-hazard loss estimation methodology earthquake model.* HAZUS-MH MR3 technical manual. Washington, DC: Federal Emergency Management Agency (FEMA).

He, R., Sneed, L. H., & Belarbi, A. (2013). Rapid repair of severely damaged RC columns with different damage conditions: An experimental study. *International Journal of Concrete Structures and Materials 7*(1), 35–50. doi:10.1007/s40069-013-0030-7.

Huang, W., & Andrawes, B. (2014). Seismic performance of SMA retrofitted multiple-frame RC bridges subjected to stong mainshock-aftershock sequences. In *10th U.S. National conference on earthquake engineering (10NCEE) Anchorage, Alaska.*

Jirawattanasomkul, T. (2013). *Ultimate shear behavior and modeling of reinforced concrete members jacketed by fiber reinforced polymer and steel.* PhD thesis. Hokkaido University, Sapporo, Japan.

Kent, D. C., & Park, R. (1973). Cyclic load behaviour of reinforcing steel. *Strain 9*(3), 98–103.

Kunnath, S. K., El-Bahy, A., Taylor, A. W., & Stone, W. C. (1997). *Cumulative seismic damage of reinforced concrete bridge piers.* In Technical report NCEER (No. 97-0006). US National Center for Earthquake Engineering Research.

Lee, D. H., Kim, D., & Lee, K. (2009). Analytical approach for the earthquake performance evaluation of repaired/retrofitted RC bridge piers using time-dependent element. *Nonlinear Dynamics 56*(4), 463–482.

Lee, D. H., Park, J., Lee, K., & Kim, B. H. (2011). Nonlinear seismic assessment for the post-repair response of RC bridge piers. *Composites Part B Engineering 42*(5), 1318–1329.

Li, Y., Song, R., & Van De Lindt, J. W. (2014). Collapse fragility of steel structures subjected to earthquake mainshock-aftershock sequences. *Journal of Structural Engineering 140*(12), 04014095.

Li, Y., Song, R., van de Lindt, J., Nazari, N., & Luco, N. (2012). Assessment of wood and steel structures subjected to earthquake mainshock-aftershock. In *15th world conference on earthquake engineering, Lisbon, Portugal.*

Lin, Z., Fakhairfar, M., Wu, C., Chen, G., Bevans, W., Gunasekaran, A. V. K., & Sedighsarvestani, S. (2013). *Design, construction and load testing of the Pat Daly Road Bridge in Washington County, MO, with internal glass fiber reinforced polymers reinforcement.* Report no. NUTC R275.

Mackie, K. R., & Stojadinovic, B. (2007). R-factor parameterized bridge damage fragility curves. *Journal of Bridge Engineering 12*(4), 500–510.

Madas, P., & Elnashai, A. S. (1992). A new passive confinement model for transient analysis of reinforced concrete structures. *Earthquake Engineering and Structural Dynamics 21*, 409–431.

Mander, J. B., Priestley, M. J. N., & Park, R. (1988). Theoretical stress–strain model for confined concrete. *Journal of Structural Engineering 114*(8), 1804–1826.

Marson, J., & Bruneau, M. (2004). Cyclic testing of concrete-filled circular steel bridge piers having encased fixed-based detail. *Journal of Bridge Engineering, ASCE 9*(1), 14–23.

Martinez-Rueda, J. E., & Elnashai, A. S. (1997). Confined concrete model under cyclic load. *Materials and Structures 30*(197), 139–147.

Menegotto, M., & Pinto, P. E. (1973). Method of analysis for cyclically loaded R.C. plane frames including changes in geometry and non-elastic behaviour of elements under combined normal force and bending. In *Symposium on the resistance and ultimate deformability of structures acted on by well defined repeated loads, international association for bridge and structural engineering, Zurich, Switzerland* (pp. 15–22).

Nazari, N., van de Lindt, J. W., & Li, Y. (2013). Effect of mainshock-aftershock sequences on woodframe building damage fragilities 1. *Journal of Performance of Constructed Facilities 29*(1), 04014036.

Nielson, B. G., & DesRoches, R. (2007). Analytical seismic fragility curves for typical bridges in the central and southeastern United States. *Earthquake Spectra 23*(3), 615–633.

PEER. (2010). Retrieved July 24, 2014, from http://nisee2.berkeley.edu/peer/prediction_contest/?page_id=25.

Pinto, A. V., Verzeletti, G., Pegon, P., Magonette, G., Negro, P., & Guedes, J., (1996). *Pseudo-dynamic testing of large-scale R/C Bridges.* Report EUR 16378, Ispra (VA), Italy.

Priestley, M. J. N., Calvi, G. M., & Kowalsky, M. J. (2007). *Displacement based seismic design of structures.* Pavia, Italy: Istituto Universitario di Studi Superiori Press.

Priestley, M. N., Seible, F., & Calvi, G. M. (1996). *Seismic design and retrofit of bridges.* New York, NY: Wiley.

Ribeiro, F. L., Barbosa, A. R., & Neves, L. C. (2014). Application of reliability-based robustness assessment of steel moment resisting frame structures under post-mainshock cascading events. *Journal of Structural Engineering 140*(8), A4014008.

Rodriguez, M. E., Botero, J. C., & Villia, J. (1999). Cyclic stress–strain behavior of reinforcing steel including effect of buckling. *Journal of Structural Engineering 125*(6), 605–612.

Saatcioglu, M., & Grira, M. (1999). Confinement of reinforced concrete columns with welded reinforced grids. *ACI Structural Journal 96*(1), 29–39.

Saatcioglu, M., & Yalcin, C. (2003). External prestressing concrete columns for improved seismic shear resistance. *Journal of Structural Engineering 129*(8), 1057–1070.

Schoettler, M. J., Restrepo, J. I., Guerrini, G., Duck, D. E., & Carrea, F. (2012). *A full-scale, single-column bridge bent tested by shake-table excitation.* Las Vegas, NV: Center for Civil Engineering Earthquake Research, Department of Civil Engineering, University of Nevada.

Scholz, C. H. (2002). *The mechanics of earthquakes and faulting.* Cambridge, MA: Cambridge University Press.

Scott, M. H., & Fenves, G. L. (2006). Plastic hinge integration methods for force-based beam–column elements. *Journal of Structural Engineering 132*(2), 244–252.

Seismosoft. (2013a). SeismoStruct—A computer program for static and dynamic nonlinear analysis of framed structures. www.seismosoft.com.

Seismosoft. (2013b). SeismoStruct ver. 6.0 and 7.0—Verification report.

Shamsabadi, A., Khalili-Tehrani, P., Stewart, J. P., & Taciroglu, E. (2009). Validated simulation models for lateral response of bridge abutments with typical backfills. *Journal of Bridge Engineering 15*(3), 302–311. doi:10.1061/(ASCE)BE.1943-5592.0000058.

Shinozuka, M., Feng, M. Q., Kim, H. K., & Kim, S. H. (2000). Nonlinear static procedure for fragility curve development. *Journal of Engineering Mechanics 126*(12), 1287–1295.

Sivaselvan, M., & Reinhorn, A. M. (1999). *Hysteretic models for cyclic behavior of deteriorating inelastic structures.* Report MCEER-99-0018, MCEER/SUNY/Buffalo.

Sivaselvan, M., & Reinhorn, A. M. (2001). Hysteretic models for deteriorating inelastic structures. *Journal of Engineering Mechanics ASCE 126*(6), 633–640, with discussion by Wang and Foliente and closure in Vol. 127, No. 11.

Spoelstra, M., & Monti, G. (1999). FRP-confined concrete model. *Journal of Composites for Construction, ASCE 3*, 143–150.

Tehrani, P., & Mitchell, D. (2013). Seismic risk assessment of four-span bridges in Montreal designed using the Canadian Bridge design code. *Journal of Bridge Engineering 19*(8), A4014002.

Terzic, V., & Stojadinovic, B. (2013). Hybrid simulation of bridge response to three-dimensional earthquake excitation followed by truck load. *Journal of Structural Engineering 140*(8), A4014010.

USGS. (2012). United States Geological Survey. http://www.usgs.gov/.

Vamvatsikos, D., & Cornell, C. A. (2002). Incremental dynamic analysis. *Earthquake Engineering and Structural Dynamics 31*(3), 491–514.

Villaverde, R. (2007). Methods to assess the seismic collapse capacity of building structures: State of the art. *Journal of Structural Engineering 133*(1), 57–66.

Vosooghi, A., & Saiidi, M. (2010). *Post-earthquake evaluation and emergency repair of damaged RC bridge columns using CFRP materials.* Rep. no. CCEER-10-05. Reno, NV:

Center for Civil Engineering Earthquake Research, Dept. of Civil Engineering, Univ. of Nevada.

Vosooghi, A., & Saiidi, M. S. (2012). Design guidelines for rapid repair of earthquake-damaged circular RC bridge columns using CFRP. *Journal of Bridge Engineering 18*(9), 827–836. doi:10.1061/(ASCE)BE.1943-5592.0000426.

White, T., & Ventura, C. E. (2004). Ground motion sensitivity of a Vancouver-style high rise. *Canadian Journal of Civil Engineering 31*, 292–307.

Xiao, Y., & Ma, R. (1997). Seismic retrofit of RC circular columns using prefabricated composite jacketing. *Journal of Structural Engineering 123*(10), 1357–1364.

Yang, J. (2009). *Nonlinear responses of high-rise buildings in giant subduction earthquakes.* PhD thesis. California Institute of Technology, CA, USA.

Yankelevsky, D. Z., & Reinhardt, H. W. (1989). Uniaxial behavior of concrete in cyclic tension. *Journal of Structural Engineering, ASCE 115*(1), 166–182.

Ying, X. F., Chen, G., Silva, P. F., LaBoube, R., & Yen, P. W. (2006). Thin steel sheet wrapping on RC columns and steel plate strengthening on beam-column joints for seismic ductility and capacity improvements. In *National conference on earthquake engineering, paper no. 513, conference proceeding, San Fransisco, CA, USA, April 18–22, 2006.*

Youm, K. S., Lee, H. E., & Choi, S. (2006). Seismic performance of repaired RC columns. *Magazine of Concrete Research 58*(5), 267–276.

5

Effect of Anchorage Number on Behavior of Reinforced Concrete Beams Strengthened with Glass Fiber Plates

Mustafa Kaya[1],*, and Zeynel Çağdaş Kankal[2]

Abstract: Reinforced concrete beams with insufficient shear reinforcement were strengthened using glass fiber reinforced polymer (GFRP) plates. In the study, the effect of the number of bolts on the load capacity, energy dissipation, and stiffness of reinforced concrete beams were investigated by using anchor bolt of different numbers. Three strengthened with GFRP specimens, one flexural reference specimen designed in accordance to Regulation on Buildings Constructed in Disaster Areas rules, and one shear reinforcement insufficient reference specimen was tested. Anchorage was made on the surfaces of the beams in strengthened specimens using 2, 3 and 4 bolts respectively. All beams were tested under monotonic loads. Results obtained from the tests of strengthened concrete beams were compared with the result of good flexural reference specimen. The beam in which 4 bolts were used in adhering GFRP plates on beam surfaces carried approximately equal loads with the beam named as a flexural reference. The amount of energy dissipated by strengthened DE5 specimen was 96 % of the amount of energy dissipated by DE1 reference specimen. Strengthened DE5 specimen initial stiffness equal to DE1 reference specimen initial stiffness, but strengthened DE5 specimen yield stiffness about 4 % lower than DE1 reference specimen yield stiffness. Also, DE5 specimen exhibited ductile behavior and was fractured due to bending fracture. Upon the increase of the number of anchorages used in a strengthening collapsing manner of test specimens changed and load capacity and ductility thereof increased.

Keywords: glass fibers, laminates, plates, adhesion, shear strengthening.

1. Introduction

In reinforced concrete beams very important shear problems are encountered due to projecting, material and application errors. Reinforced concrete specimens have the ductile behavior under bending effect. However, if these specimens have insufficient shear reinforcement, they are fractured suddenly and in a brittle manner.

In reinforced concrete beams, it is compulsory to place transverse reinforcement along the length of the beam in order to prevent shear cracks of beams (Regulation on Buildings Constructed in Disaster Areas 2007). Stirrups placed longitudinally as perpendicular to the reinforcement are used as shear reinforcement. In any beam with sufficient shear reinforcement bending cracks remain at low levels and the specimen exhibits the ductile behavior.

Some studies used steel plates or fiber reinforced polymer (FRP) sheets to improve the shear strength of reinforced concrete beams (Trianafillou 1998).

[1]Faculty of Engineering, Aksaray University, Ankara, Turkey.
*Corresponding Author;
E-mail: kaya261174@hotmail.com

[2]Republic of Turkey Ministry of Health, Ankara, Turkey.

In this study, the effect of the number of anchorages that prevent separation of GFRP plates from the beam surface on the shear strength of beams was investigated. Five reinforced concrete beams with T cross-section with a length of 4000 mm were designed. One flexural reference specimen designed according to the disaster regulation (Regulation on Buildings Constructed in Disaster Areas 2007), 1 shear deficient reference specimen with insufficient shear reinforcement, and 3 strengthened specimens with GFRP were designed. GFRP plate span and width used in all strengthened specimens were the same. The number of anchorages used on these specimens was determined as 2, 3 and 4.

All beams were tested under monotonic loads. At the end of the test load capacity, energy dissipation, stiffness, ductility and collapsing mechanism of strengthened concrete beams with the different number of anchorages were compared to flexural reference specimen.

Trianafillou (1998) tested beams with a length of 1000 mm, width of 70 mm and height of 110 mm and the shear reinforcement of which is not used by adhering their surfaces GFRP plate with different angle in the study performed. Khalifa and Nanni (2002) increased the shear strength of beams using GFRP in the different angle in the test specimens with T cross-section in the study they performed. Kachlakev and McCurry (2000) tested 4 beams with insufficient shear strength in the study they performed. One of the beams is a control specimen and another 3 of them are strengthened with GFRP and CFRP (Kachlakev and

McCurry 2000). Raghu et al. (2000) aimed at increasing the shear strength of reinforced concrete beams with T cross-section using carbon FRPs in the study they performed. For this purpose, they applied GFRP plates on all beam surfaces with and without anchorage (Raghu et al. 2000). Li et al. (2001) tested the beams with insufficient shear reinforcement in the study they performed. The effect of the amount of GFRP used to strengthened beams on beam shear strength was researched (Li et al. 2001). Ali et al. (2001) studied on separation mechanisms in the reinforcement of the beams in terms of bending and shear in the study they performed. In the study, the steel plates used for reinforcement and FRP plates were compared (Ali et al. 2001). Khalifa and Nanni (2000) increased the shear strength of beams with a rectangular cross-section using GFRP plate in the study they performed. Diagana et al. (2003) aimed at reinforcing rectangular beams with insufficient shear reinforcement against shear in the study they performed. GFRP plate was adhered on the surface of tested beams in 4 different forms. GFRP plates were adhered as perpendicular and 45° to the horizontal (Diagana et al. 2003). Wegian and Abdalla (2006) strengthened beams against shear in the study they performed. GFRP, CFRP, and FRP were used on the specimens tested in the study (Wegian and Abdalla 2006). Riyadh and Riadh (2006) aimed at reinforcing the reinforced concrete beams against shear and bending with GFRP plates in the study they performed. Anıl (2006) studied on strengthening of reinforced concrete beams against shear using GFRP plates. GFRP plate width and method of application of plates were determined as experiment parameters (Anıl 2006). Bencardino et al. (2007) strengthened beams without shear reinforcement against shear using GFRP in the study they performed. Kang et al. (2014) used carbon fibers (CF) and glass fibers (GF) combined to strengthen concrete flexural members. In their study, data of tensile tests of 94 hybrid carbon-glass FRP sheets and 47 carbon and GF rovings or sheets were thoroughly investigated in terms of tensile behavior (Kang et al. 2014). Kang and Ary (2012) used fiber-reinforced polymers (FRP) to enhance the behavior of structural components in either shear or flexure. The research focused on the shear-strengthening of reinforced and pre-stressed concrete (PC) beams using FRP (Kang and Ary 2012). Ary and Kang (2012) experimentally evaluated the impact of carbon fiber-reinforced polymers (CFRP) amount and strip spacing on the shear behavior of PC beams and evaluated the applicability of existing analytical models of FRP shear capacity of PC beams shear-strengthened with CFRP. Kang et al. (2012) reviewed the debonding failure of FRP laminates externally attached to concrete. They also discussed the influences on bond strength and failure modes as well as the existing experimental research and developed equations (Kang et al. 2012).

A review of the literature shows that there is very limited research being carried out on the effect of the number of anchorage specimens (bolt) providing anchorage of plates used for strengthening of reinforced concrete beams against shear fracture. In this study, the effect of the number of anchorages that provide connection of GFRP on the beam surface was investigated.

2. Experimental Study

2.1 General

The most important factor that determines the collapse mechanism of reinforced concrete beams is the ratio of shear span (a) to useful beam height (d). Flexural failures rather than shear failures will govern the capacity of moderately long beams a/d approximately equal to 5.

Upon the increase of load on the beam firstly bending cracks arise, and upon the increase off -tensile strengths bending cracks arise. As a result of the combination of one or several bending cracks with sloping cracks the brittle cracks occur. When brittle crack occurred test specimen breaks without significant deformation (strain), and absorb relatively little energy prior to fracture.

It was aimed to see higher load capacity, stiffness, energy dissipation, and ductile behavior from strengthened DE3, DE4, and DE5 specimen than DE1 good reference. This beam shear span determined as 1550 mm, useful height is determined as 330 mm and (a/d) determined as (4,7). This dimension approximately equal to 5 (Kankal 2011).

2.2 Detailing Test Specimens

The test specimens were detailed in 1/2 scale. Five beams were tested in the experimental program. The test specimens were designed as a T cross-section beam with a length of 4000 mm. In the beam cross-section web width was designed as 120 mm, beam height was designed as 360 mm, topping concrete width was designed as 320 mm and topping concrete depth was designed as 75 mm. 3Ø16, and 2Ø14 longitudinal reinforcement was used in all specimens as tensile reinforcement. The percentage of tensile reinforcement in the beams is $\rho = 0.0230$. Since this ratio is smaller than the balanced reinforcement ratio $\rho_b = 0.0305$ specified in TS 500 (2000). 2Ø8 longitudinal reinforcement was used as compressive reinforcement. Transverse reinforcements (stirrup) produced with Ø6 straight reinforcements were used in the beams. Stirrups were placed with a span of 75 mm in reference test specimens with sufficient shear reinforcement, shear transverse reinforcements were placed with a span of 300 mm in reference test specimens with insufficient shear reinforcement. The stirrup ratio in beams with insufficient shear reinforcement is $\rho_w = 0.00157$. With this stirrup ratio, it was aimed to keep shear strength at the low level. The geometric form and reinforcement plan test specimens are given in Fig. 1.

Total 5 beams were produced for the experimental study, namely 1 flexural reference specimen with sufficient shear reinforcement, 1 shear deficient reference specimen with insufficient shear reinforcement, and 3 strengthened specimens with GFRP plates. In the other 3 test specimens, GFRP plates were adhered to beam side surfaces. The epoxy based adherent is used in adhering GFRP plates. Plate thickness was designed as 5 mm, plate width was designed as (w_f) 90 mm and the span between the axes of plates (s_f) was designed as 100 mm. In all strengthened specimens, the anchorage was used in order to prevent separation of GFRP

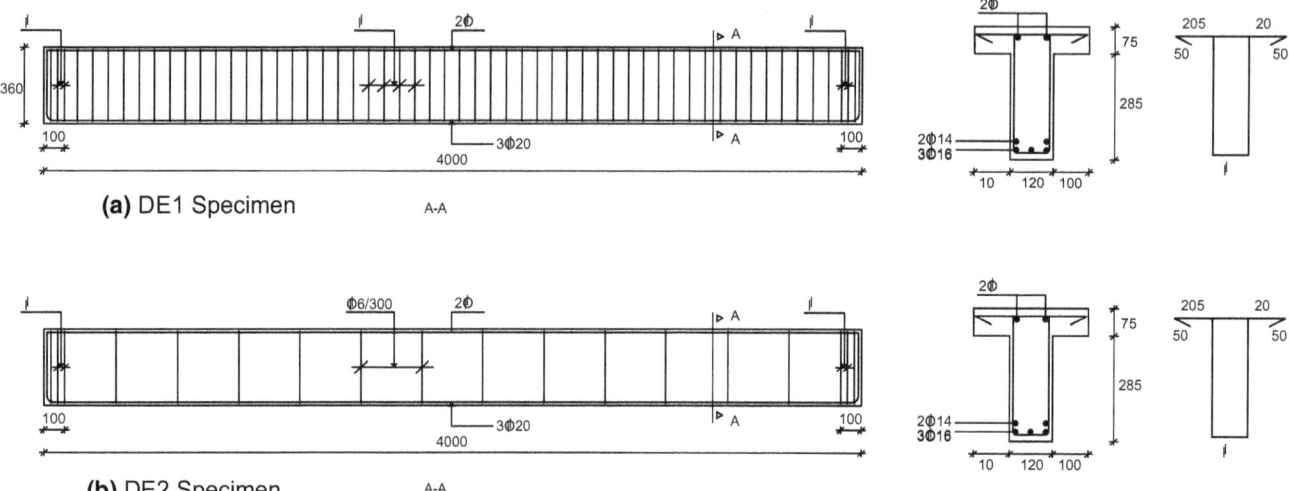

Fig. 1 DE1 and DE2 reference test specimens.

plates from the beam surface. The number of anchorages is different in each specimen. In the specimen with two anchorages, the span between anchorage axes was prepared as 180 mm, the specimen with three anchorages anchorage the span between anchorage axes was prepared as 90 mm and the specimen with four anchorages the span between anchorage axes was prepared as 60 mm. Reinforcement status of test specimens was given in Table 1. The plates were adhered to the beams to the same regions along the shear span. GFRP plate status of test specimens was given in Table 2.

In the Regulation of the Buildings to be Constructed in Disaster Regions (Trianafillou 1998), there is the condition that if FRP is used in the form of plates the span between the axes of plates (s_f) will be smaller than the sum of plate width (w_f) and one-fourth of the useful beam height (d) ($w_f + d/4$) (Trianafillou 1998). The span between plate axes to be used according to this condition should be maximum 172.5 mm ($90 + 330/4 = 172.5$). Since the span between the axes of GFRP plates is applied as 100 mm in this study, a design was made in conformity with the values in regulations. A strengthening technique performed in the beams by adhering carbon fiber and glass FRPs by Anil (2006).

In this study, the anchorage was used in order to prevent separation of GFRP plates from the beam surface. Keeping this study effect of the number of anchorages on the load capacity, stiffness, energy dissipation, ductility, and collapse mode of strengthened beams was investigated. In the experimental studies performed, the issue that the ratio

between the effective anchorage depth to anchorage diameter was considered as 5. The anchorage diameter was applied at approximately 8 mm in the experimental study mild steel bolt was used as an anchorage. Consequently, the ratio between the anchorage depth to anchorage diameter was provided. Anchorage status of test specimens was given in Table 3. GFRP plate placement used on the specimens and anchorage details was given in Fig. 2.

2.3 Properties and Strengths of Materials

For correct examination of the results of the experimental study, the test specimens were produced from materials with similar characteristics. For this purpose, the mechanical properties of the materials used in the experimental study became the same.

2.3.1 Concrete and Reinforcement

A compressive concrete strength of 16 MPa was used in this experimental study based on the average strength of an existing building collapsed during the 1999 Izmit earthquake event. Concrete samples were tested in order to determine the compressive strength for 28 days after being kept waiting in cure pool in the laboratory environment. The strength of the concrete samples is presented in Table 4. Since the difference between the compressive strength of the concrete samples did not exceed 2 %, normalization was not performed on the concrete strength. The properties of the reinforcements used in the test specimens were given in Table 5.

Table 1 Reinforcement status of test specimens.

Test specimens	(a/d)	Longitudinal reinforcement	Shear reinforcement	Shear reinforcement ratio
DE1 (reference)	4,7	3Ø16 + 2Ø14	Ø6/75	0.00628
DE2 (reference)	4,7	3Ø16 + 2Ø14	Ø6/300	0.00157
DE3 (strengthening)	4,7	3Ø16 + 2Ø14	Ø6/300	0.00157
DE4 (strengthening)	4,7	3Ø16 + 2Ø14	Ø6/300	0.00157
DE5 (strengthening)	4,7	3Ø16 + 2Ø14	Ø6/300	0.00157

Table 2 GFRP plate status of test specimens.

Test specimen	GFRP length (mm)	GFRP width (mm)	GFRP thickness (mm)	GFRP space (mm)
DE1 (reference)	–	–	–	–
DE2 (reference)	–	–	–	–
DE3 (strengthening)	280	90	5	100
DE4 (strengthening)	280	90	5	100
DE5 (strengthening)	280	90	5	100

Table 3 Anchor conditions of test specimens.

Specimen number	Anchor diameter (mm)	Anchor number	Anchor distance (mm)
DE1 (reference)	–	–	–
DE2 (reference)	–	–	–
DE3 (strengthening)	8	2	180
DE4 (strengthening)	8	3	90
DE5 (strengthening)	8	4	60

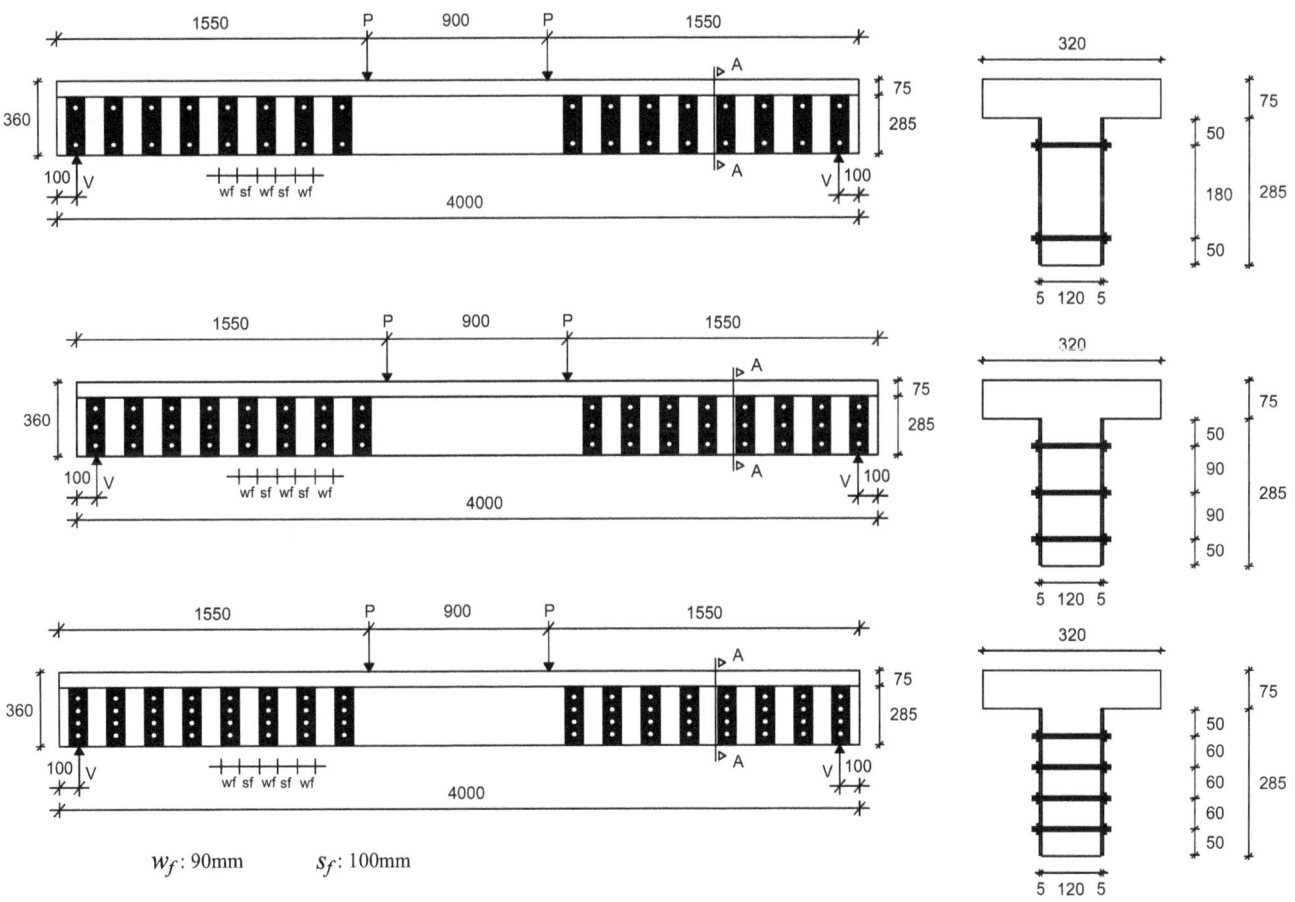

w_f: 90mm s_f: 100mm

Fig. 2 DE3, DE4 and DE5 reinforced test specimens.

2.3.2 GFRP

Glass fiber is composed of silica sand (SiO_2) which makes up the oxides and bases of sodium, calcium, aluminum, boron and iron specimens. Glass is the cheapest and most easily obtained strengthening material among the materials used in the production of fiber polymer. Glass fiber reinforcement polymer (GFRP) used in test specimens was produced by Dogus Plastic Industry (www.dogusplastiksanayi. com/index.php?pg=epoxy. 23 Haziran 2011). The properties of GFRP used in the test specimens were given in Table 6.

Table 4 Concrete specimens average compressive strengths.

Specimen number	Concrete compressive strength (MPa)
1	15.8
2	16.0
3	16.2
4	15.9
5	16.4
6	16.1

Table 5 Yielding and tensile strengths of the mild steel used at the experiments.

Reinforcement diameter	Steel class	Yielding stress (MPa)	Fracture stress (MPa)
Ø6	S220	390	630
Ø8	S420	440	670
Ø14	S420	450	680
Ø16	S420	470	695

Table 6 Properties of GFRP.

Unit weight (g/cm^3)	1.5–2.1
Tensile strength (MPa)	200–340
Impact strength (MPa)	33

2.3.3 Epoxy

Sikadur 31 (www.sika.com.tr/index.php?s=2&s2=products &s3=4. 23 Haziran 2011) was used to adhere GFRP plates on beam surfaces. Sikadur 31 is structural adhering and repair mortar with moisture tolerance, 2 components, containing epoxy resins and special filling. The mechanical/physical properties of Sikadur 31, epoxy adherent were given in Table 7.

2.4 Production of Test Specimens
2.4.1 Preparing Reinforcements

The production was started with the production of reinforcement in the beams. 3Ø16 and 2Ø14 ribbed reinforcement was used in the tensile surfaces of all beams. The tensile reinforcement ratio in beams is $\rho = 0.0230$. By designing the tensile reinforcement ratio in the beam as smaller than $\rho_b = 0.0305$, the balanced reinforcement ratio.

Table 7 Sikadur 31 epoxy mechanical and physical properties.

	Cure duration	Cure temperature	
		+20 °C	+10 °C
Compressive strength	1 days	40–45 N/mm^2	35–40 N/mm^2
	10 days	60–70 N/mm^2	50–60 N/mm^2
	Cure duration	Cure temperature	
		+10 °C ile +20 °C	
Flexural strength	10 days	30–40 N/mm^2	
Tensile strength	10 days	15–20 N/mm^2	
Adhesion strength	10 days (concrete)	3.0–3.5 N/mm^2	
	10 days (steel)	15 N/mm^2	
Modules of elasticity	4300 N/mm^2		
Density	1.65 kg/l		

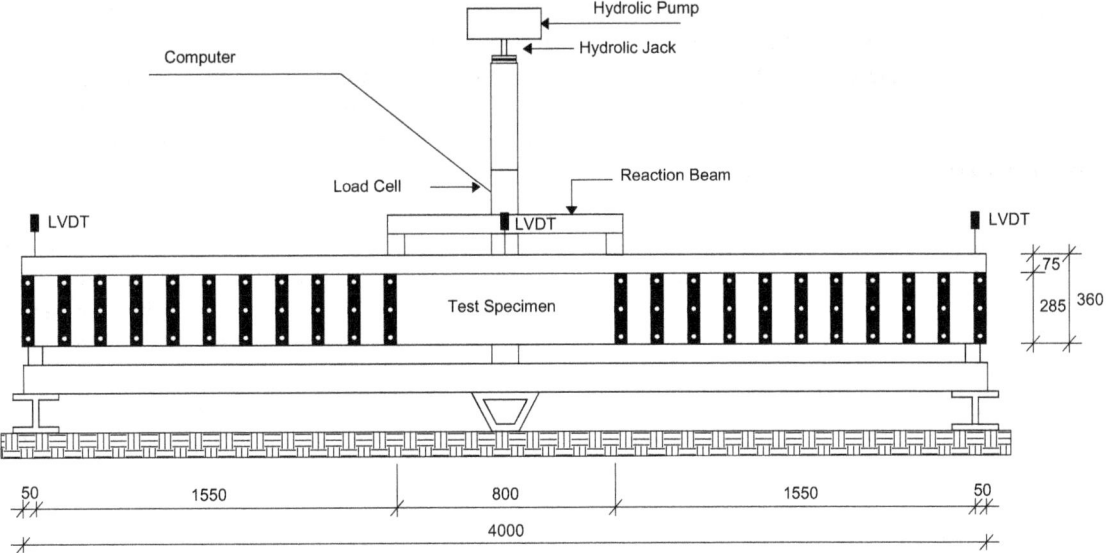

Fig. 3 Test setup and measurement system.

Shear reinforcements in DE1 were placed with spans of 75 mm. Straight closed stirrups with a diameter of 6 mm were used as shear reinforcement.

Stirrups with a diameter of 6 mm were placed with spans of 300 mm as shear reinforcement in all test specimens except for DE1. This stirrup ratio is $\rho_w = 0.00157$ and it is approximate ¼ of the stirrup ratio to be found. Using insufficient shear reinforcement in the specimens, it was aimed to create the shear crack. Three transverse reinforcements were placed in beam ends with spans of 30 mm in order to prevent the local break in beams.

2.4.2 Concrete Casting

Concrete casting was applied to test specimens which were made ready for concrete casting. C16 ready made concrete was used in the casting. The concrete vibrator was used during concrete casting in order to place the concrete homogeneously in the mold. Concrete test specimens which were removed from the mold and which took hardening of minimum 28 days were drilled from the marked points in order to fit anchorage bolts. The diameter of anchorage holes is 10 mm.

2.4.3 Preparing Anchorages

Mild steel bolts with diameters of 8 mm were used for reinforcement of test specimens. Anchorages were not used in reference test specimens numbered 1 and 2. Two anchorages were used per GFRP plate in the test specimens numbered 3. Three anchorages were used per GFRP plate in the test specimens numbered 4. Four anchorages were used per GFRP plate in the test specimens numbered 5.

2.4.4 Preparing GFRP Plates

The regions where GFRP plates will be adhered are marked inside surfaces of beams. Those regions marked were cleaned in order to create solid and clean adherence surface. Cleaning the concrete grouting from the surface area where GFRP plates will be adhered is very significant for providing adherence between the adhesive and the concrete. The points where anchorage bolts will be placed on GFRP plates were marked and the marked points were drilled in order to pass anchorage bolts in it. Drilled GFRP plates were made ready to be adhered to beam side surface. The surfaces where GFRP plates will be adhered were applied Sikadur 31 epoxy adherent with a thickness of 1 mm. After the GFRP plates were adhered to beam side surface and those plates were fixed with anchorage bolts, the strengthening process was completed and the beams to be tested were made ready to be placed in the loading system.

2.5 Loading and Measurement System

The loading program was applied to the specimens as load-controlled until the specimens collapsed. The loading program was manually applied to the specimens at the same loading velocity. The loads and displacements observed in the specimens during the loading steps were monitored via the computer display.

The load was applied to test specimens with 50 kN capacity mechanic pump connected to a hydraulic jack and the applied load was measured with 40 kN capacity load cell. The reaction was placed on the test specimens as simple support. The single load with P size transmitted to the reaction beam from the hydraulic jack was transmitted to the test specimens in the form of two equal single loads with $P/2$ sizes. The loading was started from zero and continued till collapse. Test setup and measurement system in the test specimens were given in Fig. 3.

Electronic measurement devices (LVDT) were placed on the midpoint of the beam on the sports in order to determine midpoint displacement of test specimens. To determine the midpoint displacement 200 mm capacity LVDT was used. To determine right, and left supports displacements 100 mm capacity LVDTS were used. The load, and displacement values were used to draw load–displacement curves of test specimens.

The net displacement at the end of the beams measured from the LVDT (D0) were equal to the difference of the average vertical displacements of the D1 and D2 LVDTS.

3. Experimental Test Results

The DE1 test specimen was a reference specimen with sufficient shear reinforcement prepared to be compared to strengthened test specimen.

The first bending crack in this specimen was occurred in the fixed moment region at the load of 45.4 kN. The midpoint displacement measured when the first crack occurred is 3.80 mm. Bending cracks at a load of 170.0 kN reached a width up to 1 mm. At the load of 180.24 kN tensile reinforcement started to yield. At that load, midpoint displacement was measured as 25.42 mm. While there was no definite increase in the load capacity of the specimen after that load, the midpoint displacement continued to increase. The highest load value in the test specimen was measured as 187.38 kN. The midpoint displacement at that value became 51.95 mm. At the load value of 185.96 kN, the test specimen collapsed as a result of the crash of concrete in the concrete compressive region. The midpoint displacement measured in the test specimen during the collapse is 56.08 mm. The collapse of this test specimen happened in the form of bending fracture. The view of the test specimen after a collapse is shown in Fig. 4.

The DE2 test specimen was a reference specimen with insufficient shear reinforcement prepared to be compared to strengthened test specimen. The first crack in DE2 was occurred in the fixed moment region in the form of bending crack at the load of 39.2 kN and at 3.98 mm midpoint displacement value. In parallel with the increase of loading, shear cracks arose. Two shear cracks became at the load of 130.0 kN in the right shear span. The test specimen collapsed as a result of the main shear crack at a load of 138.76 kN in the right shear span and reaching the topping concrete. When the specimen collapsed, midpoint displacement of 22.07 mm was measured in the test specimen. The collapse of this test specimen happened in the form of shear fracture. The view of DE2 after the collapse is shown in Fig. 5.

The DE3 test specimen was a strengthened with GFRP plates of which 2 anchorages were used. The first bending crack in DE3 occurred at the of 41.0 kN load and midpoint displacement of 3.64 mm. The first shear crack occurred at a load of 130.0 kN in the right shear span. The crack occurred in the plate numbered 5 and passed under the plate, and reached a thickness of 1 mm. Two main shear cracks became at the load of 168.0 kN in left shear span the plates numbered 3 and 6 passed and passed under the plates plates numbered 4 and 5. As a result of the arrival of this main crack to the topping concrete, the Specimen collapsed at a load of 169.08 kN. The midpoint displacement of the specimen measured at the load of collapse is 28.23 mm. The view of DE3 after a collapse is shown in Fig. 6. The collapse of this specimen happened in the form of shear failure.

The DE4 test specimen was a strengthened with GFRP plates of which 3 anchorages were used. The first bending crack in this specimen occurred at 43.8 kN load. At that load in the same specimen, midpoint displacement was measured as 3.79 mm. When it reached to a load level of 100.0 kN, the first shear crack occurred in the right shear span. The first shear crack that occurred in the right shear span passed under the plate numbered 12 reached to the plate numbered 13. The specimen collapsed at a load of 174.94 kN with sudden and brittle as a result of shear crack that occurred under the plates numbered 11 and 12 reaching to the topping concrete. The midpoint displacement of the specimen measured at the moment of collapse is 28.45 mm. The collapse of this specimen happened in the form of shear failure. The view of after a collapse is shown in Fig. 7.

The DE5 test specimen was a strengthened with GFRP plates of which 4 anchorages were used. The first bending crack in DE5 occurred in the maximum moment region at a load of 43.2 kN and at midpoint displacement of 3.65 mm. Upon the increase in load cracks in the specimen continued to increase. Tensile reinforcement began to yield at the load of 178.28 kN. The midpoint displacement at the moment of yield was measured as 25.44 mm. While the load applied stayed approximately the same upon the increase of the loading, the midpoint displacement continued to increase. Beam maximum load capacity was reached at a load of 186.04 kN and midpoint displacement value of 44.38 mm. Specimen collapsed upon the crush of topping concrete in

Fig. 4 View of the DE1 specimen after test finished.

Fig. 5 View of the DE2 specimen after test finished.

Fig. 6 View of the DE3 specimen after test finished.

the compressive region at a load of 185.30 kN. The midpoint displacement of 44.98 mm was measured at the load of collapse. The test specimen's collapse mechanism is in the mode of bending fracture. Tensile reinforcement in the specimen yielded and a ductile behavior was observed. The view of DE5 after a collapse is shown in Fig. 8.

4. Comparison of Experiment Results

The results obtained from the experiments were compared. In the comparison, response envelopes of test specimens were used (Fig. 9). Furthermore, test specimens' yield load, collapse load, yield and collapse displacements, stiffness, and energy dissipation capacities were compared (Table 8).

4.1 Behavior of Test Specimens

The load displacement curves of the reference test specimen DE1 with sufficient shear reinforcement, DE2 the reference test specimens with insufficient shear reinforcement, and strengthened with different numbers of anchorages DE3, DE4 and DE5 test specimens were compared in Fig. 9. The reference test specimen with sufficient shear reinforcement (DE1) is the flexural reference specimen to which strengthening was not applied. This test specimen carried 34 % more load compared to the insufficient shear reinforcement (DE2). DE1 shear reinforcement sufficient specimen carried 10 %

more load compared to the insufficient shear reinforcement DE3 specimen. DE1 test specimen carried 8 % more load compared to the insufficient shear reinforcement DE4 specimen. DE1 shear reinforcement sufficient specimen specimen carried approximately equal load with strengthened specimen (DE5) for which four anchorages were used.

The shear reinforcement sufficient reference specimen DE1 lost its load capacity at 185.96 kN. This specimen made 56.08 mm displacement at maximum load, and collapsed as a result of the flexural collapse.

The shear reinforcement insufficient reference specimen (DE2) lost its load capacity at 138.76 kN. This specimen made 22.07 mm displacement at maximum load, and collapsed as a result of the shear fracture.

The DE3 specimen for which 2 anchorages were applied to GFRP plates carried a load of 169.08 kN, and it collapsed at that load as a result of the shear fracture. The midpoint displacement of this specimen measured at the load of collapse reached to 28.23 mm. This specimen collapsed suddenly and in a brittle manner as a result of the shear crack.

The DE4 specimen for which 3 anchorages were applied to GFRP plates reached maximum load capacity at a load of 174.94 kN and the midpoint displacement was measured as 28.45 mm at that load. This specimen collapsed suddenly and in a brittle manner as a result of the shear crack.

The DE5 specimen for which 4 anchorages were applied to GFRP plates reached maximum load capacity at a load of 185.30 kN, and the midpoint displacement was measured as 44.98 mm at that load. This specimen carried approximately equal load with DE1 reference test specimen which is the flexural reference. This specimen reached higher load capacity, and midpoint displacement compared to other shear reinforcement insufficient DE2, strengthened DE3, and DE4 test specimens.

While the number of anchorages applied to GFRP plates increased, load capacity of specimens increased and DE5 specimen which was applied the highest number of anchorages (4 anchorages) carried approximately equal load with DE1 specimen This test specimen (DE5) exhibited ductile behavior and collapsed as a result of the yield of bending reinforcement (Table 9).

Fig. 7 View of the DE4 specimen after test finished.

Fig. 8 View of the DE5 specimen after test finished.

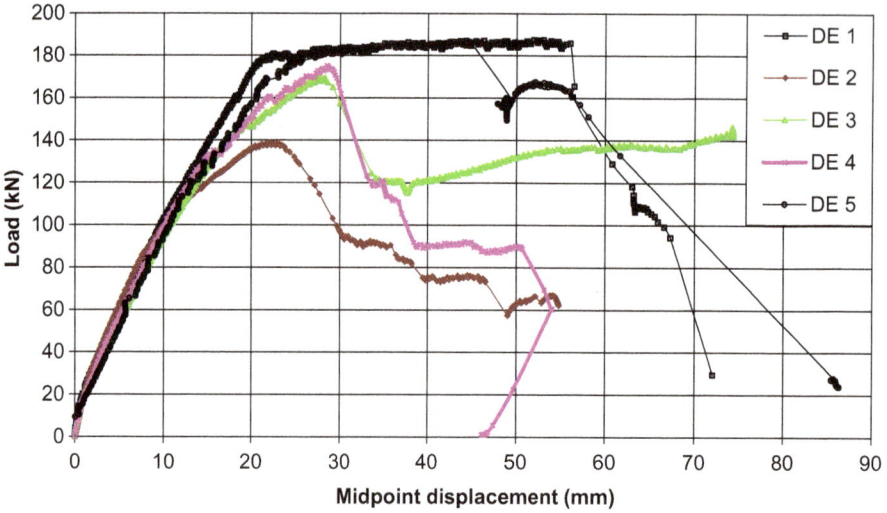

Fig. 9 Superpose load–displacement curves of the test specimens.

Table 8 Test results.

Specimen number	Yielding load (kN)	Yielding displacement (mm)	Fracture load (kN)	Fracture displacement (mm)	Fracture mechanism
DE1	180.24	25.42	185.96	56.08	Flexure
DE2	–	–	138.76	22.07	Shear
DE3	–	–	169.08	28.23	Shear
DE4	–	–	174.94	28.45	Shear
DE5	178.28	25.44	185.30	44.98	Flexure

4.2 Stiffness

Stiffness of the test specimens were assessed as initial stiffness, at the yield load, and at the collapse loads. Initial stiffness were determined as the slope at the place in the load–displacement curve where the first crack was seen. Similarly, stiffness at yield load was determined as the ratio of the load at the yield to the displacement at the yield, and stiffness at collapse was determined as the ratio of the load at the collapse to the displacement at the collapse. Initial stiffness, stiffness at yield, and stiffness at collapse values were given in Table 10.

The DE5 specimen exhibited approximately the same initial and yield stiffness with the reference shear reinforcement sufficient DE1 specimen. DE2, DE3, and DE4 strengthened specimens initial stiffness equal to DE1 DE % specimens initial stiffness. Since tensile reinforcement did not yield in DE2, DE3, and DE4 specimens, the yield stiffness could not be calculated.

5. Summary and Conclusions

In this study, strengthening of RC beams against shear fracture by using epoxy adhered GFRP plates with different anchorage number was investigated. Roughening of the concrete surfaces, cleaning of these surfaces and complying completely with the epoxy application procedures was crucial for successful bonding. Also anchorage bolts strengths, process order was very important for this study. In these type applications, after the plates adhered on the beam faces, anchorages installed in the holes before epoxy hardened. Also, 1 mm thick steel material used between plate and beam faces to provide uniform epoxy thickness. Results obtained from the experimental research are as follows:

Load capacity of strengthened test specimens has increased compared to the load capacity of the shear deficient reference specimen DE2. DE3 carried 22 % more load than DE2, DE4 carried 26 % more load than DE2, and DE5 carried 34 % more load than DE2. Increasing the number of anchorages used for strengthening, increased the load capacity of the specimen as well.

Upon the increase of the number of anchorages used in a strengthening collapsing manner of test specimens changed and load capacity thereof increased. The DE5 specimen for which the highest amount of anchorages were used was fractured by exhibiting ductile behavior as a result of the yield of bending reinforcement.

The dissipated energy amount by strengthened DE5 specimen was 4 % lower than the dissipated energy amount by strengthened DE1.

Table 9 Displacement–ductility ratio.

Specimen number	Yielding displacement	Fracture displacement	Ductility ratio
DE1	25.42	56.08	221
DE2	–	22.07	–
DE3	–	28.23	–
DE4	–	28.45	–
DE5	25.44	44.98	177

Table 10 Stiffness of test specimens.

Test specimen	Initial stiffness	Yielding stiffness	Fracture stiffness
1	11.9	71	3.31
2	9.85	–	6.29
3	11.26	–	5.99
4	11.56	–	6.15
5	11.84	68	4.12

The collapse of the DE5 test specimen happened in the form of bending fracture. While the number of anchorages used in GFRP plates increased in strengthening, ductility of specimens increases.

Upon using anchorage separation of GFRP plates from beam surface in strengthened test specimens, was prevented. No separation of GFRP plates from beam surface was observed in the test specimens.

No cracks were seen in GFRP plates, because these plates thickness and strength was enough to carry forces applied them.

References

Ali, M. S. M., Oehlers, D. J., & Park, S. M. (2001). Comparision between FRP and steel plating of reinforced concrete beams. *Composites Part A Applied Science and Manufacturing, 32*(9), 1319–1328.

Anıl, O. (2006). Improving shear capacity of RC T-beams using GFRP composites subjected to cyclic load. *Cement & Concrete Composites, 28*(7), 638–649.

Ary, M. I., & Kang, T. H.-K. (2012). Shear-strengthening of reinforced & prestressed concrete beams using FRP: Part I—Review of Previous Research. *International Journal of Concrete Structures and Materials, 6*(1), 41–48.

Bencardino, F., Spadea, G., & Swamy, R. N. (2007). The problem of shear in RC beams strengthened with GFRP laminates. *Construction and Building Materials, 21*(11), 1997–2006.

Diagana, A., Li, A., Gedalia, B., & Delmas, Y. (2003). Shear strengthening with CFF strips. *Engineering Structures, 25*(4), 507–516.

Kachlakev, D., & McCurry, D. D. (2000). The behavior of full-scale reinforced concrete beams retrofitted for shear and flexural with FRP laminates. *Composites Part B· Engineering, 31*(6–7), 445–452.

Kang, T. H.-K., & Ary, M. I. (2012). Shear-strengthening of reinforced & Prestressed concrete beams using FRP: Part II—Experimental Investigation. *International Journal of Concrete Structures and Materials, 6*(1), 49–57.

Kang, T. H.-K., Howell, J., Kim, S., & Lee, D. J. (2012). A State-of-the-Art Review on Debonding Failures of FRP Laminates Externally Adhered to Concrete. *International Journal of Concrete Structures and Materials, 6*(2), 123–134.

Kang, T. H.-K., Kim, W., Ha, S.-S., & Choi, D.-U. (2014). Hybrid Effects of Carbon-Glass FRP Sheets in Combination with or without Concrete Beams. *International Journal of Concrete Structures and Materials, 8*(1), 27–42.

Kankal, Z. Ç. (2011). Effect of Anchorage Number on Strengthening Reinforced Concrete Beams Against Shear With GFRP, Ms. Thesis. Aksaray, Turkey: Aksaray University.

Khalifa, A., & Nanni, A. (2000). Improving shear capacity of existing RC T-section beams using GFRP composites. *Cement & Concrete Composites, 22*(3), 165–174.

Khalifa, A., & Nanni, A. (2002). Rehabilitation of rectangular simply supported RC beams with shear deficiencies using GFRP composites. *Construction and Building Materials, 16*(3), 135–146.

Li, A., Diagana, C., & Delmas, Y. (2001). GFRP contribution to shear capacity of strengthened RC beams. *Engineering Structures, 23*(10), 1212–1220.

Raghu, A., Bette Meyer, M. M., Myers, J. J., & Nanni, A. (2000). An assessment of in situ FRP shear and flexural strengthening of reinforced concrete joists, ASCE Structures Congress, Philadelphia, PA, M. Elgaaly, Ed., May 8–10, CD version, 8.

Regulation on Buildings Constructed in Disaster Areas. (2007). The Ministry of Public Works and Settlement, Ankara, Turkey.

Riyadh, A., & Riadh, A. (2006). Coupled flexural-shear retrofitting of RC beams using GFRP straps. *Composite Structures, 75*(1–4), 457–464.

Trianafillou, T. C. (1998). Shear strengthening of RC beams using epoxy bonded FRP composites. *ACI Structural Journal, 95*(2), 107–115.

Wegian, F. M., & Abdalla, H. A. (2006). Shear capacity of concrete beams reinforced with fiber reinforced polymers. *Composite Structures, 71*(1), 130–138.

URL-1, www.dogusplastiksanayi.com/index.php?pg=epoxy. 23 Haziran 2011.

URL-2, www.sika.com.tr/index.php?s=2&s2=products&s3=4. 23 Haziran 2011.

6

Modeling of Mechanical Properties of Concrete Mixed with Expansive Additive

Hyeonggil Choi[1], and Takafumi Noguchi[2],*

Abstract: This study modeled the compressive strength and elastic modulus of hardened cement that had been treated with an expansive additive to reduce shrinkage, in order to determine the mechanical properties of the material. In hardened cement paste with an expansive additive, hydrates are generated as a result of the hydration between the cement and expansive additive. These hydrates then fill up the pores in the hardened cement. Consequently, a dense, compact structure is formed through the contact between the particles of the expansive additive and the cement, which leads to the manifestation of the strength and elastic modulus. Hence, in this study, the compressive strength and elastic modulus were modeled based on the concept of the mutual contact area of the particles, taking into consideration the extent of the cohesion between particles and the structure formation by the particles. The compressive strength of the material was modeled by considering the relationship between the porosity and the distributional probability of the weakest points, i.e., points that could lead to fracture, in the continuum. The approach used for modeling the elastic modulus considered the pore structure between the particles, which are responsible for transmitting the tensile force, along with the state of compaction of the hydration products, as described by the coefficient of the effective radius. The results of an experimental verification of the model showed that the values predicted by the model correlated closely with the experimental values.

Keywords: expansive additive, compressive strength, elastic modulus, pore volume, modeling.

1. Introduction

The application of an expansive additive is known to be an effective means of reducing shrinkage and increasing crack resistance. Thus, such an application is gradually becoming more common in construction projects (Choi et al. 2012a, b). In this study, we attempted to theoretically model the compressive strength and elastic modulus of hardened cement that had been treated with an expansive additive to reduce the shrinkage. The compressive strength and elastic modulus of hardened cement paste mixed with an expansive additive are closely related to the structure formation of the hardened cement paste by the hydration reaction of the cement and expansive additive (Al-Rawi 1976; Woods 1933; Taplin 1959). In other words, hydrates are generated as a result of the hydration between the cement and expansive additive; these hydrates then fill up the pores in the hardened cement. Consequently, a dense, compact structure is formed through the contact between the particles of the expansive additive and cement, which leads to the

manifestation of the strength and elastic modulus. Hence, it is important to determine the change in the organizational structure of a hardened cement paste by the progression of the hydration to estimate the strength and elastic modulus of a hardened cement paste mixed with an expansive additive.

In this study, the compressive strength and elastic modulus of a hardened cement paste mixed with an expansive additive were modeled and validated by considering the pore structure of a hardened cement paste based on the hydration, as shown in Fig. 1.

2. Modeling of Compressive Strength

2.1 Equation of Strength Development

Many studies have reported the relationship between the pore volume and strength (Maruyama 2003; Ryshkewitch 1953; Schiller 1958). In this study, it was assumed that the strength development of a cement paste mixed with an expansive additive was closely related to the pore volume based on the results of the existing studies. A typical example showing that the strength development behavior demonstrates an application possibility from the viewpoint of the pore volume and strength of a cement paste is shown in the following equations of Ryshkewitch (Ryshkewitch 1953) and Schiller (Schiller 1958).

$$f_c = f_o \cdot e^{-BP} \quad \text{(Ryshkewitch's equation)}$$
$$\text{: In the case of low porosity} \tag{1}$$

[1]Graduate School of Engineering, Muroran Institute of Technology, Hokkaido, Japan.

[2]Graduate School of Engineering, The University of Tokyo, Tokyo, Japan.

*Corresponding Author; E-mail: noguchi@bme.arch.t.u-tokyo.ac.jp

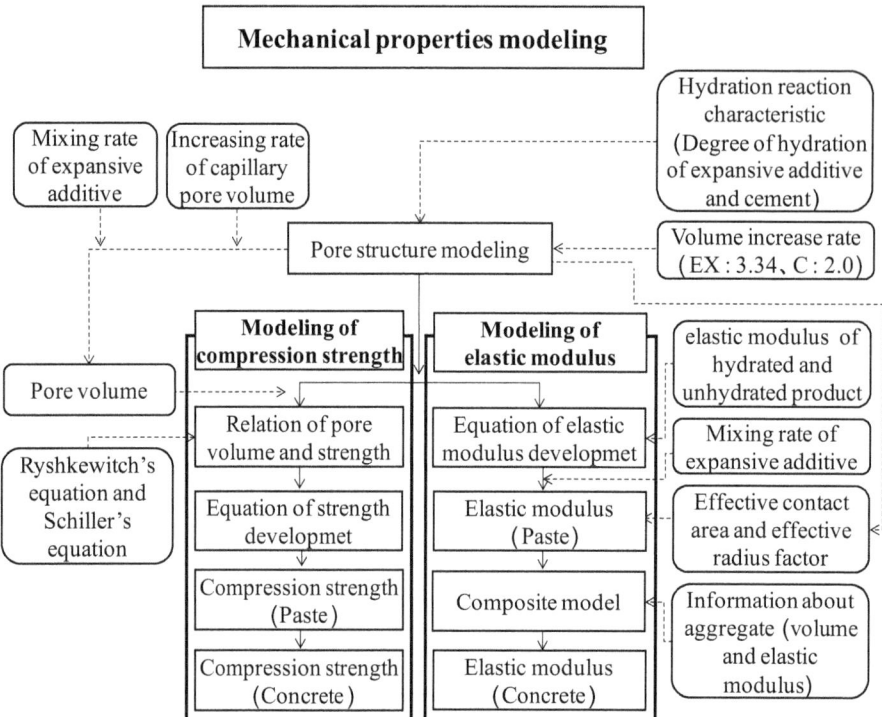

Fig. 1 Flow of mechanical properties modeling of concrete using expansive additive.

$$f_c = C \cdot ln(P_{cr}/P)$$

(Schiller's equation): In the case of high porosity   (2)

where f_c is the compressive strength of a hardened cement paste, P is the porosity, f_o is the strength of the hardened cement paste when the porosity is zero, P_{cr} is the porosity of the hardened cement paste when the strength is zero and B and C are material constants.

Similar to the above equations of Ryshkewitch (1953) and Schiller (1958), many researchers have suggested various forms for the equations describing the relationship between the porosity and strength. However, the relationships represented by these equations are rather simple, and their applicability is limited because only experimental values were used for the parameters in the equations. On the other hand, because all of the cement particles are the same size, Katsura et al. (1996) defined the strength of a cube of cement paste as the cement's original strength, where the strength of the hydrated cement paste could be expressed in terms of the contact between the cement particles within the cube produced by hydration. This strength is the strength of a cement paste with zero porosity. That is, the relative strength $R_s = f_0/f$, which is the ratio of the original strength f_o to the strength with pores f, is compared to the contact area, which has a linear relationship with the porosity, suggesting that the B value of Ryshkewitch's equation is 5.0, and the P_{cr} value of Schiller's equation is 0.523. In addition, in the C-CBM model suggested by Maruyama (2003), based on the assumption of this contact area, the contact area A_c produced by increasing the maximum radius of a hydrate R_t is modeled and applied in the following equation.

Case of $(0.5 \leq R_t < \sqrt{2}/2)$:
$$A_c = \pi(R_t^2 - 0.5^2)$$ (3)

Cas of $(\sqrt{2}/2 \leq R_t < \sqrt{3}/2)$
$$A_c = 8\left\{\frac{1}{2}\left(R_t^2 - 0.5^2\right)\left(\frac{\pi}{4} - ACOS\left(\frac{0.5}{\sqrt{R_t^2 - 0.5^2}}\right)\right) + 0.25\sqrt{R_t^2 - 0.5}\right\}$$

The suggested B value of Ryshkewitch's equation is 5.43 for a low pore volume of less than 0.3 based on the concept of the contact area in the C-CBM model (Maruyama 2003). In addition, for the high-porosity range, values of 64.5 and 0.523 are suggested for C and P_{cr} of Schiller's equation, respectively, based on the least squares method, with the data showing the relationship between the porosity and strength from Schiller's equation.

In this study, based on the equations of Ryshkewitch (Ryshkewitch 1953) and Schiller (1958), we referred to the existing studies of Maruyama (2003; Katsura et al. 1996) to calculate the coefficient of each porosity. We suggest the following equations for the compressive strength based on the relationship between the strength and porosity.

$$f_c = 182 \cdot e^{-5.215P} : 0.3 > P$$ (4)

$$f_c = 73 \cdot ln(0.523/P) : 0.3 \leq P$$ (5)

2.2 Modeling of Micro-pore Structure

The total porosity of cement paste mixed with an expansive additive was acquired by adding the porosities of the cement and expansive additive parts based on a hydration reaction. This study utilized an existing space formation model (Maruyama 2003; Park and Lee 2005; Park 2004) based on the hydration of the cement, which was used to acquire the porosities of the cement and expansive additive.

As shown in Fig. 2, the entire modeling area is represented as a 1 cm^3 cube. The density of water is denoted as ρ_w, the density of the expansive additives is denoted as ρ_e, and the water-to-binder ratio is denoted as x. Given these conditions, the volume of the expansive additive in the cube can be expressed as follows:

$$V_{c,e} = \frac{1}{x \cdot \frac{\rho_{c,e}}{\rho_w} + 1} \tag{6}$$

The radius r_o of a particle of the expansive additives in the 1 cm^3 cube is obtained as

$$r_o = \left(\frac{3V_{c,e}}{4\pi}\right)^{\frac{1}{3}} \tag{7}$$

If the particle radii of the expansive additive before and after the hydration reaction are assumed to be r_o and r_u, respectively, and if the outermost radius of the particle after the reaction is assumed to be R_t, then the hydration reaction rate can be defined as the volume of the hydration reaction divided by the volume of the expansive additive before the hydration reaction. Thus, the following equation is obtained:

$$\alpha = \frac{\frac{4}{3}\pi r_o^3 - \frac{4}{3}\pi r_u^3}{\frac{4}{3}\pi r_o^3} = 1 - \left(\frac{r_u}{r_o}\right)^3 \tag{8}$$

In addition, the volume increase rate V can be defined as the volume of the produced hydration product divided by the volume of the portion of the expansive additive that has undergone the reaction. Thus, V can be expressed as

$$V = \frac{\frac{4}{3}\pi R_t^3 - \frac{4}{3}\pi r_u^3}{\frac{4}{3}\pi r_o^3 - \frac{4}{3}\pi r_u^3} = \frac{R_t^3 - r_u^3}{r_o^3 - r_u^3} \tag{9}$$

From Eqs. (8) and (9), the following equation can be obtained:

$$R_t = (1 + (V-1)\alpha)^{\frac{1}{3}} \cdot r_o \tag{10}$$

Here, the hydration reaction rate a can be used to define the amounts of cement and expansive additive that have

reacted from their corresponding total amounts. The calories generated during hydration were measured using a multi-microcalorimeter (MMC-511 SV), and this value was used to determine the hydration reaction rate.

When both the W/C and water/expansive additive ratios are 0.50, the net calorific values are 437.53 J/g and 887.12 J/g for the cement and expansive additive, respectively. Figure 3 shows an example of the calculations for the rate of heat liberation and degree of hydration. These confirmed that at every age, the hydration reaction rate for the expansive additive was higher than that of the cement, because of an increase in the exothermic peak through the rapid reaction of the expansive additive at an early age.

On the other hand, for the cement volume that contributes to the reaction in the hydration reaction model, the volume increase rate V of the cement is obtained by adding approximately 75 % of the water volume in the total volume of gel with the volume of gel produced through the hydration reaction by chemically combining cement and water that constitute approximately 25 % of the weight of cement (Tashiro 1993). Therefore, the volume changes calculated using the cement density indicate that the appropriate volume increase rate during the complete hydration of the cement is approximately 1.59. However, the volume increase rate is expected to increase if the adsorbed water present in the hydration product or the water in the gel pores that does not contribute to the hydration is considered. For this reason, in previous studies (Powers and Brownyard 1947; van Breugel 1997), the volume increase rate of cement was defined as being in the range of 1.9–2.2. Using this range as a reference, in this study, we set the volume increase rate to 2.0 by considering the time that it converged to a certain value with regard to the hydration reaction rate, along with the adsorbed water present in the hydration product or the water in the gel pores that did not contribute to the hydration reaction.

In previous studies (Yanimoto et al. 2003) that described the expansion mechanism of an expansive additive, the volume increase rate V of the expansive additive was acquired with the assumption that the hydration products, which expand due to the expansive additive, were calcium hydroxide and ettringite. Yaniamoto et al. used the volume

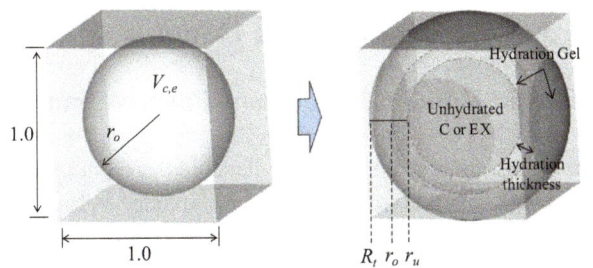

Fig. 2 Hydration reaction of cement and expansive additive particles.

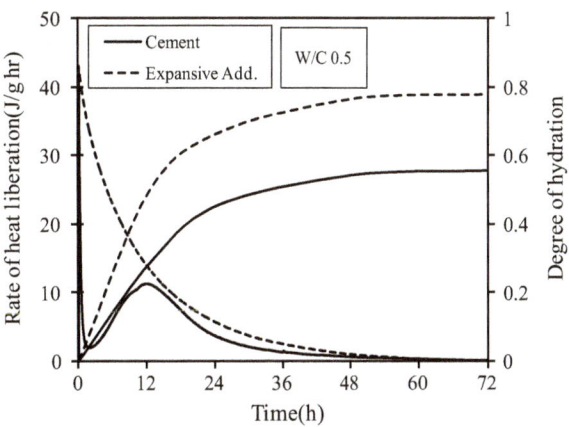

Fig. 3 Rate of heat liberation and degree of hydration.

of the layer of a hydrate product containing pores that was generated by the expansive additive for the volume of the hydration product that contributed to the expansion by the expansive additive. Here, in relation to the assumption made by Yaniamoto et al. about their model, we note that for a hydration product produced by hydration, a mixed layer containing pores of calcium hydroxide and ettringite is formed on the particle surface of the expansive additive. In addition, half of the ettringite generated by the expansive additive occurs on the surface of the C_3A in the cement. In other words, in a case where only the reaction of the expansive additive is considered, it is necessary to consider the ettringite hydration product on the surface of the C_3A in the cement. Therefore, in this study, the volume of the hydration product layer produced by the hydration of only the expansive additive was acquired by considering the volume ratio of each of the assumed hydration products (calcium hydroxide and ettringite) that were expanded as a result of the expansive additive. Figure 4 shows the volume of the ettringite hydration product created on the surface of the C_3A in the cement by the hydration of the expansive additive. This is the volume of the hydration product with the porosity created by the hydration reaction of the expansive additive. Considering the volume change of the hydrates, as previously discussed, the volume increase rate V of the expansive additive is defined as being 3.34, which is a relatively constant value.

The outermost radius R_t of the cement and expansive additive particles can be acquired using Eq. 10 by considering the volume increase rate V. Figure 5 shows a schematic view of the space formation of the cement and expansive additive particles by the outermost radius R_t. The cement particles are divided into an unlimited period, an early period of contact, and a late period of contact based on the outermost radius R_t. Then, based on the surface area and volume of each particle, the porosity can be calculated by subtracting the volumes of the particles from the volume of the cube.

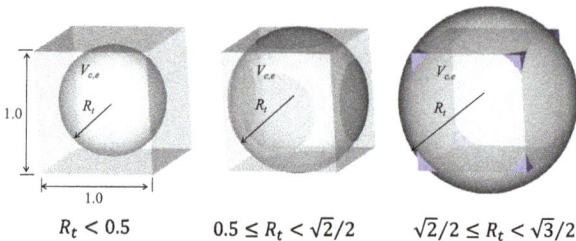

Fig. 5 Schematic view of space formation of cement and expansive additive particles (Maruyama 2003; Park 2004).

On the other hands, with the addition of an expansive additive, the hydration reaction generates capillary pores with a size range of 0.1 to several micrometers, which cause expansion. The volume increase rate of the expansive additive V already contains the pores created by the hydration reaction. Therefore, those pores need to be considered. In other words, the porosity of the expansive additive can be acquired from the values listed in Table 1, which consider the incremental factor p_{EX} of the capillary pores by the volume expansion of the expansive additive. Incremental factor p_{EX} of the capillary pores was acquired by deducing a regression equation for the porosity increase rate according to the mixing rate of the expansive additive by referring to the experimental results of existing studies (Yanimoto et al. 2003; Morioka et al. 1999).

Therefore, the total pore volume of the cement paste mixed with the expansive additive can be expressed by the following equation, depending on the mixture rate of the expansive additive according to the pore volume of the cement and expansive additive required by the hydration reaction.

$$P_{c+EX} = P_c \cdot C_{V(\%)} + P_{EX} \cdot EX_{V(\%)} \tag{11}$$

where P_{c+EX} is the porosity of the cement paste mixed with the expansive additive, P_c is the porosity of the cement part, P_{EX} is the porosity of the expansive additive part, $C_{V(\%)}$ is the volume mixing rate of the cement, and $EX_{V(\%)}$ is the volume mixing rate of the expansive additive.

The previous discussion showed that the compressive strength of cement paste mixed with an expansive additive can be acquired using Eqs. (4) and (5), which show the relations of the pore volume and strength at each range of porosity using the total pore volume acquired by Eq. (11).

3. Modeling of Elastic Modulus

3.1 Elastic Modulus of Cement Paste Without Pore

For the strength development of hardened cement, the strength and porosity seem to be highly correlated because of the stress concentration, which occurs as a result of the strength and pores of the hydration product. However, it is not easy to determine the characteristics of the elastic modulus using only its relationship with the pores because

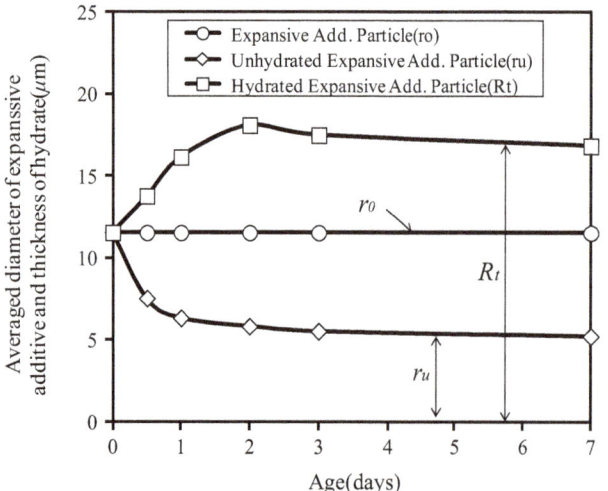

Fig. 4 Volume change of hydration products with expansive additive.

Table 1 Area of surface, and volume and porosity of particles by outermost radius R_t (Maruyama 2003).

Limitless part: $R_t < 0.5$	
Area of a surface	$S_1 = 4\pi R_t^2$
Volume	$V_1 = \frac{4}{3}\pi R_t^3$
Porosity	$P_1 = (1 - V_1) \cdot p_{EX}$
First period of contact: $0.5 \leq R_t < \sqrt{2}/2$	
Area of a surface	$S_2 = 4\pi R_t^2 - 12\pi\left(1 - \frac{0.5}{R_t}\right)$
Volume	$V_2 = \frac{4}{3}\pi R_t^3 - 6\pi\left(\frac{2}{3}R_t^3 - \frac{1}{2}R_t^2 + \frac{1}{24}\right)$
Porosity	$P_2 = (1 - V_2) \cdot p_{EX}$
Later period of contact: $\sqrt{2}/2 \leq R_t < \sqrt{3}/2$	
Area of a surface	$S_3 = \displaystyle\int\limits_{\sqrt{R_t^2 - 1/2}}^{1/2} \int\limits_{(R_t^2 - 1/2)/(4-x^2)}^{1/2} \frac{R_t}{\sqrt{R_t^2 - x^2 - y^2}}\,dx$
Volume	$V_3 = 2\sqrt{R_t^2 - 1/2} + 16\displaystyle\int\limits_{\sqrt{R_t^2-1/2}}^{1/2}\left(\frac{1}{2}\cdot 0.5 \cdot \sqrt{R_t^2 - x^2 - 1/4}\right.$ $\left. + \frac{R_t^2 - x^2}{2}\times\left[\frac{\pi}{4} - Arc\,cos\left(\frac{0.5}{\sqrt{R_t^2 - x^2}}\right)\right]\right)dx$
Porosity	$P_3 = (1 - V_3) \cdot p_{EX}$

the compressive strength and elastic modulus show the same characteristic in relation to the porosity. However, the relationship between the porosity and compressive strength is expressed as a spatial probability distribution, which is dependent on the existence probability of the weakest point in the particle connections of the hardened cement, whereas the correlation between the porosity and elastic modulus is a parameter that represents the spatial structure that shows how the stress is transferred in the state before reaching the point of destruction (Maruyama 2003). Based on these characteristics, when the paste matrix is thought of as a hardened part of the gel, non-hydrated cement, and non-hydrated expansive additive, and as a two-phase pore material, elastic modulus models for the hardened part with and without pores are needed, both of which use the concept of the contact area. On the other hand, for a hardened cement mixed with the expansive additive, two elastic moduli are needed, one for the cement part and the other for the expansive additive part. In consideration of the mix ratio of the expansive additive, each elastic modulus must have a balancing equation.

In accordance with the proportion λ_P, which is occupied by the non-hydrated cement, non-hydrated expansive additive, and hydration product, the model of the elastic modulus for the nonporous part has the shape shown in Fig. 6. To model the spatial structure, the volume ratio $V_{C,EX}$ of the non-hydrated cement to non-hydrated expansive additive is needed, based on the following equation.

$$\lambda_P = \sqrt{V_{C,EX}} \tag{12}$$

On the other hand, in the modeling of the elastic modulus, we assume that the gel, non-hydrated cement, and non-hydrated expansive additive exist in a certain ratio based on the hydration reaction ratio.

$$V_{C,EX} = \frac{(1 - \alpha)}{(1 - \alpha) + v_{gel(C,EX)}\alpha} \tag{13}$$

$$V_{gel(C,EX)} = 1 - V_{C,EX} \tag{14}$$

where α is the hydration reaction rate, $v_{gel(C,EX)}$ is the volume ratio for the produced gel of the cement and expansive additive volumes that contribute to hydration, $V_{C,EX}$ is the volume ratio of the unhydrated cement or unhydrated expansive additive, and $V_{gel(C,EX)}$ is the volume

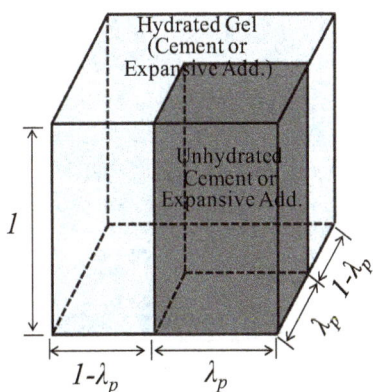

Fig. 6 Schematic view of elastic modulus of cement paste without pores.

ratio of the gel. In this case, the elastic modulus can be expressed by the following equation.

$$E_{u-paste(C,EX)} = \cfrac{1}{\cfrac{\lambda_p}{\lambda_p \cdot E_{gel(C,EX)} + (1-\lambda_p) \cdot E_{C,EX}} + \cfrac{1-\lambda_p}{E_{gel(C,EX)}}} \quad (15)$$

where $E_{u-paste(C,EX)}$ is the elastic modulus of the hardened cement paste or hardened expansive paste without pores, $E_{C,EX}$ is the elastic modulus of the unhydrated cement or unhydrated expansive additive, and $E_{gel(C,EX)}$ is the elastic modulus of the hydration product.

Here, $E_{C,EX}$ and $E_{gel(C,EX)}$ have values of 50 and 25 GPa, respectively. These values were taken from the results of the study by Maruyama (2003) on the calculation of the elastic modulus of cement. Maruyama determined these values by referring to the values of 40 and 20 GPa suggested by Hua et al. (1997) and 60 and 30 GPa suggested by Lokhorst and van Breugel (1997). In the case of the expansive additive, the cement value was applied because sufficient data could not be found.

3.2 Elastic Modulus of Cement Paste with Pore

The elastic modulus with pores was represented in the C-CBM model (Maruyama 2003) and modeled using the contact area concept, particularly the effective radius factor and effective contact area. The contact area in the C-CBM model is the characteristic value of the area that transfers stress through the connectivity of the particles. However, because the early hydration reaction occurs with the spatial expansion of an external product and becomes increasingly densified because of the hydration, it cannot be expressed using the contact area alone. That is, the early hydration reaction enlarges the radius of the external hydration product R_t, as shown in Fig. 7. Given that the thickness and density of the external product remain at a certain degree, the ratio of the actual external product to its thickness $R_t - r_0$ is the effective radius factor E_{fr} and can be applied to the effective contact area A_{ceff}, which is used for modeling. This concept is also applied to the modeling of the elastic modulus with pores.

Effective radius factor E_{fr} and effective contact area A_{ceff} of C-CBM are shown in the following equations.

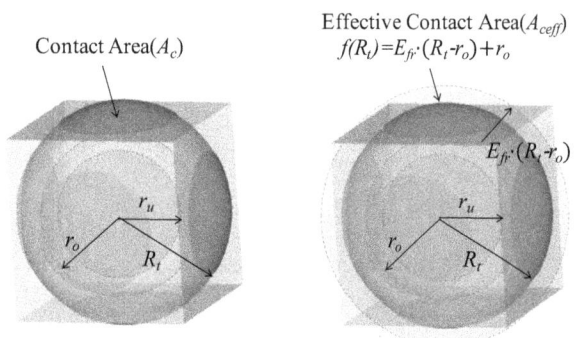

Fig. 7 Schematic view of effective radius factor and effective contact area.

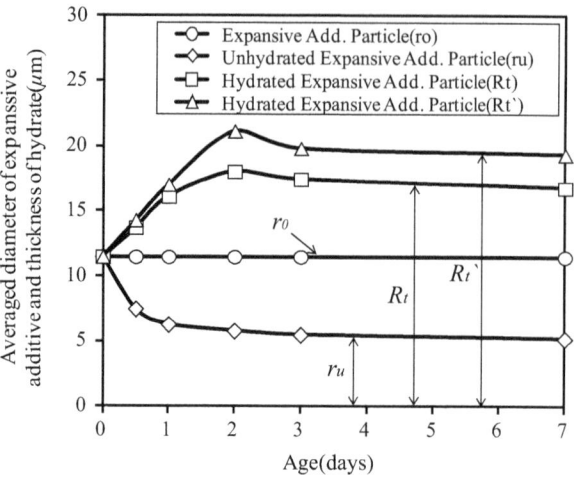

Fig. 8 Volume change of hydration layer of expansive additive by considering hydration at early age.

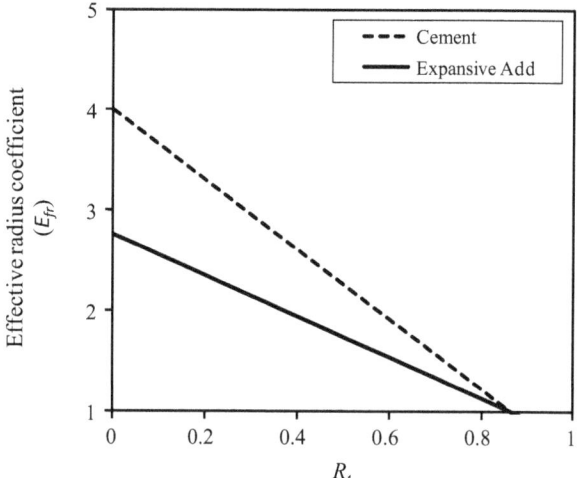

Fig. 9 Change in E_{fr} by R_t.

$$E_{fr} = E_{fro} \cdot \left(1 - \frac{2}{\sqrt{3}} R_t\right) + 1 \quad (16)$$

$$A_{ceff} = f(R_t), \quad f(R_t) = E_{fr} \cdot (R_t - r_o) + r_o \quad (17)$$

where E_{fro} is the initial value of the effective radius factor. In the case of cement, a value of 3.0 was applied, which was selected by giving equal consideration to the development of the elastic modulus in previous studies using the C-CBM model. In the case of the expansive additive, it was necessary to obtain the increase in the radius of external products by the hydration reaction of the expansive additive at an early age, as shown in Fig. 8, by referring to previous studies (Yanimoto et al. 2003). This was used to derive the effective radius factor of the expansive additive, which here has an initial value of 1.759. Figure 9 shows the change in the effective radius factor E_{fr} by the outermost particle radius R_t for the cement and expansive additive. The initial effective radius factors of the cement and expansive additive are different because the volume increase rate V for the expansive additive already considers the pores generated as a result of the hydration reaction. Function $f(R_t)$ can be acquired using Eq. (3), which is a function to express the relation of the outermost particle radius R_t and apparent effective contact area A_{ceff}.

Therefore, the elastic modulus of the cement paste part and expensive additive paste part $E_{paste(C,EX)}$ can be obtained by the following equation using the external contact area A_{ceff}.

$$E_{paste(C,EX)} = E_{u-paste(C,EX)} \cdot A_{ceff} \qquad (18)$$

In addition, the elastic modulus of the paste mixed with the expansive additive can be expressed by the following equation, depending on the mixture rate of the expansive additive according to the balance between the required cement-part elastic modulus and expansive additive-part elastic modulus.

$$E_{paste(C+EX)} = E_{paste(C)} \cdot C_{V(\%)} + E_{paste(EX)} \cdot EX_{V(\%)} \qquad (19)$$

where $E_{paste(C+EX)}$ is the elastic modulus of the cement paste mixed with the expansive additive, $E_{paste(C)}$ is the elastic modulus of the cement part, $E_{paste(EX)}$ is the elastic modulus of the expansive additive part, $C_{V(\%)}$ is the volume mixing rate of the cement, and $EX_{V(\%)}$ is the volume mixing rate of the expansive additive.

3.3 Elastic Modulus of Concrete

In order to extend the elastic modulus to concrete, the concrete is considered to consist of two materials, the aggregate and paste. It is also assumed that there is no behavior by the aggregate. Hence, the behavior of the paste dominates in the behavior of the concrete. Therefore, the aggregate has a resistor function without affecting the behavior of the paste.

Assuming that a composite model for the aggregate and paste is established based on the aforementioned assumptions and ideas, the elastic modulus of the concrete can be estimated from the elastic modulus of the paste, as shown in Fig. 10.

The elastic modulus of the concrete can be expressed by the following equation without considering Poisson's ratio.

$$E_{concrete(C+EX)} = \cfrac{1}{\cfrac{\lambda_c}{\lambda_c \cdot E_{paste(C+EX)} + (1-\lambda_c) \cdot E_{agg}} + \cfrac{1-\lambda_c}{E_{paste(C+EX)}}} \qquad (20)$$

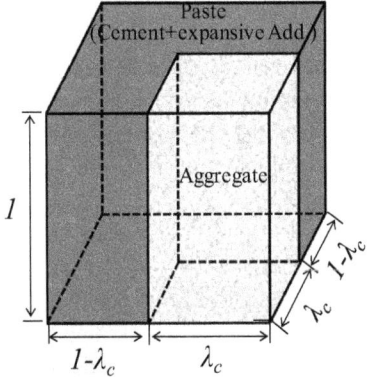

Fig. 10 Composite model of elastic modulus for concrete.

where $E_{concrete(C+EX)}$ is the elastic modulus of the concrete mixed with the expansive additive, $E_{paste(C+EX)}$ is the elastic modulus of the cement paste mixed with the expansive additive, and E_{agg} is the elastic modulus of the aggregate.

In addition, λ_c is a constant of the volume of the aggregate V_{agg} in the unit volume, and can be expressed by the following equation.

$$\lambda_c = \sqrt{V_{agg}} \qquad (21)$$

4. Verification of Model

4.1 Compressive Strength

An experiment was performed to verify the compressive strength model. Water and binder at a ratio of 0.50 was mixed with 0, 5, and 10 % expansive additive (Ettringite-gypsum type, Density 3.05 g/cm^3). The increase in the coefficient p_{EX} of the capillary pores generated by the volume increase with this expansive additive admixture was set by referring to the studies of Yaniamoto and Morioka (Yanimoto et al. 2003; Morioka et al. 1999), in which a mercury porosimeter test measured the change in a specimen's porosity.

For the increase rate of the capillary pores following the mixing of the expansive additive, a regression equation was derived for the admixture of the expansive additive and the increase rate of the pores, as shown in Fig. 11. By considering this, the total porosity was found for the case of mixing in the expansive additive.

Figure 12 shows the change in the capillary pore volume of the paste. We confirmed that the capillary pore volume increased with the admixture of the expansive additive and decreased with the progress of the hydration. Figure 13 shows the compressive strength estimates found using the model and the values measured in an experiment, based on the total porosity. Because our water/binder ratio was 0.50, the porosity found by the model was in the range of a high porosity of 0.3 or greater, and Eq. (5) from Schiller was used for the estimated values. The figure also shows Morioka's

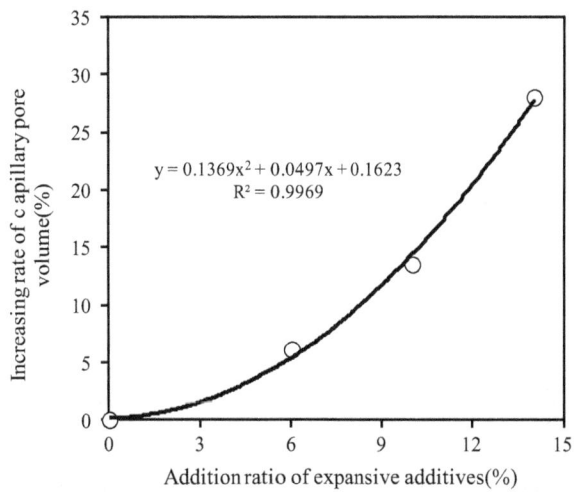

Fig. 11 Increasing rate of capillary pores by addition ratio of expansive additive.

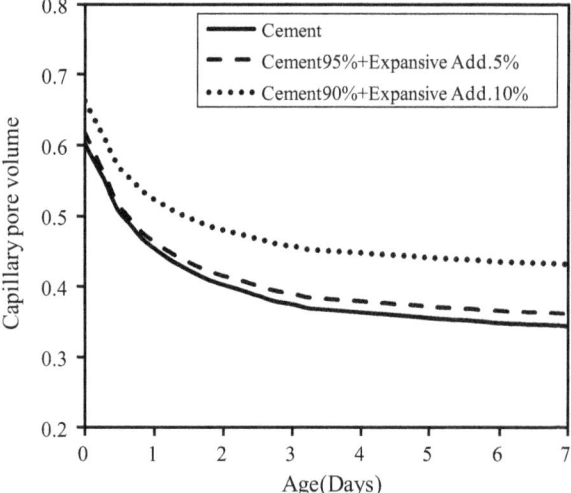

Fig. 12 Change in capillary pore volume.

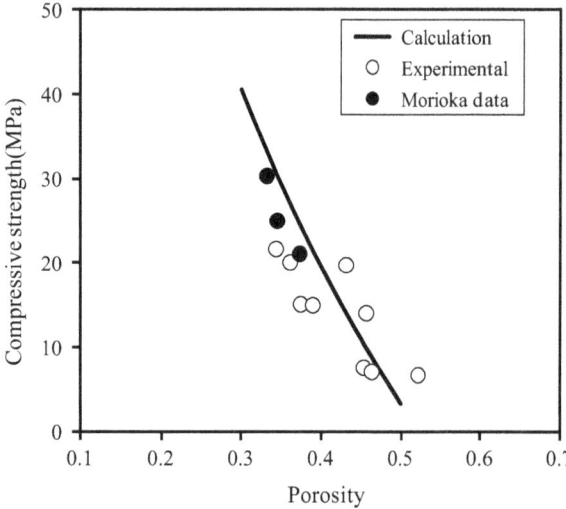

Fig. 13 Compressive strength development by model compared with experimental data.

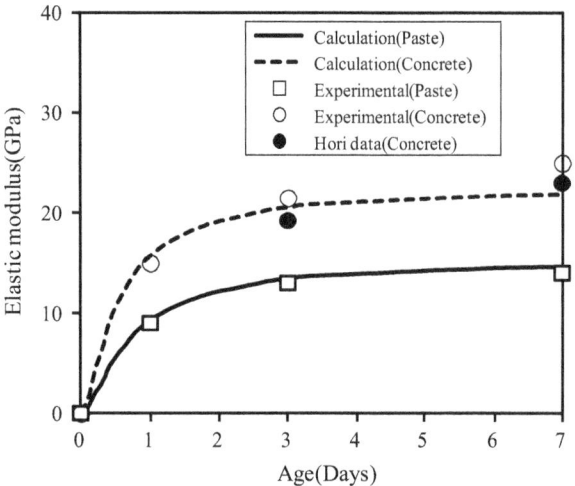

Fig. 14 Elastic modulus development by model compared with experimental data.

data (Morioka et al. 1999), in which the expansive additive was mixed at 10 and 12 % and aged for 7 days, with a water/binder ratio of 0.50. As shown in the figure, we confirmed

that the compressive strength estimates were positively correlated with the experiment values.

4.2 Elastic Modulus

To verify the elastic modulus model, we measured the elastic modulus of the cement paste after 1, 3, and 7 days, in which 10 % expansive additive(Ettringite-gypsum type, Density 3.05 g/cm^3) was mixed as a binder, where the water/binder ratio was 0.50. The actual values for the paste measured in the experiment and the model's estimated values are shown in Fig. 14. In addition, this figure shows the estimated elastic modulus of the concrete and Hori's data for the elastic modulus of the concrete after 3 and 7 days, where the concrete included 30 kg (approximately 10 % of the cement) of the expansive additive, with a water/binder ratio of 0.50 (Hori et al. 2000). Elastic modulus of aggregates were used 60 GPa for the calculation on elastic modulus of concrete. As shown in the figure, we confirmed that the elastic modulus estimates were positively correlated with the experiment values. It was confirmed that the aging of the elastic modulus of the concrete followed the estimate suggested in Hori's data and experimental data, indicating that the composite model can estimate the elastic modulus of concrete based on the behavior of the paste.

5. Conclusions

In this study, the compressive strength and elastic modulus of a hardened cement paste mixed with an expansive additive to control cracking were modeled, and the models were verified. The results of the study are summarized below.

(1) In the modeling of the mechanical properties of the concrete using the expansive additive, in the hardened cement paste with an expansive additive, hydrates are generated as a result of the hydration between the cement and expansive additive; these hydrates then fill up the pores in the hardened cement. Consequently, a dense, compact structure is formed through the contact between the particles of the expansive additive and the cement, leading to the manifestation of the strength and elastic modulus. Hence, modeling was performed to evaluate the compressive strength by assuming that the strength development of the cement paste was closely related to the pore volume.

(2) On the other hand, the elastic modulus of the hardened part without pores was modeled by assuming that the paste matrix consisted of the gel, non-hydrated cement, and hardened part of the non-hydrated expansive additive, with two-phase pores in the case where it contained pores. The elastic modulus was modeled using the concept of the effective radius factor and effective contact area.

(3) The estimates of the models were positively correlated with experimental values, which verified the compressive strength model and elastic modulus model. In addition, the elastic modulus of the concrete could

effectively be estimated based on the composite model of the aggregate and paste.

References

Al-Rawi, R. S. (1976). Choice of curing temperature for accelerated cured concrete. *Cement and Concrete Research, 16*(5), 603–611.

Choi, H. G., Tsujino, M., Kitagaki, R. & Noguchi, T. (2012a). Expansion–contraction behaviors and cracking control effects of expansion concrete in buildings. In *Proceedings of the 5th ACF international conference*, Pattaya, Thailand. [ACF2012-0096 (Session 3–2)].

Choi, H. G., Tsujino, M., Noguchi, T., & Kitagaki, R. (2012b). Expansion/contraction behavior and cracking control effect of expansive concrete in building structure. *Proceedings of the Japan Concrete Institute, 34*(1), 424–429.

Hori, A., Tamaki, T., & Hagiwara, H. (2000). Cracking behavior of expansive concrete in homoaxial tension. *Proceedings of the Japan Concrete Institute, 22*(2), 511–516.

Hua, C., Ehrlacher, A., & Acker, P. (1997). Analyses and models of the autogenous shrinkage of hardening cement paste II modeling at scale of hydration grains. *Cement and Concrete Research, 27*(2), 245–258.

Katsura, O., Morimoto, J., & Nawa, T. (1996). A model of strength development considering hydration of cement. *Proceedings of the Japan Concrete Institute, 39*, 109–114.

Lokhorst, S. J., & van Breugel, K. (1997). Simulation of the effect of geometrical changes of the microstructure on the deformational behavior of hardening concrete. *Cement and Concrete Research, 27*(10), 1465–1479.

Maruyama, I. (2003). Time dependent property of cement based materials on the basis of micro-mechanics. PhD Dissertation, The University of Tokyo, Tokyo, Japan.

Morioka, M., Hagiwara, H., Sakai, E., & Daimon, M. (1999). Chemical shrinkage and autogenous volume change of cement paste with expansive additive. *Proceedings of the Japan Concrete Institute, 21*(2), 157–162.

Park, S. G. (2004). A study on the reduction of autogenous shrinkage of high performance concrete using shrinkage reducing admixture and expansive additive at early age. PhD Dissertation, The University of Tokyo, Tokyo, Japan.

Park, K. B., & Lee, H. S. (2005). Prediction of temperature profile in high strength concrete structures using a hydration model. *Journal of the Architectural Institute of Korea Structure & Construction, 21*(10), 111–118.

Powers, T. C., & Brownyard, T. L. (1947). Studies of the physical properties of hardened Portland cement paste. *Journal of American Concrete Institute (Proceedings), 43*.

Ryshkewitch, E. (1953). Composition and strength of porous sintered alumina and zirconia. *Journal of the American Ceramic Society, 36*, 65–68.

Schiller, K. K. (1958). *Mechanical properties of non-metallic materials* (pp. 35–50). London, UK: Butterworths.

Taplin, J. H. (1959). A method for following the hydration reaction in Portland cement paste. *Australian Journal of Applied Sciences, 10*, 329–345.

Tashiro, C. (1993). Water behavior in cement & concrete. Report of water subcommittee, research committee of cement & concrete.

van Breugel, K. (1997). Simulation of hydration and formation of structure in hardening cement-based materials. PhD Dissertation, Delft University, Delft, Netherlands.

Woods, H. (1933). Heat evolved by cement in relation to strength. *Engineering News-Record, 1933*, 431–433.

Yanimoto, K., Morioka, M., Sakai, E., & Daimon, M. (2003). Expansion mechanism of cement added with expansive additive. *Concrete research and technology, Japan Concrete Institute, 14*(3), 23–31.

7

Estimation of Friction Coefficient Using Smart Strand

Se-Jin Jeon[1],*, Sung Yong Park[2], Sang-Hyun Kim[1], Sung Tae Kim[2], and YoungHwan Park[2]

Abstract: Friction in a post-tensioning system has a significant effect on the distribution of the prestressing force of tendons in prestressed concrete structures. However, attempts to derive friction coefficients using conventional electrical resistance strain gauges do not usually lead to reliable results, mainly due to the damage of sensors and lead wires during the insertion of strands into the sheath and during tensioning. In order to overcome these drawbacks of the existing measurement system, the Smart Strand was developed in this study to accurately measure the strain and prestressing force along the strand. In the Smart Strand, the core wire of a 7-wire strand is replaced with carbon fiber reinforced polymer in which the fiber Bragg grating sensors are embedded. As one of the applications of the Smart Strand, friction coefficients were evaluated using a full-scale test of a 20 m long beam. The test variables were the curvature, diameter, and filling ratio of the sheath. The analysis results showed the average wobble and curvature friction coefficients of 0.0038/m and 0.21/radian, respectively, which correspond to the middle of the range specified in ACI 318-08 in the U.S. and Structural Concrete Design Code in Korea. Also, the accuracy of the coefficients was improved by reducing the effective range specified in these codes by 27–34 %. This study shows the wide range of applicability of the developed Smart Strand system.

Keywords: friction coefficient, fiber Bragg grating sensor, prestressing tendon, strand, sheath, duct, prestressed concrete structure.

1. Introduction

The calculation and control of elongation and the prestressing force during tensioning of tendons are of primary importance in post-tensioned concrete structures. In this respect, the friction that occurs through the interaction between strands and a sheath during tensioning in a post-tensioning system has a significant effect on the distribution of prestressing force and elongation of tendons. Underestimation or overestimation of the friction coefficients can lead to unexpected structural behavior in terms of camber, deflection, and stress distribution (ACI 2014). Although the relevant design codes and specifications recommend that the friction coefficients be experimentally determined (ACI 2014; KCI 2012), the set-up of test specimens and measurement of forces or strains of tendons required to obtain the coefficients are not easy to carry out. Furthermore, the accuracy of the coefficients is not always guaranteed because of a number of variables affecting the coefficients while testing. Therefore, the friction coefficients that are

specified in design codes and specifications are still referred to frequently. However, the coefficients show a wide range of differences depending on the provisions, and are sometimes expressed as a range rather than as a specific value. This has caused some confusion and trial-and-error practices for designers and constructors, and has led to the inconsistent use of friction coefficients. An acceptable error limit of ±5 or ±7 % of the jacking force between the measured value in a jack and the calculated value from the elongation of tendons (AASHTO 2014; ACI 2014) may still provide a source of discrepancy from the original calculation sheet in the stress distribution of concrete as well as tendons.

In order to reasonably determine the friction coefficients, a number of studies have been performed, but a standard method has not yet been established (Gupta 2005; Jeon et al. 2009; Jeung et al. 2000; Kitani and Shimizu 2009; Moon and Lee 1997). It is found that in each method, some assumptions have been made and that each method depends on inaccurate or incomplete data. In particular, attempts made to derive friction coefficients using conventional electrical resistance strain gauges do not seem to lead to reliable results, mainly due to the damage of sensors and lead wires during the insertion of strands into a sheath and during tensioning as well as the difficulty of gauge installation on a strand. Although a load cell can be installed on the dead end of the test specimen in the opposite side of the live end that is subjected to jacking, the load cell can only provide additional information on the prestressing force at

[1]Department of Civil Systems Engineering, Ajou University, Suwon-si, Gyeonggi-do 443-749, Korea.
*Corresponding Author; E-mail: conc@ajou.ac.kr
[2]Structural Engineering Research Institute, Korea Institute of Civil Engineering and Building Technology, Goyang-si, Gyeonggi-do 411-712, Korea.

the dead end, which is not sufficient to determine the exact distribution of prestressing force required to derive reliable friction coefficients.

In order to overcome these drawbacks of the existing measurement system, the Smart Strand with the embedded fiber Bragg grating sensors was developed in this study to accurately measure the strain and prestressing force along the strand (KICT 2013; Kim et al. 2015). As one of the applications of the Smart Strand, friction coefficients were evaluated using a full-scale test of a 20 m long beam. The obtained friction coefficients were compared with those specified in current provisions for verification and, as a result, several improvements were proposed.

2. Friction Coefficients

2.1 Friction in Post-tensioning System

While the prestressing tendons are tensioned using a jack, the loss of prestress occurs along the tendons due to the friction between the tendons and the sheath in a post-tensioning system. Figure 1 demonstrates two major types of friction. Curvature friction is induced at the curved section of a sheath, where the tendons come into contact with the sheath during tensioning. On the other hand, wobble friction occurs even in a straight sheath due to the unintended deformation of the sheath during handling or casting of concrete, although the sheath is supported at a certain interval before casting.

The predictive equation of the prestressing force as affected by the friction can be derived as shown in Eq. (1) (Nilson 1987).

$$P_{x1} = P_{x2}e^{-(\mu\alpha+kl)} \tag{1}$$

where P_{x1} and P_{x2} are the prestressing forces at the points of $x1$ and $x2$ on a tendon, respectively, with $x2$ closer to the tensioning point, μ is the curvature friction coefficient, k is the wobble friction coefficient, α is the variation of angle between $x1$ and $x2$, and l is the distance between $x1$ and $x2$.

2.2 Friction Coefficients in Provisions

The friction coefficients specified in various codes, specifications, and manuals, etc. are summarized and compared in Table 1. It is noted that some provisions provide different friction coefficients according to the type and surface condition of the prestressing steel and sheath. Although Table 1 represents the case of strands in a galvanized metal sheath that is most frequently used worldwide, it still shows a wide range of variation of the values from provision to provision, and even in a single provision. For the wobble friction

coefficient, the lowest value is 0.00066 as specified in several American provisions (AASHTO 2002; AASHTO 2014; Caltrans 2005; PCI 2011), whereas the highest value is 0.0066 as presented in the Korean design codes (KCI 2012; KRTA 2010) and the previous ACI 318 code (ACI 2008). The difference is as much as ten times. On the other hand, most of the curvature friction coefficients fall into the range of 0.15–0.25, with the recommended value of around 0.20, except for the exceptionally high value specified in the Japanese specifications (JRA 2012; JSCE 2007).

Therefore, a number of attempts have been made to develop more reasonable friction coefficients. However, in several studies, one of the two types of friction coefficients was assumed, while the other friction coefficient was evaluated (Kitani and Shimizu 2009; Moon and Lee 1997); this involves an intrinsic inaccuracy that is strongly affected by the initial assumption of the value of a friction coefficient. The errors that may be induced by this type of methodology were analyzed in some studies (Park and Gil 2004; Park and Kang 2003). Some studies referred to the strains measured by the conventional electrical resistance strain gauges attached to the surface of a strand (Jeung et al. 2000). It is generally accepted, however, that the reliability of the strains obtained by this method is somewhat questionable due to a number of sources of uncertainty and inaccuracy. Gupta (2005) developed a technique to measure the prestressing force at any point of a strand using a tension tester based on the relationship between the lateral deflection and tension of the strand. However, it may be regarded that this method uses an indirect measurement of tension, which possibly involves some errors. Therefore, a more reliable methodology is required to derive realistic friction coefficients in terms of acquirement of the actual strain distribution of a strand and evaluation of the friction coefficients using the measured data.

3. Principle of Smart Strand

Usually, 7-wire strands with a nominal diameter of 12.7 or 15.2 mm are used as prestressing tendons in prestressed concrete (PSC) structures (ASTM 2012; KATS 011). These prestressing tendons consist of one core wire and six outer helical wires made of steel as shown in Fig. 2a. When the strain distribution along a strand needs to be measured, in most cases, the only possibility has been to attach the usual electrical resistance strain gauges on the helical wires exposed outside; although it is desirable to install sensors in the straight core wire to minimize the damage caused by the contact between the strands and between the strand and the sheath. Moreover, the strain measured from this gauge does

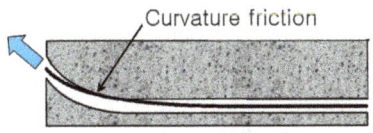
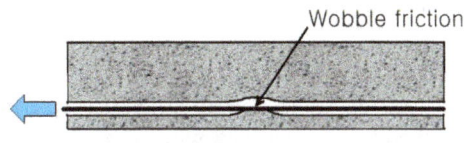

Fig. 1 Friction in a post-tensioning system.

Table 1 Recommended friction coefficients.

Provisions	Wobble friction coefficient, k (/m)	Curvature friction coefficient, μ (/radian)
Structural concrete design code (KCI 2012)	0.0015–0.0066	0.15–0.25
Design code for highway bridges (KRTA 2010)	0.0015–0.0066	0.15–0.25
ACI 318-08 (ACI 2008)	0.0016–0.0066	0.16–0.25
ACI 318-14 (ACI 2014)	Not specified	
Standard specifications for highway bridges (AASHTO 2002)	0.00066	0.15–0.25
AASHTO LRFD bridge design specifications (AASHTO 2014)	0.00066	0.15–0.25
Bridge design manual (PCI 2011)	0.00066	0.20
Post-tensioning manual (PTI 2006)	0.0010–0.0023 (Recommended value: 0.0016)	0.14–0.22 (Recommended value: 0.18)
Prestress manual (Caltrans 2005)	0.00066	0.15, 0.20, 0.25, etc. (Related to the length of a strand)
Canadian highway bridge design code (CSA 2006)	0.003, 0.005 (Related to the diameter of a sheath)	0.20
BS 8110 (BSI 1997)	Not less than 0.0033	0.20, 0.25, 0.30 (Related to rust)
Eurocode 2 (CEN 2002)	0.00095–0.0019	0.19
CEB-FIP model code (CEB 1993)	0.00095–0.0019	0.19
fib model code for concrete structures (fib 2013)	0.0008–0.002	0.16–0.20
Standard specifications for concrete structures (JSCE 2007)	0.004	0.30
Specifications for highway bridges (JRA 2012)	0.004	0.30

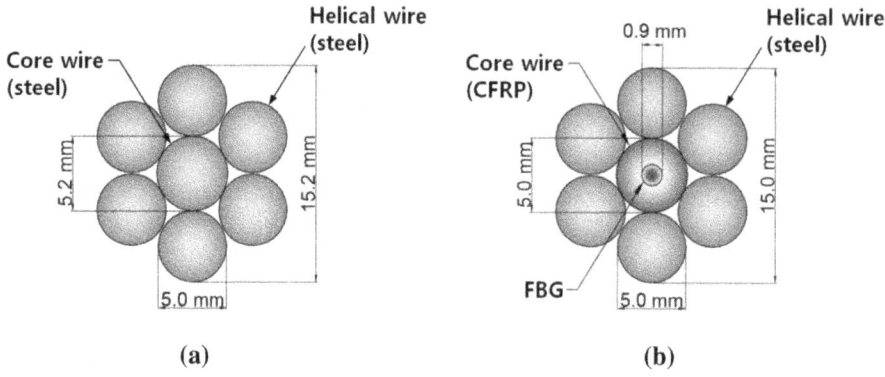

Fig. 2 Comparison of strands. **a** General strand, **b** smart strand.

not represent the actual axial strain because of the inclination of the helical wires and the difference of the length between the helical wire and the core wire. Damage of the lead wires required for this type of gauge during the insertion of strands into a sheath and during tensioning is another anticipated problem.

In order to address the aforementioned conventional problems, the Smart Strand with the embedded fiber Bragg grating (FBG) sensors was developed in this study as shown in Fig. 2b to accurately measure the strain and prestressing force along the strand (KICT 2013, 2014; Kim 2015; Kim et al. 2015). In the Smart Strand, the steel core wire of a general strand is replaced with carbon fiber reinforced polymer (CFRP) to contain the optical fiber and Bragg grating sensors at the center of the core wire section. Among the various possible ways to fabricate the CFRP, the braidtrusion method was adopted in the developed Smart Strand to prevent the galvanic corrosion that may occur due to the contact with the outer steel helical wires, by virtue of the coated nylon fiber (Kim et al. 2015). In comparison, some researchers developed FBG sensors embedded in an ordinary steel core wire (Kim et al. 2012). However, it was demonstrated that the CFRP core wire developed in this study is more advantageous than the steel core wire in terms of mechanical property and the convenience in the fabrication and embedment of the optical fibers (KICT 2013). On the other hand, a different type of FRP and FBG sensing technique to that used in this study was employed in another study (Zhou et al. 2009).

Detail of the principle of the FBG sensor can be found in many references (Jang and Yun 2009; Kim et al. 2012; Nellen et al. 1999). FBG sensors have widely been used recently due to a number of advantages over the conventional sensing technique using electrical resistance, such as non-sensitivity to electromagnetic interference and tolerance for extremely low or high temperatures, etc. When light penetrates into an optical fiber, each Bragg grating embedded in the optical fiber reflects light waves that have a particular wavelength and transmits all other light waves. By analyzing the reflected light waves, the strain at the point of each Bragg grating can be obtained.

The mechanical properties of the CFRP core wire and the developed Smart Strand were verified through a number of specimen tests. Based on the stress–strain relationship curves of the Smart Strand, in addition to the sensing purposes, it was confirmed that the Smart Strand can be used even for structural purposes under service load and ultimate load conditions in most cases (KICT 2014; Kim 2015).

4. Full-Scale Test for Friction Coefficient

4.1 Test Specimen and Variables

Test variables were established to include various cases of PSC beams or girders constructed in practice. Test variables

of the full-scale test of a PSC beam for evaluating friction coefficients are the diameter, curvature, and filling ratio of sheaths as shown in Table 2. Seven, twelve and nineteen strands are inserted into the sheath with 66, 85, and 100 mm diameters, respectively, in usual cases. However, the effect of the filling ratio on the friction coefficients was also taken into consideration by intentionally reducing the number of strands. In addition to the parabolically curved sheaths, straight sheaths were also arranged to separate the curvature effect from the wobble effect. The curvature shown in Table 2 does not refer to a mathematical curvature, but is defined by the sag of a curved sheath divided by the specimen length. Minor lateral curvature, which is inevitable when arranging many sheaths at mid-span, was ignored. The degree of the curvature of sheaths was determined by taking into account the ordinary curvature in PSC girder bridges, including box girders, and the intentionally increased curvature. Similar to an ordinary bonded post-tensioning system, the strands were not lubricated.

Figure 3 shows the 20 m long full-scale test specimen, where the height varies from 2.0 to 2.5 m to realize the largest curvature with an economical use of concrete. The specified compressive strength of the specimen is 40 MPa, which is the same as that of the standard PSC girders used in Korea. Also, self-consolidating concrete with a slump flow of 600 mm was used to accommodate the casting work and to ensure complete compaction of concrete, even in the case of congested sheaths and reinforcements.

The tendons were tensioned at only one end with the opposite end remaining as the dead end in order to increase friction loss and thus to purposely highlight the friction

Table 2 Test variables.

Category	Test variables	Values
Sheath	Diameter (mm)	66, 85, 100
	Curvature (sag/length)	0 (straight), 0.0295, 0.0490, 0.0785
	Material	Galvanized metal
Strand	Number	1, 7, 12, 13, 19
	Nominal diameter (mm)	15.2
	Ultimate strength (MPa)	1860
	Lubrication	Not applied

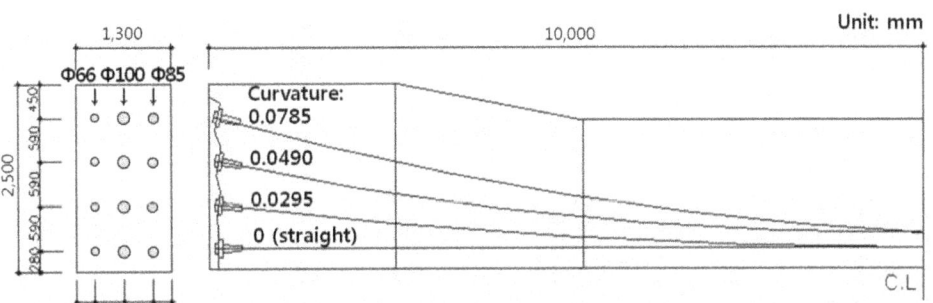

Fig. 3 Full-scale test specimen.

effect, although in practice, the tendons can also be tensioned at both ends. The jacking force per strand was increased by 20–30 MPa until it reached 180–200 kN, which nearly corresponds to the allowable tensile stress during jacking ($0.80f_{pu}$ or $0.94f_{py}$, where f_{pu}: Ultimate strength, and f_{py}: Yield strength) of the prestressing steel with an ultimate strength of 1860 MPa (ACI 2014; KCI 2012). A total of 13 Smart Strands were fabricated, where six strands have seven Bragg gratings (G1–G7 from dead end to live end) and the remaining seven strands have five Bragg gratings (G1–G5 from dead end to live end) that are equally spaced along the length. Among the total number of strands in one sheath, Smart Strands occupied some portion and normal strands occupied the remaining portion. The locations of the Smart Strands in the holes of the anchor head were determined so that various interactions related to the friction (between strands and between the strands and a sheath) can be accounted for. Figure 4 shows the arrangement of Smart Strands and normal strands in the case of the anchor head with 19 holes, corresponding to the sheath with a diameter of 100 mm and curvature of 0.0785 in Fig. 3. The figure also shows the sequence of work performed to investigate the effect of the filling ratio on friction coefficients.

In addition to the FBG sensors in the Smart Strands, extra measures were taken to complement or compare with the data obtained by the FBG sensors. First, a load cell was installed at the dead end to measure the total prestressing force at that location. Second, electro-magnetic (EM) sensors were installed immediately inside the jack, as shown in Fig. 5a, to measure the individual jacking force of the

strands that reflects the jack loss (Cho et al. 2015). Third, electrical resistance gauges were attached to the strand surface corresponding to the selective locations of the FBG sensors. The dead end, i.e. the passive side of the strands, was realized by applying compressed grips as shown in Fig. 5b instead of normal wedges, to ensure reuse of the Smart Strands by minimizing the damage that may possibly be caused by the wedges. The optical fibers were connected to lead wires and a data logger by using a fusion splice.

4.2 Test Results

Figure 6 shows an example of the relationship between jacking force and strains obtained in one of the 19 strands (strand A in Fig. 4) inserted in a sheath with a diameter of 100 mm and curvature of 0.0785 from in Fig. 3. Strand A is an example of a Smart Strand with five Bragg gratings, while seven gratings were also applied for some other Smart Strands. Because the strands were tensioned up to less than the allowable stress, the strains are within an elastic range and thus are almost linearly proportional to the jacking force. It can also be identified in Fig. 6 that the Smart Strands can be utilized to check that the jacking force is correctly transmitted through the tendons.

The distribution of the prestressing force obtained at the same Smart Strand as that shown in Fig. 6 is presented in Fig. 7, where the values measured at the load cell, EM sensor, and jack are also indicated. The strains measured at the Bragg gratings of a Smart Strand can be converted to the prestress at each grating by multiplying the modulus of elasticity of the Smart Strand, and can further be converted to the prestressing force by multiplying the cross sectional

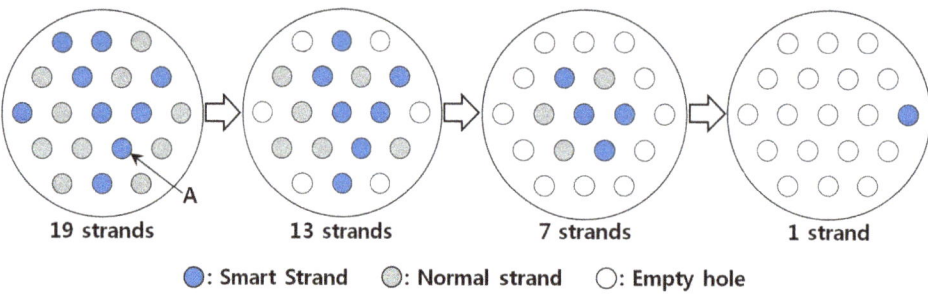

Fig. 4 Arrangement of strands in an anchor head with 19 holes.

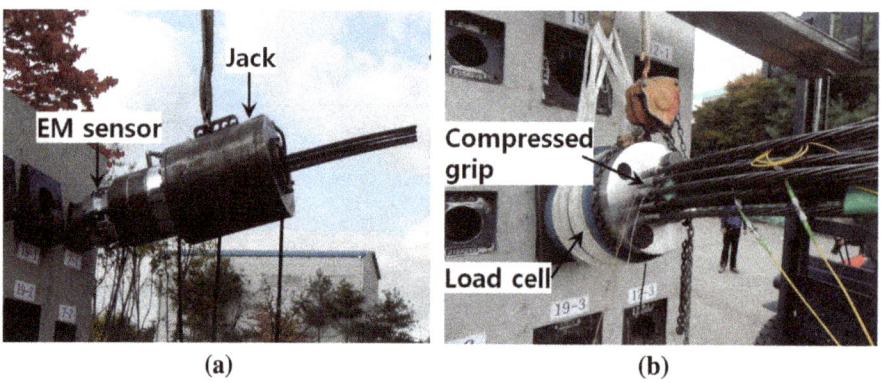

Fig. 5 Live and dead ends of test specimen. **a** Live end, **b** dead end.

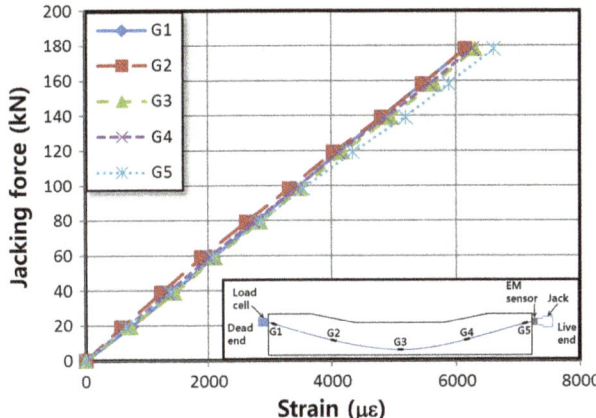

Fig. 6 Jacking force–strain relationship.

area of the Smart Strand. However, the computation of the prestressing force from the strain of the core wire of a Smart Strand can further be refined by accounting for the twist of helical wires and the difference of the cross sectional area and material property between the helical wires and the core wire as proposed in Eq. (2) (Cho et al. 2013; KICT 2013) in an analytical way. In a usual approximate calculation for a normal strand, $E_p = 2 \times 10^5$ MPa and $A_p = 138.7$ mm^2 are used to produce $E_p A_p = 27,740$ kN. On the other hand, the refined calculation for a Smart Strand using Eq. (2) results in $(E_p A_p)_{smart} = 26,007$ kN.

$$P = \left(f_c E_{p,c} A_{p,c} + f_h E_{p,h} A_{p,h}\right)\varepsilon_p = \left(E_p A_p\right)_{smart}\varepsilon_p \qquad (2)$$

where P is the prestressing force at the location of a Bragg grating, ε_p is the strain measured at a Bragg grating of a Smart Strand, f_c and f_h are the correction factors for core wire and helical wires, respectively, $E_{p,c}$ and $E_{p,h}$ are the modulus of elasticity of core wire and helical wires, respectively, $A_{p,c}$ and $A_{p,h}$ are the cross sectional area of core wire and helical wires, respectively, and $(E_p A_p)_{smart}$ is the equivalent $E_p A_p$ for a Smart Strand.

It should be noted from Fig. 7 that the prestressing force of each strand can be separately determined in the Smart

Strands and EM sensors in each load level (Lv. 1–Lv. 9), while only the averaged prestressing force in each strand can be obtained in the jack and load cell system by dividing the total force by the number of strands. This implies another advantage of the Smart Strands system; it can individually predict the distribution of the prestressing force of a specific strand. Therefore, the difference of the friction coefficients of each strand can be evaluated in the Smart Strands system, which can be used to investigate the variation of friction coefficients depending on the location of a strand inside the sheath. This shows a clear contrast to the conventional method where only the average friction coefficients can be derived inside a sheath using the elongation and jacking force measured at a jack, and the force measured at a load cell installed at the dead end, if available (Jeon et al. 2009). Also, it can be seen that the difference in prestressing forces between the EM sensor and the jack refers to the amount of jack loss that occurs due to the friction inside the jack.

5. Evaluation of Friction Coefficients

5.1 Methodology

Friction coefficients can be determined by applying the basic equation shown in Eq. (1) and the distribution of prestressing force as presented, for example, in Fig. 7. In this study, the friction coefficients were evaluated in two steps for sheaths with a specific diameter. First, the wobble friction coefficient was evaluated in the straight sheath. Since the variation of angle (α) does not exist in the straight sheath, the wobble friction coefficient (k) can be obtained from two prestressing forces (P_{x1} and P_{x2}) that were arbitrarily selected in a Smart Strand, judging from the form of Eq. (1), with the term of $\mu\alpha$ removed. The curvature friction coefficient (μ) can then be evaluated from Eq. (1) by applying two prestressing forces on a curved Smart Strand within a curved sheath with the wobble friction coefficient maintained as the previously obtained value for the straight sheath of the same diameter.

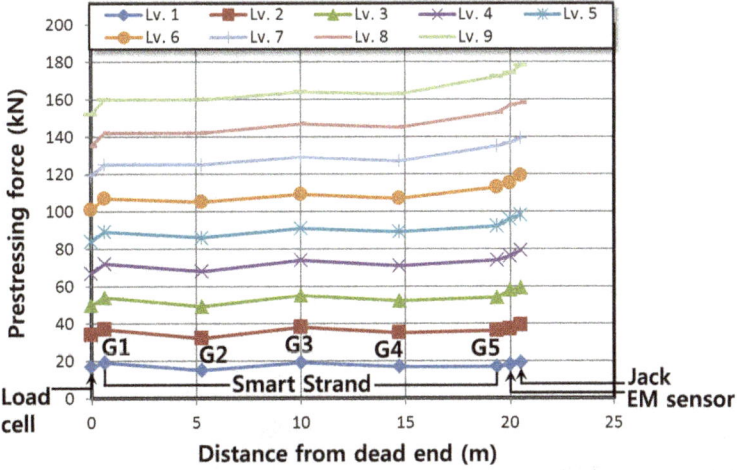

Fig. 7 Distribution of prestressing force.

As can be expected, the friction coefficients obtained in such a way vary depending on the two prestressing forces chosen. Therefore, a statistical approach is required to derive friction coefficients that are more reliable. During the statistical process, some of the friction coefficients may exhibit exceptionally high or low values when compared to the ordinary range of the coefficients shown in Table 1. This behavior can be attributed to the abnormal distribution of prestressing force that can occur in a local region due to an excessive twist of strands while inserting or jacking, or the inevitable irregularity of alignment of a sheath caused by insufficient support combined with the casting pressure of concrete. Therefore, data filtering has been performed for a minority of these exceptional values based on the upper or lower limits of the friction coefficients shown in Table 1. The filtering was performed in two different ways and the results are compared. The first case is based on the two Korean design codes; Structural Concrete Design Code (KCI 2012) and Design Code for Highway Bridges (KRTA 2010). Therefore, the wobble and curvature friction coefficients that were calculated outside the range of 0.0015–0.0066/m and 0.15–0.25/radian, respectively, have been excluded from the statistics. It can be identified that the values of friction coefficients of ACI 318-08 (ACI 2008) are almost identical to those of the Korean design codes. In the second case, the entire provisions in Table 1 were accounted for and, as a result, the effective range was extended to 0.00066–0.0066/m and 0.14–0.30/radian for the wobble and curvature friction coefficients, respectively.

When Eq. (1) is used to evaluate the friction coefficients, any two arbitrary prestressing forces measured at different points can be adopted, regardless of where they are measured among the Smart Strand, load cell, EM sensor, and jack. In this respect, two different approaches were employed in this study. First, two prestressing forces corresponding to the jack and one of the gratings in a Smart Strand were referred to. As mentioned previously, however, the prestressing force measured at the jack is only an average value and does not represent the exact prestressing force of the strand under consideration. Furthermore, although friction loss may also occur inside the jack and at the anchorage devices, the jacking force does not include these losses. These are the sources that may lower the accuracy of the resulting friction coefficients. In order to cope with these problems, in the second method, two prestressing forces obtained purely in two gratings of a Smart Strand were employed.

In the above evaluation process of friction coefficients, the accurate calculation of the distance and the variation of angle between two points is needed. The distance between two arbitrary points of Bragg grating can be easily calculated, since the gratings are embedded at equal spaces along the strands. The variation of angle can be calculated by measuring the angle formed between two tangent lines drawn from the two grating points. In order to perform this calculation, the following mathematical equation, Eq. (3) (Kreyszig 2011), for calculating the length of a curve is required to inversely obtain the horizontal distance, i.e. x coordinate, corresponding to the

grating point. The curved form of a sheath is assumed to be a parabola, as has frequently been assumed in design practice.

$$l = \int_a^b \sqrt{1 + (f'(x))^2}\, dx \tag{3}$$

where l is the length of the partial curve of $f(x)$ between $x = a$ and $x = b$, and $f(x)$ is the shape of a sheath assumed as a parabola.

5.2 Analysis Results

The results of the statistical analyses are presented in Figs. 8 and 9 for wobble and curvature friction coefficients, respectively. In these figures, 'jack-grating' implies that the jacking force and prestressing force at a grating were used for analysis, while 'grating-grating' indicates that two prestressing forces obtained at two gratings were adopted. Also, 'Korean provisions' and 'entire provisions' imply that the statistical data were filtered based on the limit of the two Korean design codes (KCI 2012; KRTA 2010) and entire provisions shown in Table 1, respectively.

The average wobble and curvature friction coefficients of the cases shown in Figs. 8 and 9 were evaluated as 0.0038/m and 0.21/radian, respectively. Therefore, the wobble friction coefficient was slightly smaller than the average value of 0.0041/m in the Korean design codes (KCI 2012; KRTA 2010), while the curvature friction coefficient was a little larger than the average value of 0.2/radian in the Korean design codes. In general, however, the evaluated values were close to the average values specified in the Korean design codes. Also, it can be observed that, in each pair of the wobble and curvature friction coefficients, if the wobble friction coefficient is increased, the corresponding curvature friction coefficient decreases, and vice versa. This can be expected as a matter of course because the two coefficients are interrelated in Eq. (1). The difference of the values in each group, i.e. jack-grating or grating–grating, is due to the difference of the range used for data filtering.

In most of the previous studies using the strands with FBG sensors, only the distribution of prestressing force considering prestress losses was estimated, and the friction coefficients were not derived (Kim et al. 2012; Xuan et al. 2009; Zhou et al. 2009). At most, the distribution of prestressing force obtained by assuming different friction coefficients was compared with the measured data (Kim et al. 2012). The research significance of this study can be found in a direct evaluation of the friction coefficients by utilizing the advanced sensing technology using FBG embedded in a strand. When the scope is extended to the previous studies for proposing friction coefficients, regardless of which method is adopted, the coefficients show a wide range of variation depending on the methodology used (Gupta 2005; Jeon et al. 2009; Kim et al. 2012; Kitani and Shimizu 2009; Moon and Lee 1997) and consistent coefficients have not yet been established. Another factor for this large variation may be the difference in material and workmanship in each study. For example, the degree of wobble friction sensitively varies according to the supporting interval, stiffness, and surface condition of a sheath and to the workmanship dedicated to maintain the original shape of a sheath during the installation of the sheath and the casting of concrete.

The confidence level of each friction coefficient was also investigated as shown in Figs. 8 and 9. The 95 % confidence interval was calculated using the corresponding mathematical equation (Kreyszig 2011) for each method, by assuming normal distribution of the data. Although each method has a narrower band of the confidence interval, the 95 % confidence interval marked with dotted lines in Figs. 8 and 9 only presents the absolute lower and upper limits that can cover all cases with sufficient reliability. Through this type of statistical method, the wide range of the friction coefficients specified in a specific provision can be reduced to enhance the accuracy and reliability. For example, while the range of the vertical axes shown in Figs. 8 and 9 corresponds to that of ACI 318-08 (ACI 2008) and Korean design codes (KCI 2012; KRTA 2010), the range can be narrowed to

0.0021–0.0058/m and 0.178–0.244/radian for wobble and curvature friction coefficients, respectively, by applying a 95 % confidence level. This means that the range was reduced by 27 and 34 % for the wobble and curvature friction coefficients in this study, respectively, which may accommodate the choice of friction coefficients for field engineers and designers.

5.3 Discussion

As mentioned earlier, a number of conventional electrical resistance strain gauges were also implemented in this study for comparison with the Smart Strand system. As has been frequently reported, however, a large number of electrical resistance strain gauges were made unavailable due to the damage of sensors and lead wires during insertion of strands into a sheath and during tensioning. The strains measured by Smart Strands were compared with those measured by the remaining electrical resistance gauges attached to the surface

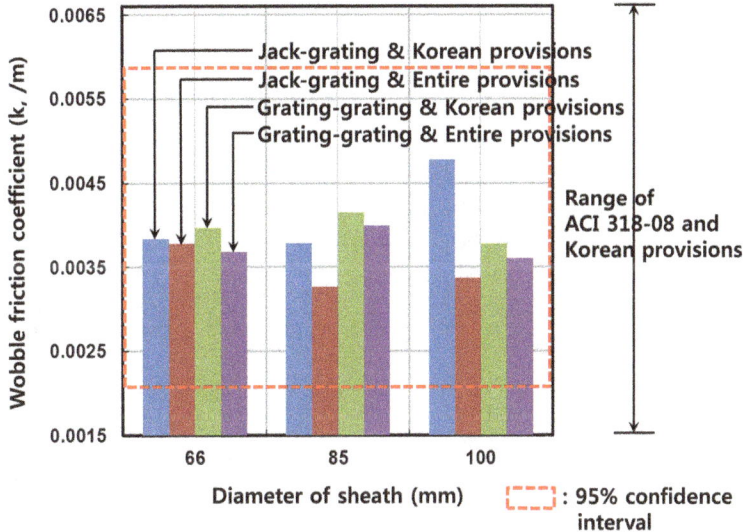

Fig. 8 Wobble friction coefficient.

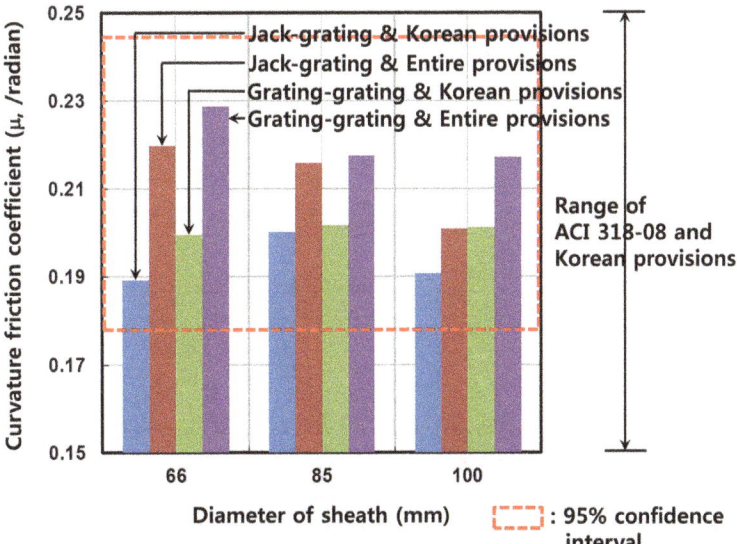

Fig. 9 Curvature friction coefficient.

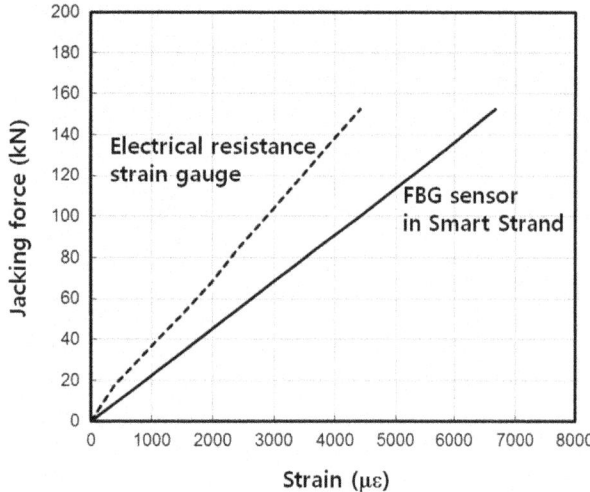

Fig. 10 Comparison of strains of a strand.

of helical wires, corresponding to the same grating points. The data obtained using electrical resistance gauges showed a great amount of difference from those of Smart Strands, since the electrical resistance gauges may be sensitively affected by such factors as the alignment of the gauges on the helical wires and the workmanship related to the proper bonding of the gauges. In the overall trend, the strains of the conventional gauges were smaller than those of Smart Strands by 20–30 % as representatively shown in Fig. 10. The difference between the strains can be attributed to the difference in the length between the core wire and helical wire, and the slope of the helical wire with respect to the core wire (8.2°), etc. For example, the longer length of the helical wire wound around the core wire can result in a smaller strain. The difference in the actual stresses between the core wire and helical wire has also been investigated by Cho et al. (2013). Therefore, the strains measured using electrical resistance gauges should be interpreted with special care, especially when applied to the strands, although they are still widely used.

In this study, the effect of curvature, diameter, and filling ratio of a sheath, and the effect of the location of a strand in a sheath on the friction coefficients have also been investigated. However, these topics will be dealt with in another paper since they involved extensive analyses. This study presented the general average friction coefficients in terms of wobble and curvature, taking into consideration all the test variables, since the friction coefficients are specified without any limited condition in most provisions as shown in Table 1.

6. Conclusions

In order to overcome the various drawbacks of the existing measurement system of the prestressing force of strands in PSC structures, the Smart Strand system was developed in this study by devising a core wire made of carbon fiber reinforced polymer with the fiber Bragg grating sensors embedded. As one of the applications of the Smart Strand, friction coefficients were evaluated using the strain distribution of strands obtained from the full-scale test of a 20 m long beam. Based on the results of the above investigation, the following conclusions can be drawn:

(1)   The tests were performed for various curvatures, diameters, and filling ratios of a sheath and for various strand locations inside a sheath to include general cases. The analysis results showed the wobble and curvature friction coefficients of 0.0038/m and 0.21/radian on average, respectively, for general galvanized metal sheaths. These values correspond to the values within the middle of the range specified in ACI 318-08 in the U.S. and Structural Concrete Design Code in Korea.

(2)   A wide range of friction coefficients specified in the general provisions may cause some difficulty and trial-and-error in choosing an appropriate coefficient for design purposes. Through statistical analyses using a confidence interval, the accuracy of the coefficients was improved by reducing the effective range. For example, the ranges in ACI 318-08 in the U.S. and Structural Concrete Design Code in Korea were reduced by 27–34 %.

(3)   The strains measured using conventional electrical resistance strain gauges showed a great amount of difference from those of Smart Strands. In the overall trend, the strains of conventional gauges were smaller than those of Smart Strands by 20–30 %. The differences in the strains may originate from the differences in the length between the core wire and helical wire, and the slope of the helical wire with respect to the core wire, etc. Therefore, the strains measured using electrical resistance gauges should be interpreted with special care in terms of the strains of the strands.

(4)   The effect of curvature, diameter, and filling ratio of a sheath, and the effect of the location of a strand in a sheath on the friction coefficients were also investigated, but these will be dealt with in another paper as further study. By applying the Smart Strand system to reliably estimate the distribution of the prestressing force of strands, as has been demonstrated in this study, relevant provisions regarding the design of PSC structures can be verified and improved, if necessary. It is also expected that the Smart Strand system will be used for the maintenance or management of PSC structures by the long-term monitoring of prestressing force.

Acknowledgments

This research was supported by a grant from a Strategic Research Project (Development of Smart Prestressing and Monitoring Technologies for Prestressed Concrete Bridges) funded by the Korea Institute of Civil Engineering and Building Technology.

References

ACI Committee 318. (2008). *Building code requirements for structural concrete (ACI 318-08)*. Farmington Hills, MI, US: American Concrete Institute (ACI).

ACI Committee 318. (2014). *Building code requirements for structural concrete (ACI 318-14)*. Farmington Hills, MI, US: American Concrete Institute (ACI).

American Association of State Highway and Transportation Officials (AASHTO). (2002). *Standard specifications for highway bridges* (17th ed.). Washington, D.C., US: AASHTO.

American Association of State Highway and Transportation Officials (AASHTO). (2014). *AASHTO LRFD bridge design specifications* (7th ed.). Washington, D.C., US: AASHTO.

American Society for Testing and Materials (ASTM). (2012). *Standard specification for steel strand, uncoated seven-wire for prestressed concrete (ASTM A416/A416M-12a)*. West Conshohocken, PA, US: ASTM.

British Standards Institution (BSI). (1997). *Structural use of concrete (BS 8110)*. London, UK: BSI.

Canadian Standards Association (CSA). (2006). *Canadian highway bridge design code, CAN/CSA-S6* (10th ed.). Mississauga, ON, Canada: CSA.

Cho, K. H., Kim, S. T., Park, S. Y., & Park, Y. H. (2013). Computation of the strand resistance using the core wire strain measurement. *Engineering, 5*, 850–855.

Cho, K. H., Park, S. Y., Cho, J. R., Kim, S. T., & Park, Y. H. (2015). Estimation of prestress force distribution in the multi-strand system of prestressed concrete structures. *Sensors, 15*, 14079–14092.

Euro-International Committee for Concrete (CEB). (1993). *CEB-FIP model code*. Lausanne, Switzerland: Thomas Telford Services Ltd.

European Committee for Standardization (CEN). (2002). *Design of concrete structures (Eurocode 2)*. Brussels, Belgium: CEN.

Gupta, P. R. (2005). *Rational determination of friction losses in post-tensioned construction* (pp. 129–144). SP-231, American Concrete Institute (ACI).

Jang, I. Y., & Yun, Y. W. (2009). Study on stress transfer property for embedded FBG strain sensors in concrete monitoring. *International Journal of Concrete Structures and Materials, 3*(1), 33–37.

Japan Road Association (JRA). (2012). *Specifications for highway bridges-Part III. Concrete bridges*. Tokyo, Japan: JRA.

Japan Society of Civil Engineers (JSCE). (2007). *Standard specifications for concrete structures-design*. Tokyo, Japan: JSCE.

Jeon, S. J., Park, J. C., Park, I. K., & Shim, B. (2009). Estimation of friction coefficients based on field data. *Journal of the Korean Society of Civil Engineers, 29*(5A), 487–494.

Jeung, B. K., Han, K. B., & Park, S. K. (2000). An experimental study on the frictional loss of stress in the prestressing tendons. *Journal of the Korean Society of Civil Engineers, 20*(5-A), 797–804.

Kim, S. H. (2015). *An experimental study on measurement of friction coefficient using smart strand*. Master's thesis, Ajou University, Suwon-si, Korea.

Kim, J. M., Kim, H. W., Park, Y. H., Yang, I. H., & Kim, Y. S. (2012). FBG sensors encapsulated into 7-wire steel strand for tension monitoring of a prestressing tendon. *Advances in Structural Engineering, 15*(6), 907–917.

Kim, S. T., Park, Y. H., Park, S. Y., Cho, K. H., & Cho, J. R. (2015). A sensor-type PC strand with an embedded FBG sensor for monitoring prestress forces. *Sensors, 15*, 1060–1070.

Kitani, T., & Shimizu, A. (2009). Friction coefficient measurement test on 13MN class tendon of PC strands for prestressed concrete containment vessel (PCCV). In: *Proceedings of 20th international conference on structural mechanics in reactor technology (SMiRT 20)*, Paper 1825.

Korea Concrete Institute (KCI). (2012). *Structural concrete design code*. Seoul, Korea: KCI.

Korea Institute of Civil Engineering and Building Technology (KICT). (2013). *Development of smart prestressing and monitoring technologies for prestressed concrete bridges, KICT 2013-167*. Goyang-si, Korea: KICT.

Korea Institute of Civil Engineering and Building Technology (KICT). (2014). *Development of smart prestressing and monitoring technologies for prestressed concrete bridges, KICT 2014-171*. Goyang-si, Korea: KICT.

Korea Road and Transportation Association (KRTA). (2010). *Design code for highway bridges*. Seoul, Korea: KRTA.

Korean Agency for Technology and Standards (KATS). (2011). *Uncoated stress-relieved steel wires and strands for prestressed concrete (KS D 7002)*. Seoul, Korea: Korean Standards Association (KSA).

Kreyszig, E. (2011). *Advanced engineering mathematics* (10th ed.). Hoboken, NJ, US: Wiley.

Moon, J. K., & Lee, J. H. (1997). A study on the determination of prestressing force considering frictional loss in PS concrete structures. *Journal of the Korean Society of Civil Engineers, 17*(I-1), 89–99.

Nellen, P. M., Frank, A., Broennimann, R., Meier, U., & Sennhauser, U. J. (1999). Fiber optical Bragg grating sensors embedded in CFRP wires. *SPIE Proceedings, 3670*, 440–449.

Nilson, A. H. (1987). *Design of prestressed concrete* (2nd ed.). Hoboken, NJ, US: Wiley.

Park, Y. H., & Gil, H. B. (2004). An error analysis of the friction assessment method for PS tensioning. In: *Proceedings of Korean Society of Civil Engineers (KSCE) conference*, pp. 106–111.

Park, Y. H., & Kang, H. T. (2003). A criterion for application of friction assessment methods in PS tensioning management. In: *Proceedings of Korean Society of Civil Engineers (KSCE) conference*, pp. 594–599.

Post-Tensioning Institute (PTI). (2006). *Post-tensioning manual* (6th ed.). Phoenix, AZ, US: PTI.

Precast/Prestressed Concrete Institute (PCI). (2011). *Bridge design manual* (3rd ed.). Chicago, IL, US: PCI.

State of California Department of Transportation (Caltrans) Engineering Services. (2005). *Prestress manual*. Sacramento, CA, US: Caltrans.

The International Federation for Structural Concrete (fib). (2013). *fib model code for concrete structures 2010*. Lausanne, Switzerland: fib.

Xuan, F. Z., Tang, H., & Tu, S. T. (2009). In situ monitoring on prestress losses in the reinforced structure with fiber-optic sensors. *Measurement, 42,* 107–111.

Zhou, Z., He, J., Chen, G., & Ou, J. (2009). A smart steel strand for the evaluation of prestress loss distribution in post-tensioned concrete structures. *Journal of Intelligent Material Systems and Structures, 20,* 1901–1912.

8

Heavy Metal Leaching, CO_2 Uptake and Mechanical Characteristics of Carbonated Porous Concrete with Alkali-Activated Slag and Bottom Ash

G. M. Kim, J. G. Jang, Faizan Naeem, and H. K. Lee*

Abstract: In the present study, a porous concrete with alkali activated slag (AAS) and coal bottom ash was developed and the effect of carbonation on the physical property, microstructural characteristic, and heavy metal leaching behavior of the porous concrete were investigated. Independent variables, such as the type of the alkali activator and binder, the amount of paste, and CO_2 concentration, were considered. The experimental test results showed that the measured void ratio and compressive strength of the carbonated porous concrete exceeded minimum level stated in ACI 522 for general porous concrete. A new quantitative TG analysis for evaluating CO_2 uptake in AAS was proposed, and the result showed that the CO_2 uptake in AAS paste was approximately twice as high as that in OPC paste. The leached concentrations of heavy metals from carbonated porous concrete were below the relevant environmental criteria.

Keywords: porous concrete, coal bottom ash, alkali activated slag, carbonation, heavy metal leaching.

1. Introduction

Porous concrete is defined as a type of concrete with continuous open voids (Sriravindrarajah et al. 2012). In general, porous concrete is fabricated with a uniform size of coarse aggregates without fine aggregates to create continuous voids into concrete intentionally (Sriravindrarajah et al. 2012). Since the development of porous concrete in Japan in the 1980s, it has been applied in various ways to restore groundwater supplies and to reduce the runoff from a site (Sriravindrarajah et al. 2012; Bhutta et al. 2012). In recent years, industrial by-products such as coal ash and slag have attracted several studies to utilize the materials in the fabrication of eco-friendly porous concretes (Jang et al. 2015a; Park and Tia 2004).

Since the voids in porous concrete are mainly formed by coarse aggregates, a morphological characteristic of the aggregate is deemed most important, and need to be of uniform size, rough surface, and low specific gravity (Jang et al. 2015a). Coal bottom ash (BA) is generated by the combustion of pulverized coal, and many features of BA largely satisfy the aforementioned conditions to fabricate a porous concrete (Singh and Siddique 2014; Kim et al. 2014a). In particular, the highly porous structure and the rough surface property of BA is suitable for forming sufficient voids in the matrix of porous concrete (Kuo et al. 2013; Kim et al. 2014b). However, BA contains and may leach a variety of heavy metals such as cadmium, copper, and lead, inhibiting a broad range of applications in a construction industry (Jang et al. 2015a).

Several attempts have been made to immobilize the heavy metals in a cementitious matrix. Li et al. reported that a cement based matrix could be effective to immobilize Cu, Pb, and Zn (Li et al. 2012). Such heavy metals could be encapsulated in the calcium silicate hydrate (C–S–H) phase of a Portland cement matrix (Li et al. 2012). Zhang et al. found that fly ash based geopolymer can immobilize Cr(VI), Cd(II), and Pb(II), relying on the chemical binding and precipitation of the heavy metals (Zhang et al. 2008). Alkali-activated slag (AAS) paste is also known as an effective material to immobilize heavy metals. Deja found that AAS paste can immobilize Cd, Zn and Pb up to 99.9 % (Deja 2002).

In order to fabricate a porous concrete, AAS paste using blast furnace slag (BFS) was considered in the present study. The immobilization effectiveness of AAS paste has been reported in several studies, and details can be found in Shi and Fernandez-Jimenez (2006), Qian et al. (2003a, 2003b). However, the microstructure and pH in AAS paste could be significantly affected by the carbonation process (Kar et al. 2014). In particular, pH in carbonated AAS paste can be decreased to 9 or lower, and the solubility of some heavy metals may be increased (Puertas et al. 2006; Bernal et al. 2014). Thus, the effect of carbonation on the leaching characteristics of heavy metals in porous concrete should be investigated.

In addition to immobilization of heavy metals, AAS paste can be used for uptake of CO_2 which is a typical greenhouse

Department of Civil and Environmental Engineering, Korea Advanced Institute of Science and Technology, Daejeon 305-701, South Korea.

*Corresponding Author; E-mail: haengki@kaist.ac.kr

gas. In most studies on carbonation of AAS paste, calcium carbonate was only considered as a carbonated product (Park and Tia 2004; Kim et al. 2013; Deja 2002). Eloneva et al. reported that AAS paste can be used for CO_2 uptake via the carbonation of calcium components (Eloneva et al. 2008). However, sodium carbonates in previous studies were found in AAS paste when NaOH or water glass was used as an alkali activator (Puertas et al. 2006; Bernal et al. 2014), thus the sodium carbonates should be considered to evaluate the CO_2 uptake of AAS paste (Bertos et al. 2004).

In the present study, a porous concrete with alkali activated slag and coal bottom ash was developed. BFS and BA were used as a binder and an aggregate material, respectively, for the fabrication of the porous concrete. Since carbonation is known to induce not only the change in physicochemical properties of AAS but also the change in the leaching behavior of heavy metal from BA, the present study aims to investigate the effect of carbonation on heavy metal leaching, mechanical property and CO_2 uptake of porous concrete through a series of characterization tests. Independent variables, such as the type of the alkali activator and binder, the amount of paste, and CO_2 concentration, were considered.

2. Experimental Program

2.1 Materials

The chemical compositions of BFS and BA used in this study, are listed in Table 1. The BA was obtained from the Seo-Cheon thermoelectric power plant, sieved in the range of 2.5–5.0 mm, and was then washed to remove impurities from the surface. The absolute volume ratio of the BA to design the mixing proportion was calculated in accordance with the procedure described in JIS A 1104 (2006), and the result was approximately 39 %. Inductively coupled plasma (ICP, iCPA-6300 Duo ICP-OES and ICP/MS 7700X) spectrometers were used to measure the total amount of heavy metals in the BA, and the results are listed in Table 2. Detailed test procedures for the ICP analysis of BA can be found in Jang et al. (2015a).

Sodium hydroxide (NaOH), and a mixture of water glass and sodium hydroxide were used as alkali activators. The sodium hydroxide with the specific gravity of 2.13 was used in the form of pellets. The chemical composition of the waterglass produced by Duksan Chemicals Co. Ltd., was

SiO_2 of 28–38 wt% and Na_2O of 9–19 wt%, and its specific gravity was 1.38.

2.2 Mix Proportion and Fabrication Process of Porous Concrete

The mix proportion of the porous concretes is listed in Table 3. Three different types of porous concretes were fabricated in this study. The AAS-based porous concretes were categorized according to the type of alkali activator, and were compared with the OPC-based porous concretes. The effect of alkali activator on the AAS-based porous concretes was investigated by using two different types of alkali activator, i.e., one sodium hydroxide and both sodium hydroxide and waterglass. It has been reported in many studies that silicate modulus (Ms: SiO_2/Na_2O) of 1.0–1.5 and Na_2O to slag ratio of 4 % were appropriate to ensure an appropriate compressive strength of AAS paste (Wang and Scrivener 1995). Therefore, in the present study, the silicate modulus was chosen to be 1.2 and controlled by an addition of sodium hydroxide and water (Ravikumar and Neithalath 2013; Chi et al. 2012). The Na_2O/slag was fixed at 4 % for all the alkali activators. The liquid to binder ratio of AAS-based porous concretes and the water to cement ratio of OPC-based porous concretes were fixed at 0.5 and 0.3, respectively. Paste to aggregate (P/G) ratios were varied from 25 to 35 % to investigate the effect of P/Gs on the physical properties of porous concrete. The P/G of PC2 was kept identical to that of NS2 and SS2, i.e., 30 %, to examine the effect of the binder type on the properties of the porous concrete.

The specimens were cast into the cylindrical molds with a diameter of 10 cm and a height of 20 cm. The mixing procedure was as follows: BA and BFS were mixed for 1 min and alkali activator was then added to the dry mixture. Then, mixing was conducted for further 3 min. Since the compaction process can significantly affect the mechanical properties of porous concretes, the process was carefully controlled as follows: Half of the mold was filled with the mixture, and a tamping bar was used to compact the mixture 10 times. Then, the rest of the mold was fully filled, and a vibrator was used to compact the mixture for 10 s. The specimens were demolded after 2 days and cured for 28 days. The specimens were cured under temperature of 25 ± 5 °C and ambient humidity in laboratory condition.

The mix proportion of the paste specimens is shown in Table 4. The same mixing proportion as the part of paste for

Table 1 Chemical compositions of BFS, OPC and BA.

Chemical composition (%)	Blast-furnace slag (BFS)	Ordinary Portland cement (OPC)	Coal bottom ash (BA)
CaO	56.10	65.9	3.95
SiO_2	21.00	10.6	50.10
Al_2O_3	17.00	3.8	26.90
Fe_2O_3	0.62	3.89	10.80
SO_3	0.77	2.26	–

Table 2 Measured heavy metal contents of BA by ICP analysis (in ppm).

Cr	As	Cu	Cd	Hg	Pb
121.952	3.416	276.220	139.524	ND[a]	355.578

[a] ND: non detected.

Table 3 Mix proportion of porous concretes (unit: kg/m^3).

Series	Binder material	BA	Water	NaOH	Waterglass	W/B	P/G (%)
NS1	248.1	662.3	110.9	13.1	–	0.5	25
NS2	285.6	636.8	127.7	15.1	–	0.5	30
NS3	319.6	613.7	142.9	16.9	–	0.5	35
SS1	259.5	661.7	79.2	8.4	42.2	0.5	25
SS2	297.6	637.0	90.9	9.6	48.4	0.5	30
SS3	333.5	613.6	101.8	10.8	54.2	0.5	35
PC2	365.0	642.4	109.5	–	–	0.3	30

P/G: ratio of paste by the weight of aggregates. NS: porous concrete fabricated with slag and NaOH (Na$_2$O/Slag: 4 %). SS: porous concrete fabricated with slag, waterglass and NaOH (Na$_2$O/Slag: 4 %, silicate modulus (Ms): 1.2). PC2: porous concrete fabricated with Portland cement and water.

Table 4 Mix proportion of paste specimens.

Series	Binder material	water	NaOH	Waterglass	W/B
NS(P*)	6.37	2.85	0.336	–	0.5
SS(P)	6.67	2.03	0.215	1.083	0.5
PC(P)	9.66	2.90	–	–	0.3

*P: paste specimen.

porous concrete was adopted to fabricate pastes in an effort to investigate the carbonation effect on the microstructure. The size of these specimens was 50 mm × 50 mm × 50 mm, and the curing condition was identical to that of the porous concretes.

2.3 Experimental Details
2.3.1 Accelerated Carbonation Test

An accelerated carbonation test was conducted on the porous concretes and pastes at a temperature of 20 °C and a relative humidity of 65 %. All specimens were exposed to a CO$_2$ concentration of 5 or 10 % in a carbonation chamber for 2 weeks. Non-carbonated specimens were prepared as references in a sealed condition to prevent natural carbonation. The carbonated specimens were cut at the center, and a phenolphthalein spraying test was performed on the cut surface of the specimens to measure the carbonation depth.

2.3.2 Microstructural Analysis of AAS Pastes

A mercury intrusion porosimetry (MIP) test was conducted on the pastes, using a porosimeter (Auto pore IV 9500, Micromeritics Instrument Corporation). The pressure in the test increased up to 414 MPa to detect a pore diameter in the range of 0.003–1000 μm (Ravikumar and Neithalath 2013). The surface tension and the contact angle in the experiment were set at 0.485 N/m and 130°, respectively

(Ravikumar and Neithalath 2013). A thermogravimetric/differential thermal analysis (TG/DTA) was performed to analyze the hydration and carbonation products in the pastes. The heat rate and the flux of nitrogen gas in the test were set at 30 °C min and 100 ml min, respectively (Kim et al. 2013; Ravikumar and Neithalath 2013). The temperature was varied from 30 to 1000 °C (Kim et al. 2013).

2.3.3 Void Ratio Test

The open-void ratio (V_{open}) and closed-void ratio (V_{close}) of the porous concretes were calculated with Eqs. (1) and (2), respectively (Jang et al. 2015a; Kim and Lee 2010).

$$V_{open} = \left(1 - \frac{(W_2 - W_1)\rho_w}{V_1}\right) \times 100\,\% \qquad (1)$$

$$V_{close} = \left(1 - \frac{(W_3 - W_1)\rho_w}{V_1}\right) \times 100\,\% - V_{open} \qquad (2)$$

Here, W_1 is the weight of the specimen submerged in water, W_2 is the weight of the specimen dried in air for 24 h, W_3 is the weight of the specimen dried in an oven, V_1 is the volume of the specimens, and ρ_w is the density of water (Jang et al. 2015a; Kim and Lee 2010). The total void ratio (V_{total}) was calculated as the sum of V_{open} and V_{close} (Jang et al. 2015a; Kim and Lee 2010).

2.3.4 Compressive Strength Test

Compressive strength tests of the porous concretes were conducted in accordance with the procedure described in ASTM C39 (American Society for Testing and Materials 2012). The top and bottom surfaces of each specimen were ground to prevent any concentrated loading. The specimens were tested at an age of 28 days using a 3000 kN universal testing machine (UTM). A displacement control method was adopted, and the speed of the cross-head of the UTM was held at 0.01 mm/s.

2.3.5 Heavy Metal Leaching Test

A leaching test on the porous concretes was conducted to analyze the leaching behavior of the heavy metals in accordance with the NSF/ANSI 61-2007a (2007) (Drinking water system components-Health effects) (NSF International standard/American National Standard, NSF/ANSI 61-2007a 2007). The test solution of leaching test was changed after first, second and fourth day, and the solution was collected on each day. The pH of the solution, at a level of 4–5, was adjusted with nitric acid, and ICP-MS was used to measure the leached concentrations of the heavy metals from the testing solution including chromium, arsenic, copper, lead, mercury, and cadmium. In previous studies, toxicity characteristic leaching procedure (TCLP) method has been adopted to characterize the leaching behavior of heavy metals (e.g., Dermatas and Meng 2003; Perera et al. 2005). TCLP method requires the sample to be equal or greater than 3.1 cm^2/g or smaller than 1 cm in its narrowest dimension, otherwise it necessitates crushing, cutting or grinding of the specimen (Environment, Health and Safety Online 2008). Since the porous concrete does not fit into the size specified in TCLP method, grinding of the porous concrete is mandatory that could expose the BA covered by the AAS paste. In contrast, since the NSF/ANSI 61-2007a does not specify the dimension of a sample for the test procedure, the original form of the porous concrete can be maintained in the course of the test (NSF International standard/American National Standard, NSF/ANSI 61-2007a 2007).

3. Results and Discussion

3.1 Carbonation Behavior of AAS Pastes

The measured carbonation depths of AAS and OPC pastes after exposure to CO_2 are shown in Fig. 1 and summarized in Table 5. The extent of the carbonation in the OPC paste (PC(P)_C5) at a 5 % CO_2 concentration was slight, and the carbonation depth was no greater than 3 mm at all sides. In contrast, the AAS pastes underwent deeper and more intense carbonation than OPC paste did. The carbonation depth of AAS pastes at a 5 % CO_2 concentration varied according to the type of the alkali activator. The carbonation depth in the AAS paste incorporating the waterglass-based alkali activator (SS(P)_C5 specimen) was slightly lower than that of the AAS paste using NaOH (NS(P)_C5 specimen).

The total porosity and the average pore diameter of the pastes are shown in Table 6. The average pore diameter of the non-carbonated AAS pastes (NS(P)_C0 and SS(P)_C0) specimens varied according to the type of alkali activator. SS(P)_C0 specimen showed a smaller average pore diameter than NS(P)_C0 specimen. Such differences in the average pore diameters according to the type of the alkali activator presumably affected the carbonation rate (see Table 5), since high porosity can cause an increase in the diffusivity of CO_2 (Bertos et al. 2004). In the phenolphthalein spraying test, the high porosity of NS(P)_C5 could cause deeper carbonation depth than SS(P)_C5 specimen.

NS(P)_C0 and SS(P)_C0 specimens showed lower porosity in comparison with PC(P)_C0 specimen. After exposure to accelerated carbonation, the porosity of NS(P)_C10 and SS(P)_C10 specimens increased, whereas that of PC(P)_C10 specimen decreased. In general, OPC paste is known to have more calcium components in the residues of C_3S, C_2S, and $Ca(OH)_2$ in the pore solution than AAS paste (Bakharev et al. 2001). This condition may act as a buffer for the decalcification of C–S–H phases in the OPC paste (Bakharev et al. 2001). In addition, the carbonated products in an OPC matrix can create a protective area that can interfere with the diffusion of CO_2, filling the pores with the carbonated products (Bernal et al. 2014; Ylmén and Jäglid 2013). Thus, the formation of carbonates under the rich calcium condition can reduce the porosity of an OPC matrix (Bernal et al. 2014; Bakharev et al. 2001). In contrast, AAS paste has lower buffer in the pore solution such as $Ca(OH)_2$ and calcium components than OPC paste (Bernal et al. 2014; Bakharev et al. 2001). Hence, the carbonation of AAS pastes in the test was mainly caused by the decalcification from C–S–H phases, leading to an increase in the porosity (Bakharev et al. 2001).

The pore size distribution of the pastes is summarized in Table 7. The percentage of pores in the range of 0.1–10 µm increased slightly in PC(P)_C10 sample after carbonation, while the percentage of pores in the range of 0.01–0.1 µm declined slightly. In NS(P)_C10 and SS(P)_C10 samples, the percentage of pores in the range of 0.1–100 µm increased after carbonation, while the percentage of pores in the range of 0.01–0.1 µm decreased. In particular, the percentage of pores in NS(P)_C10 and SS(P)_C10 samples ranging from 0.1 to 1.0 µm significantly increased. It can be inferred from the results that the decalcification of C–S–H phases in AAS pastes induced a change in the percentage of pores ranging from 0.1 to 1.0 µm.

Figure 2 shows the TG/DTA curves of the paste samples. The weight loss of C–S–H phases and carbonated products in the TG/DTA curves is summarized in Table 8. The weight loss from 50 to 200 °C was used as an indicator of the interlayer water weight of the C–S–H phases (Kim et al. 2013). Before carbonation, the weight loss of NS(P)_C0 and SS(P)_C0 samples in the temperature range was much higher than that of PC(P)_C0 sample. After carbonation, the weight loss of the C–S–H phase in SS(P)_C10 and NS(P)_C10 samples decreased by more than 30 % compared to SS(P)_C0 and NS(P)_C0 samples, whereas the reduction in the weight loss of PC(P)_C10 was slight compared to PC(P)_C0. These results can support that

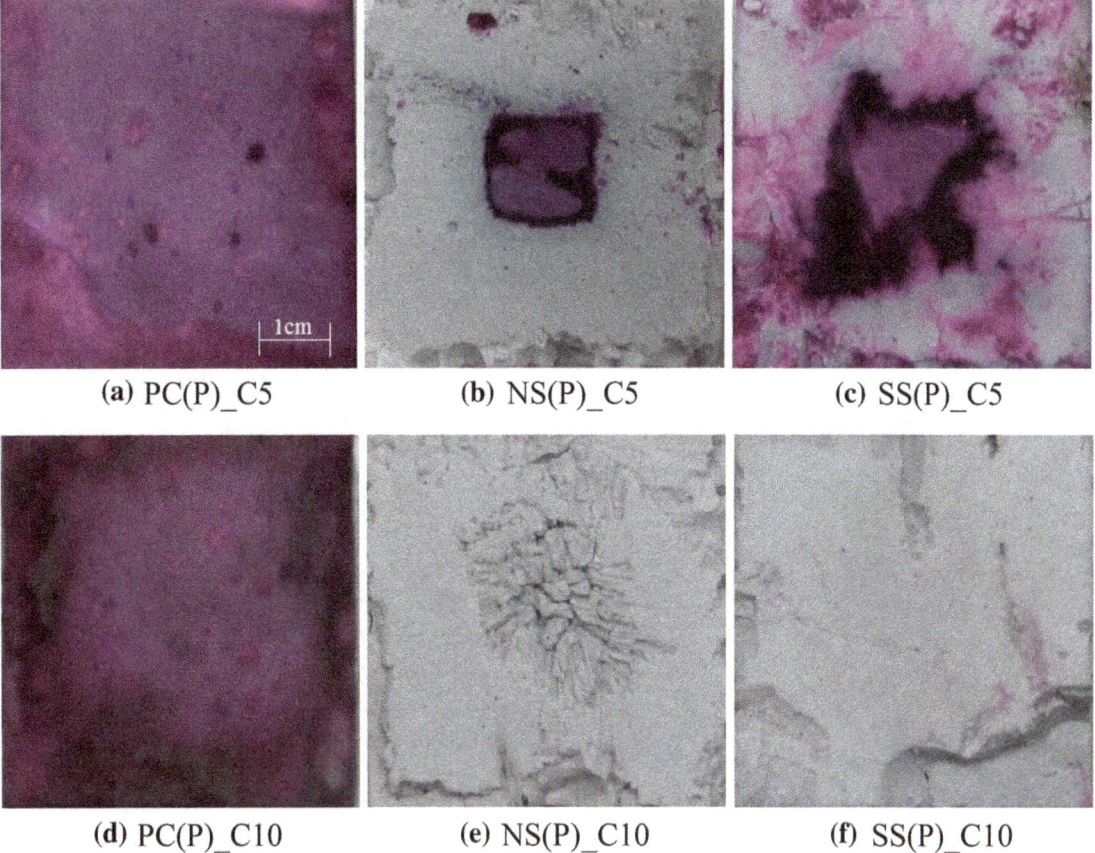

(a) PC(P)_C5   (b) NS(P)_C5   (c) SS(P)_C5

(d) PC(P)_C10   (e) NS(P)_C10   (f) SS(P)_C10

Fig. 1 Carbonation depth of AAS and OPC pastes according to CO_2 concentration: C5 and C10 indicate 5 and 10 % CO_2 concentration, respectively.

Table 5 Measured carbonation depth of paste specimens.

Specimen	PC(P)_C5*	PC(P)_C10	NS(P)_C5	NS(P)_C10	SS(P)C5	SS(P)_C10
Carbonation depth (mm)	3.33	5.33	17.67	25	14.33	25

Table 6 Total porosity and average pore diameter of paste specimens.

Specimen	PC(P)_C0	PC(P)_C10	NS(P)_C0	NS(P)_C10	SS(P)_C0	SS(P)_C10
Total porosity (%)	19.59	14.12	21.54	22.90	18.20	20.56
Average pore diameter (µm)	0.0270	0.0207	0.0167	0.0152	0.0103	0.0105

Table 7 Pore size distribution (PSD) of paste specimens.

PSD (%)	Pore diameter (µm)					
	>100	100–10	10–1.0	1.0–0.1	0.1–0.01	<0.01
PC(P)_C0	0.9	6.4	4.8	10.3	68.0	12.2
PC(P)_C10	1.1	2.1	7.9	16.1	58.1	17.4
NS(P)_C0	1.4	3.8	2.2	1.8	73.6	17.4
NS(P)_C10	0.6	7.2	4.0	27.5	29.9	31.1
SS(P)_C0	1.2	3.2	1.6	1.3	47.2	45.9
SS(P)_C10	0.8	7.6	8.3	10.8	32.3	43.0

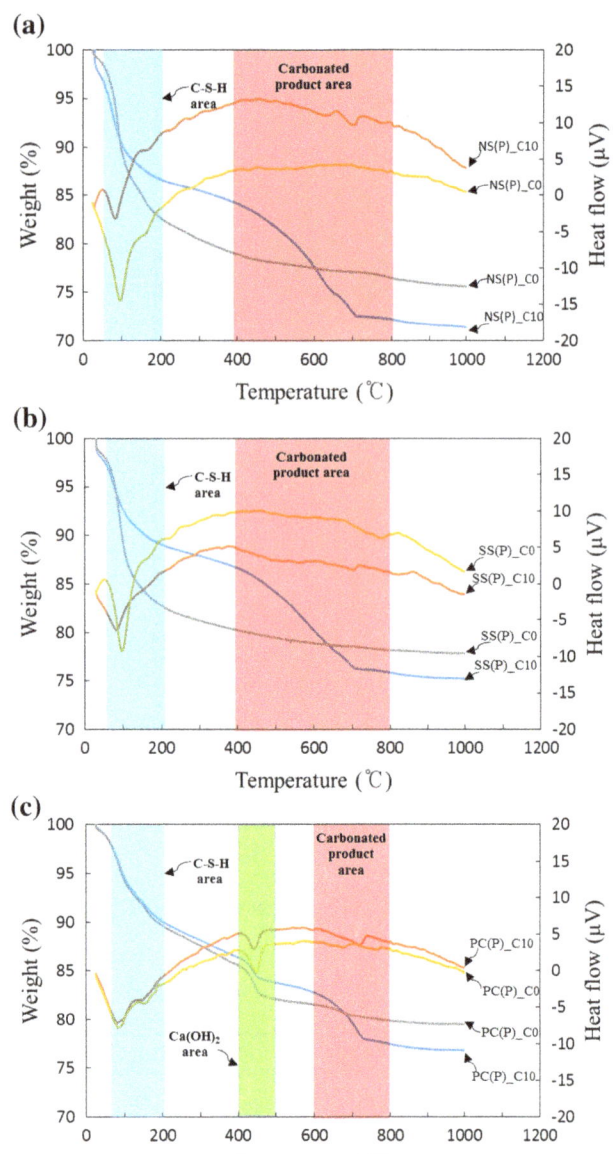

Fig. 2 TG/DTA test results of AAS and OPC pastes: **a** AAS pastes with NaOH (NS(P)), **b** AAS pastes with waterglass (SS(P)) and **c** OPC pastes (PC(P)).

carbonation of the AAS pastes in the present study was mainly caused by the decalcification from the C–S–H phases (Bernal et al. 2014; Bakharev et al. 2001).

The decomposition temperature of calcium carbonates such as calcite and vaterite was defined at the peak range of 600–800 °C in many studies (Kim et al. 2013). In the present study, the weight loss of SS(P)_C10 and NS(P)_C10 samples in the temperature range was similar to that of PC(P)_10 sample, whereas the carbonation depth of SS(P)_C10 and NS(P)_C10 specimens was much deeper than that of PC(P)_C10 specimen. This may be due that BFS has a relatively low calcium content compared to Portland cement (Puertas et al. 2006; Bakharev et al. 2001) and the aqueous concentration of calcium leached from the AAS pastes becomes low under a highly alkaline condition (Song and Jennings 1999). In addition, porosity of AAS paste increases as carbonation progressed. That is, relatively small amount of calcium component in AAS paste can form calcium carbonates and an increase in porosity due to carbonation can improve the diffusivity of CO_2 under those conditions. Consequently, the carbonation depth of AAS paste can be deeper than that of OPC paste, while the weight loss indicating calcium carbonates between them can be similar.

In the present study, an amended temperature range of carbonated products was proposed in an effort to accurately evaluate the uptake of CO_2 in AAS pastes. In general, calcium carbonates are mainly considered as a carbonated product in AAS pastes. However, sodium carbonates such as natron or trona can also exist in AAS pastes, when using a sodium incorporated alkali activator (Puertas et al. 2006; Bernal et al. 2014). Natron is a compound with sodium decahydrate ($Na_2CO_3 \cdot 10H_2O$) and sodium bicarbonate ($NaHCO_3$), and more than 80 % of natron consists of sodium decahydrate (Edwards et al. 2007). The release of CO_2 from the sodium decahydrate initiates at a temperature of 400 °C, exhibiting an endothermic reaction (Guo et al. 2011). Since trona loses water at the temperature below 100 °C and forms sodium carbonate, a similar reaction could

Table 8 Weight loss of C–S–H and carbonated products in the TG/DTA curves.

Phases	Temperature range (°C)	Weight loss (%)					
		PC(P)_C0	NS(P)_C0	SS(P)_C0	PC(P)_C10	NS(P)_C10	SS(P)_C10
C–S–H	50–200	9.27	15.90	15.44	8.90	10.26	8.61
Calcium carbonates (calcite, vaterite)	600–800	1.67	1.00	0.72	5.31	5.49	4.35
Calcium carbonates (calcite, vaterite) & sodium carbonates (natron, trona)	400–800	1.67	2.40	2.08	5.31	11.95	10.79

also begin at 400 °C (Ekmekyapar et al. 1996; Jang et al. 2015b). The downward slope of the DTA curves of SS(P)_C10 and NS(P)_C10 samples indicated an occurrence of the endothermic reaction that was observed at 400 °C (see Fig. 2a, b), and a dramatic change in the weight loss in the TG curve was initiated at 400 °C, simultaneously. This temperature is the identical point that the release of CO_2 from sodium carbonates is initiated (Guo et al. 2011; Ekmekyapar et al. 1996). Thus, it seems that the temperature range of 400–800 °C in the TG/DTA curves is more suited to define the uptake of CO_2 in AAS pastes.

In PC(P)_C10 sample, the TG/DTA curves within the proposed temperature range also showed a trend similar to that of SS(P)_C10 and NS(P)_C10 samples. However, it is well known that the weight loss and the trend of the DTA curve in PC(P)_C10 sample is caused by the decomposition of $Ca(OH)_2$ (Kim et al. 2013). That is, the weight loss within the proposed temperature range in PC(P)_C10 sample indicated the decomposition of $Ca(OH)_2$. In contrast, $Ca(OH)_2$ is not exist in AAS paste (Shi and Fernandez-Jimenez 2006). The CO_2 uptake in SS(P)_C10 and NS(P)_C10 samples in the proposed temperature range was approximately twice as high as that in PC(P)_C10 sample (see Table 8). It can be inferred from the results that porous concrete fabricated with an AAS paste can be a more effective CO_2 absorber in comparison with that fabricated with an OPC paste.

3.2 Physical Properties of Carbonated Porous Concrete

The measured total void ratios and the targeted void ratios of the porous concretes are listed in Table 9. The target void ratio was calculated based on the P/G ratio. The test results showed that the measured total void ratios of all specimens were much higher than the target void ratios. The fractional excess level between the measured total void ratio and the target void ratio was in the range of 11–31 %. The rough surface and porous internal structure of BA and compaction process for fabrication of the porous concretes may have caused such differences (Singh and Siddique 2014).

Carbonation affected the total void ratios in all the specimens as shown in Table 9. After exposure to accelerated carbonation, the total void ratios of the porous concretes reduced, and the reduction at a 10 % CO_2 concentration was as much as 10 %. In the present study, In spite of the reduction in void ratios after carbonation, the total void ratios of the porous concretes exceeded minimum void ratio

of 15 % stated in ACI 522 (American Concrete Institute 2010).

The open and closed void ratios of porous concretes according to the CO_2 concentration are given in Table 10. Most of the voids in the porous concretes consisted of open-voids ranging from 20 to 39 %, while the closed-void was ranged from 5 to 15 %. After the carbonation, the reduction in the closed-void ratios in SS and NS series specimens was greater than that in the open-void ratios. The closed-void ratios were reduced in the range of 1–8 %, while the open-void ratios were reduced in the range of 0–4 %. The residues in the pore solution of the closed-voids can react with CO_2, and the reaction products such as calcium and sodium carbonates may be filled in the closed-voids. Bernal et al. reported that carbonation of AAS pastes mainly occurs in the pore solutions (Bernal et al. 2014). In contrast, the surfaces of the open-voids may be covered with a relatively small amount of solution due to the continuity of the open-voids. Consequently, the reduction of the closed-void ratios in SS and NS series specimens after carbonation was greater than that of open-void ratios.

The reductions in the closed-void ratios of PC2 specimen were relatively lower than that in the SS and NS series specimens. As mentioned above, the porosity of the OPC paste shown in the MIP test result decreased after carbonation, while that of the AAS paste increased. It can be inferred from the results that the reduction in the porosity of the OPC paste interrupted the filling of the closed-voids as carbonation progressed, whereas the formation of carbonated products in the closed-voids of the AAS pastes accelerated (see Table 6).

The test results of compressive strength on porous concretes are shown in Fig. 3. The compressive strength of NS(P), PC(P), and SS(P) pastes in the present study were 23, 45 and 78 MPa, respectively, showing significant difference. The difference in the compressive strength between NS(P) and SS(P) is due that the hydration products in SS(P) is more densified by soluble silicate in waterglass than that in NS(P) (Phoo-ngernkham et al. 2015). However, the compressive strength of NS2_C0, PC2_C0, and SS2_C0 exhibited comparable regardless of the strength of the pastes, when the total void ratios of the porous concretes was similar. Furthermore, when NS series and SS series had comparable total void ratios, the compressive strengths were almost similar. It has been reported in previous studies that the main factor which affects the compressive strength of porous concrete is the total void ratio (Sriravindrarajah et al. 2012; Jang et al. 2015a; Lian et al. 2011; Bhutta et al. 2013).

Table 9 Measured total void ratios and targeted void ratios of porous concretes.

	Exposure condition (CO_2, %)	NS1	NS2	NS3	SS1	SS2	SS3	PC2
V_{target} (%)	–	36	31	26	36	31	26	31
V_{total} (%)	–	40	39	35	46	38	32	37
	5	37	39	29	45	41	33	37
	10	34	33	25	39	36	25	32

Table 10 Open and closed void ratios of porous concretes.

	Exposure condition (CO_2, %)	NS1	NS2	NS3	SS1	SS2	SS3	PC2
V_{open} (%)	–	31	30	20	39	32	24	32
	5	33	35	23	42	37	29	33
	10	29	28	18	36	32	20	28
V_{close} (%)	–	9	9	15	7	6	8	5
	5	4	4	6	3	4	4	5
	10	5	5	7	3	4	5	4

(a)

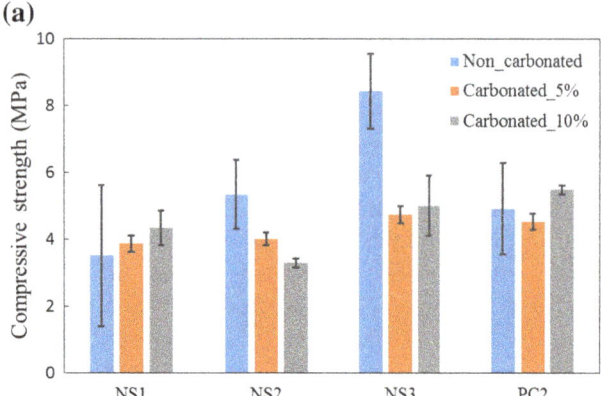

(b)

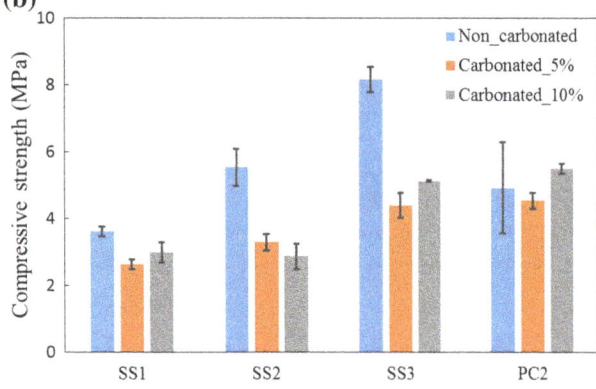

Fig. 3 Compressive strength of porous concretes: **a** fabricated with AAS pastes with NaOH (NS series) and **b** fabricated with AAS pastes with waterglass (SS series).

The compressive strength of the carbonated NS series and SS series decreased compared to non-carbonated specimens, whereas the opposite trend was observed for the PC2 specimen. It can be implied with the results of TG analysis and MIP test that the reduction in compressive strength of NS and SS series after carbonation was responsible for a decrease in the C–S–H phases due to decalcification and an increase in the porosity. However, the reduced compressive strength satisfied the minimum value noted in ACI 522 (2.8 MPa) (American Concrete Institute 2010).

3.3 Heavy Metal Leaching Characteristics of Carbonated Porous Concrete

The test result of an accumulated leaching concentration of heavy metals with exposure sequence in accordance with NSF/ANSI 61-2007a is shown in Fig. 4. The leached concentrations of Cr, Pb and Cd from the OPC-based porous concrete were observed to be higher than those from the AAS-based porous concretes. Bhutta et al. reported that Cr can be evenly distributed in the C–S–H phases of AAS paste, whereas the hydrated calcium aluminate phases in the OPC paste primarily bound Cr ions (Bhutta et al. 2013). It seems that a large amount of Cr in the test was immobilized in NS2 and SS2 specimens, since the C–S–H phases in AAS is a major reaction products. Shi and Fernandez-Jimenez reported that Pb in alkali activated cementitious material can be precipitated as Pb_3SiO_5, which is a highly insoluble silicate (Shi and Fernandez-Jimenez 2006). In the test, it seemed that Pb component in NS2 and SS2 be immobilized by the formation of Pb_3SiO_5.

The leached concentrations of Cr and Pb were not affected by the carbonation. Otherwise, the effect of carbonation on leachability of Cd was differed from that of Cr and Pb. The leached concentration of Cd from PC2_C10 and SS2_C10 specimens was higher than that of non-carbonated PC2_C0 and SS2_C0 specimens. The increased concentration of Cd may be due to a decrease of pH in carbonated specimens (Fleischer et al. 1974; Halim et al. 2004). In previous studies, solubility of Cd increased with the lowering of pH due to carbonation (Fleischer et al. 1974; Halim et al. 2004).

Arsenic is known to be soluble at a wide range of pH levels, creating oxyanions in solutions (Ravikumar and Neithalath 2013). For that reason, the predominant leaching mechanism of As may take place by washing off of leachate from the surface of the porous concretes. However, the leached concentration of As in PC2 specimen was relatively lower than that from the NS2 and SS2 specimens. Vandecasteele et al. reported that As can be reacted with $Ca(OH)_2$ in OPC paste to form the insoluble $Ca_3(AsO_4)$ (Vandecasteele et al. 2002). In contrast, $Ca(OH)_2$ is not present in the AAS matrix (Shi and Fernandez-Jimenez 2006). Therefore, it can be inferred that As in PC2 specimen was immobilized

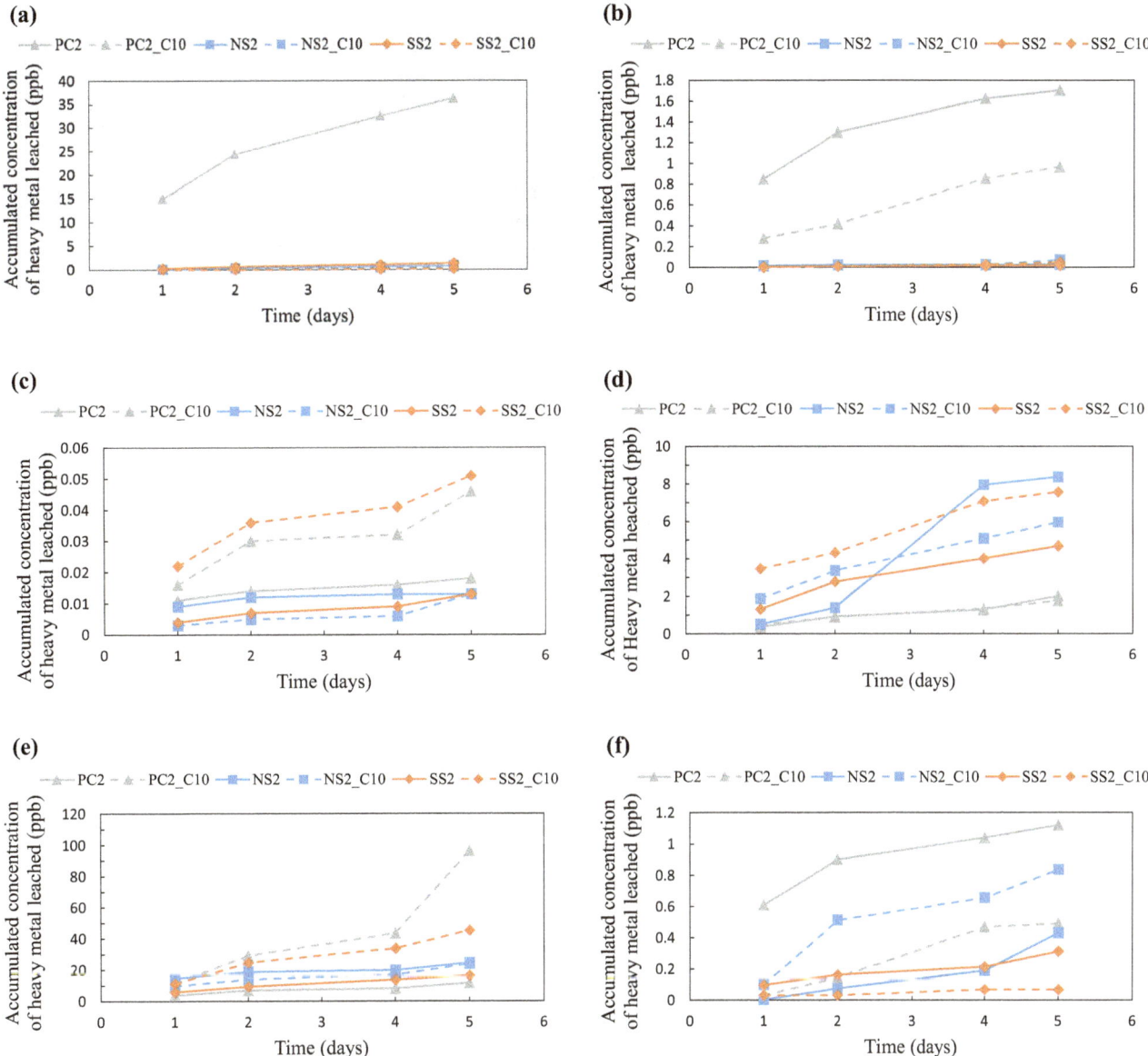

Fig. 4 Accumulated concentration of heavy metals leached from porous concretes: **a** chromium, **b** lead, **c** cadmium, **d** arsenic, **e** copper and **f** mercury.

Table 11 Single time-point concentration of the leached heavy metal according to NSF/ANSI 61-2007a, the drinking water regulatory level criteria (MCL/MAC) and the single-product allowable concentration (SPAC) (NSF International standard/American National Standard, NSF/ANSI 61- 2007a 2007).

(mg/L)	Cr	Cu	As	Pb	Cd	Hg
PC2_C0	0.003747	0.003361	0.00074	0.00008	2E−0.6	0.00008
NS2_C0	0.000168	0.004682	0.00041	3E−06	ND[a]	0.00024
SS2_C0	0.000248	0.002800	0.00066	6E−06	4E−06	0.00009
PC2_C10	0.00011	0.05289	0.00041	0.00015	0.00001	0.00002
NS2_C10	0.00011	0.00681	0.00088	0.00004	7E−06	0.00018
SS2_C10	0.00005	0.01154	0.00049	0.00002	0.00001	ND
MCL/MAC	0.1	1.3	0.01	0.015	0.005	0.002
SPAC	0.01	0.13	0.001	0.0015	0.0005	0.0002

[a] ND: Non detected.

by the formation of $Ca_3(AsO_4)$, whereas As in NS2 and SS2 specimens was washing off from the surface of them.

The leached concentrations of Cu from PC2_C0 specimen were lower than those from NS2_C0 and SS2_C0 specimens, as shown in Fig. 4e. In addition, the leached Cu concentrations in all specimens increased after carbonation. The increased Cu concentrations may be caused by a decrease of pH due to carbonation. Bochenczyk reported that solubility of Cu can increase by lowering of pH due to carbonation (Zhang et al. 2007; Bochenczyk 2010). That is, the solubility of Cu in all carbonated porous concretes may increase, since the carbonation can be lower the pH of all specimens (Puertas et al. 2006; Bernal et al. 2014). The immobilization of Hg in PC2 specimen was not effective in comparison with that in the NS2 and SS2 specimens, as shown in Fig. 4f. Qian et al. reported that the immobilization of Hg in AAS matrix is highly effective, since physical encapsulation and chemical fixation are contributed to immobilization of Hg (Bakharev et al. 2001; Qian et al. 2003c). Thus, it seems that the two mechanisms in the test lowered leached concentration of Hg from NS2 and SS2 specimens.

Lastly, the test results of the single-point concentration of the leached heavy metals, the drinking water regulatory level criteria (MCL/MAC), and the single-product allowable concentration (SPAC) are shown in Table 11. The concentrations of the leached heavy metals from all the specimens, except for Hg in NS2, satisfied the criteria specified in MCL/MAC and SPAC. The exceeding amount of Hg in NS2 was not significant.

4. Concluding Remarks

In the present study, a porous concrete with alkali activated slag and coal bottom ash was developed and the effect of carbonation on physical property, microstructural characteristic, and heavy metal leaching behavior of the porous concrete were investigated. Independent variables, such as the type of the alkali activator and binder, the amount of paste, and CO_2 concentration, were considered. The main conclusions can be summarized below:

(1) Void ratio and compressive strength of AAS-based porous concretes were reduced by carbonation, and the measured void ratio and compressive strength of the carbonated porous concretes in the test exceeded the minimum level stated in ACI 522 for general porous concrete.

(2) Reduction of void ratio in carbonated AAS-based porous concrete mainly occurred in closed-voids. Reduction of compressive strength in the carbonated porous concrete may be attributed by an increase in porosity and decomposition of C–S–H phases.

(3) An amended temperature range in TG/DTA analysis was proposed to more accurately evaluate CO_2 uptake in AAS pastes, considering sodium and calcium carbonates. The CO_2 uptake of AAS pastes in the proposed temperature range was approximately twice as high as that in OPC paste.

(4) The Immobilization of chromium, lead, cadmium and mercury in the AAS-based porous concretes was more effective in comparison with that in the OPC-based porous concrete, and carbonation of the AAS based-porous concretes could cause an increase in the leached concentration of cadmium and copper. However, the leached concentrations of heavy metals in porous concretes utilizing AAS and BA in the test were all below the SPAC and MCL/MAC criteria as described in the NSF/ANSI 61-2007a.

Acknowledgments

This research was supported by a grant from the Energy Technology Development Program (grant No. 2013T100100021) funded by the Ministry of Trade Industrial and Energy of the Korea government, and by Saudi Aramco-KAIST CO2 Management Center.

References

American Concrete Institute, ACI 522R-10. Report on pervious concrete. ACI Committee 522 2010.

American Society for Testing and Materials, ASTM C39. (2012). *Standard test method for compressive strength of cylindrical concrete specimens*. West Conshohocken, PA: ASTM International.

Bakharev, T., Sanjayan, J. G., & Cheng, Y.-B. (2001). Resistance of alkali-activated slag concrete to carbonation. *Cement and Concrete Research, 31*, 1277–1283.

Bernal, S. A., Nicolas, R., Provis, J. L., De Gutierrez, R. M., & van Deventer, J. S. (2014). Natural carbonation of aged alkali-activated slag concretes. *Materials and Structures, 47*, 693–707.

Bertos, M. F., Simons, S. J. R., Hills, C. D., & Carey, P. J. (2004). A review of accelerated carbonation technology in the treatment of cement-based materials and sequestration of CO_2. *Journal of Hazardous Materials, 112*(2), 193–205.

Bhutta, M. A. R., Hasanah, N., Farhayu, N., Hussin, M. W., Tahir, M. B. M., & Mirza, J. (2013). Properties of porous concrete from waste crushed concrete (recycled aggregate). *Construction and Building Materials, 47*, 1243–1248.

Bhutta, M. A. R., Tsuruta, K., & Mirza, J. (2012). Evaluation of high-performance porous concrete properties. *Construction and Building Materials, 31*, 67–73.

Bochenczyk, A. U. (2010). Mineral sequestration of CO2 in suspensions containing mixtures of fly ashes and desulphurization waste. *Gosposarka Surowcami Mineralnymi, 26*, 109–118.

Chi, M.C., Chang, J.J., & Huang, R. (2012). Strength and drying shrinkage of alkali-activated slag paste and mortar. *Advances in Civil Engineering, 2012*.

Deja, J. (2002). Immobilization of Cr^{6+}, $Cd^{2+,}$ Zn^{2+} and Pb^{2+} in alkali-activated slag binders. *Cement and Concrete Research, 32*, 1971–1979.

Dermatas, D., & Meng, X. (2003). Utilization of fly ash for stabilization/solidification of heavy metal contaminated soil. *Engineering Geology, 70*, 337–394.

Edwards, H. G., Currie, K. J., Ali, H. R., Villar, S. E. J., David, A. R., & Denton, J. (2007). Raman spectroscopy of natron: shedding light on ancient Egyptian mummification. *Analytical and Bioanalytical Chemistry, 388*(3), 683–689.

Ekmekyapar, A., Ersahan, H., & Yapici, S. (1996). Non-isothermal decomposition kinetics of trona. *Industrial and Engineering Chemistry Research, 35*, 258–262.

Eloneva, S., Teir, S., Salminen, J., Fogelholm, C. J., & Zevenhoven, R. (2008). Fixation of CO2 by carbonating calcium derived from blast furnace slag. *Energy, 33*, 1461–1467.

Environment, Health and Safety Online. (2008). The EPA TCLP: Toxicity characteristic leaching procedure and characteristic wastes (D-codes). Environment, Health and Safety Online.

Fleischer, M., Sarofim, A. F., Fassett, D. W., Hammond, P., Shacklette, H. T., Nisbet, I. C., & Epstein, S. (1974). Environmental impact of cadmium: a review by the Panel on Hazardous Trace Substances. *Environmental Health Perspectives, 7*, 253.

Guo, Q., Qu, J., Qi, T., Wei, G., & Han, B. (2011). Activation pretreatment of limonitic laterite ores by alkali-roasting method using sodium carbonate. *Minerals Engineering, 24*, 825–832.

Halim, C. E., Acott, J. A., Natawardaya, H., Amal, R., Beydoun, D., & Low, G. (2004). Comparison between acetic acid and landfill leachates for the leaching of Pb(II), Cd(II), As(V), and Cr(VI) from cementitious wastes. *Environmental Science and Technology, 38*, 3977–3983.

Jang, J. G., Ahn, Y. B., Souri, H., & Lee, H. K. (2015a). A novel eco-friendly porous concrete fabricated with coal ash and geopolymeric binder: Heavy metal leaching characteristics and compressive strength. *Construction and Building Materials, 79*, 173–181.

Jang, J. G., Kim, H. J., Park, S. M., & Lee, H. K. (2015b). The influence of sodium hydrogen carbonate on the hydration of cement. *Construction and Building Materials, 94*, 746–749.

Japanese Standard Association, JIS A 1104. (2006). Methods of test for bulk density of aggregates and solid content in aggregates. JSA

Kar, A., Ray, I., Halabe, U. B., Unnikrishnan, A., & Dawson-Andoh, B. (2014). Characterizations and quantitative estimation of alkali-activated binder paste from microstructures. *International Journal of Concrete Structures and Materials, 8*, 213–228.

Kim, H. K., Ha, K. A., Jang, J. G., & Lee, H. K. (2014a). Mechanical and chemical characteristics of bottom ash aggregates cold-bonded with fly ash. *Journal of Korean Ceramic Society, 51*, 57–63.

Kim, H. K., Jang, J. G., Choi, Y. C., & Lee, H. K. (2014b). Improved chloride resistance of high-strength concrete amended with coal bottom ash for internal curing. *Construction and Building Materials, 71*, 334–343.

Kim, M. S., Jun, Y., Lee, C., & Oh, J. E. (2013). Use of CaO as an activator for producing a price-competitive non-cement structural binder using ground granulated blast furnace slag. *Cement and Concrete Research, 54*, 208–214.

Kim, H. K., & Lee, H. K. (2010). Influence of cement flow and aggregate type on the mechanical and acoustic characteristics of porous concrete. *Applied Acoustics, 71*, 607–615.

Kuo, W. T., Liu, C. C., & Su, D. S. (2013). Use of washed municipal solid waste incinerator bottom ash in pervious concrete. *Cement & Concrete Composites, 37*, 328–335.

Li, X. D., Poon, C. S., Sun, H., Lo, I. M. C., & Kirk, D. W. (2012). Heavy metal speciation and leaching behaviors in cement based solidified/stabilized waste materials. *Journal of Hazardous Materials, 82*(3), 215–230.

Lian, C., Zhuge, Y., & Beecham, S. (2011). The relationship between porosity and strength of porous concrete. *Construction and Building Materials, 25*, 4292–4298.

NSF International standard/American National Standard, NSF/ANSI 61-2007a. (2007). *Drinking water system components—health effects*. Oxfordshire: NSF International.

Park, S. B., & Tia, M. (2004). An experimental study on the water purification properties of porous concrete. *Cement and Concrete Research, 34*, 177–184.

Perera, D. S., Aly, Z., Vance, E. R., & Mizumo, M. (2005). Immobilization of Pb in a geopolymer matrix. *Journal of the American Ceramic Society, 88*, 2586–2588.

Phoo-ngernkham, T., Maegawa, A., Mishima, N., Hatanaka, S., & Chindaprasirt, P. (2015). Effects of sodium hydroxide and sodium silicate solutions on compressive and shear bond strengths of FA-GBFS geopolymer. *Construction and Building Materials, 91*, 1–8.

Puertas, F., Palacious, M., & Vazquez, T. (2006). Carbonation process of alkali-activated slag mortars. *Journal of Materials Science, 41*, 3071–3082.

Qian, G., Sun, D. D., & Tay, J. H. (2003a). Immobilization of mercury and zinc in an alkali-activated slag matrix. *Journal of Hazardous Materials, 101*(2), 65–77.

Qian, G., Sun, D. D., & Tay, J. H. (2003b). Characterization of mercury- and zinc-doped alkali-activated slag matrix Part II. Zinc. *Cement and Concrete Research, 33*, 1257–1262.

Qian, G., Sun, D. D., & Tay, J. H. (2003c). Characterization of mercury- and zinc-doped alkali-activated slag matrix Part I. Mercury. *Cement and Concrete Research, 33*, 1251–1256.

Ravikumar, D., & Neithalath, N. (2013). Electrically induced chloride ion transport in alkali activated slag concretes and the influence of microstructure. *Cement and Concrete Research, 47*, 31–42.

Shi, C., & Fernandez-Jimenez, A. (2006). Stabilization/solidification of hazardous and radioactive wastes with alkali-activated cements. *Journal of Hazardous Materials, 137*(3), 1656–1663.

Singh, M., & Siddique, R. (2014). Strength properties and micro-structural properties of concrete containing coal bottom ash as partial replacement of fine aggregate. *Construction and Building Materials, 50*, 246–256.

Song, S., & Jennings, H. M. (1999). Pore solution chemistry of alkali-activated ground granulated blast-furnace slag. *Cement and Concrete Research, 29*, 159–170.

Sriravindrarajah, R., Wang, N. D. H., & Ervin, L. J. W. (2012). Mix design for pervious recycled aggregate concrete. *International Journal of Concrete Structures and Materials, 6*(4), 239–246.

Vandecasteele, C., Dutre, V., Geysen, D., & Wauters, G. (2002). Solidification/stabilization of arsenic bearing fly ash from the metallurgical industry. Immobilization mechanism of arsenic. *Waste Management, 22*(2), 143–146.

Wang, S. D., & Scrivener, K. L. (1995). Hydration products of alkali activated slag cement. *Cement and Concrete Research, 25*, 561–571.

Ylmén, R., & Jäglid, U. (2013). Carbonation of Portland cement studied by diffuse reflection fourier transform infrared spectroscopy. *International Journal of Concrete Structures and Materials, 7*, 119–125.

Zhang, J., Provis, J. L., Feng, D., & van Deventer, J. S. J. (2008). Geopolymers for immobilization of Cr^{6+}, Cd^{2+}, and Pb^{2+}. *Journal of Hazardous Materials, 157*, 587–598.

Zhang, Y., Sun, W., Chen, Q., & Chen, L. (2007). Synthesis and heavy metal immobilization behaviors of slag based geopolymer. *Journal of Hazardous Materials, 143*, 206–213.

9

Mode II Fracture Toughness of Hybrid FRCs

H. S. S. Abou El-Mal[1], A. S. Sherbini[2], and H. E. M. Sallam[3],*

Abstract: Mode II fracture toughness (K_{IIc}) of fiber reinforced concrete (FRC) has been widely investigated under various patterns of test specimen geometries. Most of these studies were focused on single type fiber reinforced concrete. There is a lack in such studies for hybrid fiber reinforced concrete. In the current study, an experimental investigation of evaluating mode II fracture toughness (K_{IIc}) of hybrid fiber embedded in high strength concrete matrix has been reported. Three different types of fibers; namely steel (S), glass (G), and polypropylene (PP) fibers were mixed together in four hybridization patterns (S/G), (S/PP), (G/PP), (S/G/PP) with constant cumulative volume fraction (V_f) of 1.5 %. The concrete matrix properties were kept the same for all hybrid FRC patterns. In an attempt to estimate a fairly accepted value of fracture toughness K_{IIc}, four testing geometries and loading types are employed in this investigation. Three different ratios of notch depth to specimen width (a/w) 0.3, 0.4, and 0.5 were implemented in this study. Mode II fracture toughness of concrete K_{IIc} was found to decrease with the increment of a/w ratio for all concretes and test geometries. Mode II fracture toughness K_{IIc} was sensitive to the hybridization patterns of fiber. The (S/PP) hybridization pattern showed higher values than all other patterns, while the (S/G/PP) showed insignificant enhancement on mode II fracture toughness (K_{IIc}). The four point shear test set up reflected the lowest values of mode II fracture toughness K_{IIc} of concrete. The non damage defect concept proved that, double edge notch prism test setup is the most reliable test to measure pure mode II of concrete.

Keywords: fiber reinforced concrete, hybrid fiber, mode II fracture toughness.

1. Introduction

Almost all FRCs used today commercially involve the use of a single fiber type. The decision to mix two fibers may be based on the properties that they may individually provide or simply based on economics (ACI committee 544 2011). Clearly, a given type of fiber can only be effective in a limited range of crack opening and deflection. The benefits of combining organic and inorganic fibers to achieve superior tensile strength and fracture toughness were recognized nearly 40 years ago by Walton and Majumdar (1975). After a long period of relative inactivity there appears to be a second wave of interest in hybrid fiber composites and efforts are underway to develop the science and rationale behind fiber hybridization.

In well-designed hybrid composites, there is a positive interaction between the fibers and the resulting hybrid performance exceeds the sum of individual fiber performances (Bentur and Mindess 1990; Xu et al. 1998). This phenomenon is often termed as "synergy". This might be due to any of the following mechanisms.

1.1 Hybrids Based on Fiber Constitutive Response

One type of fiber is stronger, stiffer and provides reasonable first crack strength and ultimate strength, while the second type of fiber is relatively flexible and leads to improved toughness and strain capacity in the post-crack zone.

1.2 Hybrids Based on Fiber Dimensions

One type of fiber is smaller, so that it bridges micro-cracks controlling their growth and delays coalescence. This leads to a higher tensile strength of the composite. The second fiber is larger and is intended to arrest the propagation of macro-cracks and therefore results in a substantial improvement in the fracture toughness of the composite. Fibers of small size (often called micro-fibers) delay crack coalescence in the cement paste and mortar phases and increase the apparent tensile strength of these phases (Banthia et al. 1995; Shah 1991).

1.3 Hybrids Based on Fiber Function

One type of fiber is intended to improve the fresh and early age properties such as ease of production and plastic

[1]Civil Engineering Department, Menofia University, Shibin El-Kom 32511, Egypt.

[2]Civil Engineering Department, Suez Canal University, Ismailya 41522, Egypt.

[3]Department of Civil Engineering, Faculty of Engineering, Jazan University, Jazan 82822-6694, Saudi Arabia.

*Corresponding Author; E-mail: hem_sallam@yahoo.com

shrinkage, while the second fiber leads to improved mechanical properties.

In the past, many attempts have been made at identifying fiber combinations that produce the maximum synergy (Larson and Krenchel 1991; Feldman and Zheng 1993; Kamlos et al. 1995; Qian and Stroeven 2000; Kim et al. 1999; Banthia and Sheng 1991; Mobasher and Li 1996; Lawler et al. 2002). More recently, Banthia and Soleimani (2005) investigated three-fiber hybrids with carbon and polypropylene microfibers added to macro-steel fibers. Their results showed that, steel macro-fibers with highly deformed geometry produce better three-fiber hybrids than those with a less deformed geometry. Finally, Banthia and Gupta (2004) showed that the strength of the matrix plays a major role in the optimization of hybrid composites.

On the other hand, many researchers are now looking at the sliding plane deformation state other than opening mode which may be associated with crack propagation and fracture known as mode II fracture. In considering this, there are two major and interrelated problems: (1) determination of fracture parameters for mode II and (2) verification both analytically and experimentally that a crack can propagate due to mode II deformation. Analytical models cannot function successfully without valid mode II data such as values of K_{IIc} (Swartz et al. 1988). The authors' knowledge to the parameters controlling the concrete fracture toughness and the fiber/matrix interface should enhance the development of concrete technology.

To study Mode II fracture toughness, various approaches have attempted to define testing geometries where self-similar crack propagation occurs with only mode II deformations (Sherbini 2014). Although there is a violent debate around the validity of such a test in driving cracks under pure mode II, the proposed test geometries briefed in Table 1 are considered the most important techniques in isolating shear parameters (Reinhardt et al. 1997; Watkins 1983; Prokopski 1991; Irobe and Pen 1992; Iosipescu 1967).

In a quick comparison between advantages and disadvantages of proposed Mode II fracture toughness test approaches, Sherbini (2014) concluded in an earlier study that, double notched cube (DNC) test setup showed higher values than all other tests due to the crack propagation miss alignment opposing sliding of crack surfaces. Regarding Brazilian notched disc (BND) test setup, the addition of fibers decreased the calculated values of K_{IIc} for all single fiber types. Finally, four point shear (4PS) test set up reflects the most reliable values of mode II fracture toughness K_{IIc} of concrete. The biases of the various concrete toughness tests developed is still unknown. Sufficient data should be gathered and sufficient research conclusions should be collected in order to define a reliable test standard (Lee and Lopez 2014).

The aim of this experimental investigation is to study the effect of adding different combinations of fibers to concrete on its mode II fracture toughness K_{IIc}. A comparison between the estimated values of K_{IIc} of concrete according to the proposed four different test techniques is reported in this investigation attempting to find an answer for the confusing argument, "which test set up is the most convenient to evaluate mode II fracture toughness in case of hybrid fiber reinforced concrete?"

2. Experimental Work

The present experimental program included Three different types of fibers; namely steel (S), glass (G), and polypropylene (PP) fibers were mixed together in four hybridization groups, (S/G), (S/PP), (G/PP), (S/G/PP) with constant cumulative volume fraction (V_f) of 1.5 %. The concrete matrix properties were kept the same for all hybrid fiber reinforced concrete (FRC) patterns. The chosen types of fibers, cumulative volume fractions (V_f), properties of raw materials, mix proportions, matrix properties, and all other laboratory conditions (specimen preparation, casting and compaction, curing, temperature, test setup, and day of testing) were kept the same as reported from previous work of Sherbini (2014) to achieve a solid comparison with his earlier study for single fiber type. Each group contains, standard cubes and cylinders to determine the mechanical properties, in addition to four different mode II fracture toughness test specimens (Reinhardt et al. 1997; Watkins 1983; Prokopski 1991; Irobe and Pen 1992; Iosipescu 1967). In the current study, three different (a/w) ratios 0.3, 0.4, and 0.5 were used in agreement with the conclusion of Lee and Lopez (2014) that, the accuracy of the size effect fracture energy determined using one size of notched beam has recently been brought into question. As a further study, a comparison of the size effect fracture energy as determined using multiple sizes of notched beams is recommended. Five specimens per sample were used for each tested parameter.

The cement used in all concrete mixes was ordinary Portland cement of 450 kg/m^3. Light gray silica fume with specific surface area (SSA) of 18 m^2/gm supplied from the Ferro silicon alloys plant in Edfo zone, Egypt, was used with 10 % added percentage to the cement content to produce HSC. The sand used was local natural siliceous sand with specific gravity of 2.55, fineness modulus (FM) of 2.51, and SSA of 50.47 cm^2/gm. The coarse aggregate was dolomite with nominal maximum aggregate size (NMAS) of 10 mm, specific gravity of 2.6, FM of 6.69, and SSA of 6.54 cm^2/gm. A superplasticizer called Adecrite PVF (naphthalene sulphonated compound) was added to the mixing water to improve the workability and to keep the slump almost constant. The mixing, casting, and compaction recommendations suggested by ACI Committee 544 (2011) were adopted in the present work to prepare all mixes.

Plain mild steel, high zirconia alkali resistance glass (NEG ARG) fibers and MC polypropylene fibers were used with different combinations in this investigation. Table 2 shows the properties of different fiber types used in the current study as reported by manufacturers. Galvanized steel fiber with a new shape was used in this work; two straight steel

Table 1 Mode II fracture toughness test geometries.

Test geometries	Calculation equation	Dimensions mm
double-edge notched prism (DENP)	If $h \geq 2a$, $w \geq \pi a$. $K_{IIc} = \frac{\sigma}{4}(\pi a)^{1/2}$ If $h \geq 2a$, $w \leq \pi a$ $K_{IIc} = \frac{\sigma}{4}w^{1/2}$ Proposed by Reinhardt et al. (1997)	$2h = 200$ $2a = 140, 120, 100$ $w = 100$ Thickness $= 100$
Double notched cube (DNC)	$K_{IIc} = \frac{5.11P_Q}{2BW}(\pi a)^{1/2}$ Proposed by Watkins (1983) and Prokopski (1991)	Cube 150 mm $a = 45, 60, 75$ $w = 150$ $B = w - a$
Brazilian Disc Specimen with inclined centered notch (BND)	$K_{IIc} = -\frac{2P}{t}\sqrt{\frac{\lambda}{\pi R}} \left\{ \begin{array}{l} B_0 - \lambda^2\left(B_0 + \frac{1}{2}B_2\right) + \lambda^4\left(-\frac{1}{8}B_0 + \frac{1}{4}B_2 + \frac{3}{8}B_4\right) \\ + \lambda^6\left(B_0 - \frac{1}{16}B_2 + \frac{1}{8}B_4 + \frac{5}{16}B_6\right) \\ + \lambda^8\left(-\frac{17}{64}B_0 + \frac{3}{8}B_2 - \frac{5}{128}B_4 + \frac{5}{64}B_6 + \frac{35}{128}B_8\right) \end{array} \right\}$ $\lambda = a/R$, and β: the notch inclination angle $= 30°$ $B_0 = \sin 2\beta$, $B_2 = 2\left[\sin 4\beta - \sin 2\beta\right]$, $B_4 = 3\left[\sin 6\beta - 2\sin 4\beta\right]$. $B_6 = 4\left[\sin 8\beta - 3\sin 6\beta\right]$, $B_8 = 5\left[\sin 10\beta - 4\sin 8\beta\right]$. Proposed by Irobe and Pen (1992)	$\beta = 30°$ $R = 75$ Thickness $= 60$ $2a = 45, 60, 75$
Four-Point Shear Beam (4PS)	$K_{IIc} = Y_{II}\sigma\sqrt{\pi a}$ Proposed by Iosipescu (1967)	Prism $100 \times 100 \times 500$ $a = 30, 40, 50$ Loaded span $= 400$

Table 2 Properties of the used steel, glass and polypropylene fiber.

Properties	Fibers type		
	Plain mild steel	NEG ARG glass	MC polypropylene*
Fiber length (mm)	25	25	15
Fiber diameter	0.5 mm/filament 1 mm/bi-filament	10–12 μm/filament 1–1.2 mm/strand	0.0965 ± 10 %**
Specific gravity (t/m^3)	7.8	2.7	0.90
Tensile strength (MPa)	3600	1400	550–600
Young's modulus (MPa)	200,000	74,000	3600–3900
Strain at failure (%)	6–9	2	14–25
Geometry	Bi-filament	Chopped strands	Monofilament

* Master Chemicals Technology Company.
** Fiber thickness (mm).

Table 3 Fiber combination percentages.

	S/G	S/PP	G/P	S/G/P
Volume fraction (V_f)	1 % steel & 0.5 % glass	1 % steel & 0.5 % PP	0.75 % glass & 0.75 % PP	0.5 % steel & 0.5 % glass & 0.5 % PP

fibers of 265 MPa yield strength were twisted around each other to form a bi-filament fiber of 25 mm length. This new shape of fiber produces a good bond between the matrix and the fiber due to the development of interlock mechanical bond depending on the fiber geometry. Chopped strands alkaline resistance glass fiber (NEG ARG) achieves its high alkali resistance from the high zirconia's content in its glass composition. "MC" polypropylene Synthetic fiber meets the requirements of ASTM C 1116 and C 1399.

The mix proportion by weight for all mixes was 1:1.92:2.00:0.38 [cement:sand:dolomite:water/(cementations materials)] as reported by Sherbini (2014). The fiber combination percentages are illustrated in Table 3.

A vertical mixer of revolving blades type was used in mixing. Materials of the specified mix were weighed first and then mixed in the following procedures. Mixing different fiber types in hybrid combinations was really a challenge. The used fibers with widely varied aspect ratios are hard to blend together due to their different behavior during mixing. Polypropylene fibers representing high aspect ratio fiber ($L/d = 167$) should be mixed with the fine dry components.

First of all: sand, cement, silica fume, and polypropylene fibers were dryly mixed together for about 3 min to achieve uniform distribution of fibers through the mix.

Then, the coarse aggregate is added gradually during dry mixing. In the second step, one-third of the water content is added to the mixture. In the following step, the admixtures are added to the residual two-thirds of the water content then added to the mixture to achieve a slump greater than the final desired slump by 50 mm. Finally, Chopped glass and steel fibers (representing low aspect ratio fiber ($L/d = 25$) are added in small increments by sprinkling them onto the surface of the mix until all the fibers were absorbed into the

matrix. This technique was performed to prevent balling or interlocking of the fibers and achieve homogeneous dispersion of the fibers through the matrix. The freshly mixed concrete was tested for slump as a quality control test; the desired slump was (100 mm) to avoid segregation during casting and compaction. The mixed materials were then placed in the molds, compacted using external vibration, leveled, and cured in water for 28 days before testing according to the recommendations of ACI committee 544 (2011). Figure 1 represents the uniform distribution of fibers along the cross section of the tested specimens reflecting that, the fiber segregation was avoided and the desired homogeneity was achieved.

Cubes of 150 × 150 × 150 mm dimensions were prepared to be tested under static compression. Cylinders of 150 mm diameter and 300 mm height were prepared to be tested under indirect tension. The mean values and the standard deviations of compressive and tensile strengths of the hybrid FRCs are listed in Table 4.

For comparison, the mean values of compressive and tensile strengths for high strength concrete matrix with single fiber type addition tested by Sherbini (2014), with the same constituents properties and mix proportions as the current study, are listed in Table 5. It is clear that, the strengths of hybrid FRCs are higher than those of individual FRC. The compressive strength increased by (2–13 %), and the tensile strength increased (up to 14.8 %) in comparison with single fiber addition to concrete at the same fiber volume fraction. The keyword explaining that behavior is "synergy", i.e. synergistic effect.

Rao and Rao (2009) and Boulekbache et al. (2012), studied the effect of steel fiber geometry (fiber aspect ratio = 47 (Rao and Rao 2009) and 65 & 80 (Boulekbache et al. 2012)) and matrix strength, i.e. $f_c \approx 20$ MPa (Rao and

Fig. 1 Distribution of fibers along the cross section of the tested specimens.

Table 4 Mechanical properties of hybrid FRCs in MPa (Mean ± SD).

	S/G	S/P	G/P	S/G/P
Compressive strength	54.9 ± 1.92	56.2 ± 2.04	53 ± 2.00	54.5 ± 1.67
Tensile strength	5.8 ± 0.2	6.2 ± 0.18	5.2 ± 0.16	5.1 ± 0.16
Shear strength (predicted)	16.1	16.6	15.5	16

Table 5 Compressive and tensile strengths in MPa (Sherbini 2014).

	SRC	GRC	PRC
Compressive strength	49.7	50.3	51.8
Tensile strength	5.4	5.2	5.1

Rao 2009) and $f_c \approx 29, 60$, and 82 MPa (Boulekbache et al. 2012), on shear behavior of fiber reinforced concrete. They concluded that (Rao and Rao 2009; Boulekbache et al. 2012), the ultimate shear strength of FRC (τ_f) is a function of $V_f\%$ and the ultimate shear strength of concrete matrix (τ_0), $\tau_0 = k\sqrt{f_{co}}$ where k is constant and f_{co} is compressive strength of concrete matrix, i.e., $\tau_f = \tau_0 + c\ (V_f\%)^n$ where c and n are constants.

Rao and Rao (2009), measured compressive, tensile, and shear strengths of concrete, while, Boulekbache et al. (2012), measured only compressive and shear strengths of the reported three types of concrete. Furthermore, the mechanical property of concrete that is designed or controlled is typically its compressive strength, since this is the most important material characteristic in concrete specification and in building codes (Li 2012). Therefore, in the present work regression analysis was carried out on their experimental data points of τ_f and compressive strength of FRC (f_c) (Rao and Rao 2009; Boulekbache et al. 2012). Through regression analysis, the empirical relation obtained can be expressed.

$$\tau_f = 0.153 f_c^{1.163} \qquad (1)$$

Coefficient of determination (R^2) of this proposed relation is 0.89, suggesting a strong correlation between these two mechanical properties. In the present study, steel fiber

represents the main fiber type due to its higher strength and stiffness, and for all hybrid FRCs except S/G/PP, the ration of V_f of steel fiber to V_f of other fibers is 2.0. Therefore, the current regression analysis of the results reported by Rao and Rao (2009) and Boulekbache et al. (2012) might herein be acceptable. According to the above equation, the predicted values of the present hybrid FRCs are tabulated in Table 4.

Concerning mode II of fracture (sliding mode), four test methods have been investigated; four point shear (4PS), Brazilian notched disc (BND), double notched cube (DNC), and double edge notched specimens (DENP), in a trial to avoid the limitations and sensitivity of each test. Test setups, layout, loading conditions, and specimens' dimensions are illustrated in Table 1. The main parameter affecting K_{IIc} obtained from previous test setups are specimen geometry, size effect, constraint condition, and the notch depth to specimen width ratio (a/w). Looking closer at these methods, it turned out that, most of them produce a mixed state of normal and shear stress mainly due to unavoidable load eccentricities, and hence bending moments which occur either from the beginning or after some deformation of the specimen.

To examine the reliability of these four mode II fracture toughness tests, the maximum undamaged defect size (d_{max}) suggested by Sallam (2003), Al Hazmi et al. (2012), Sallam et al. (2014), Sallam and Mubaraki (2015) will be compared

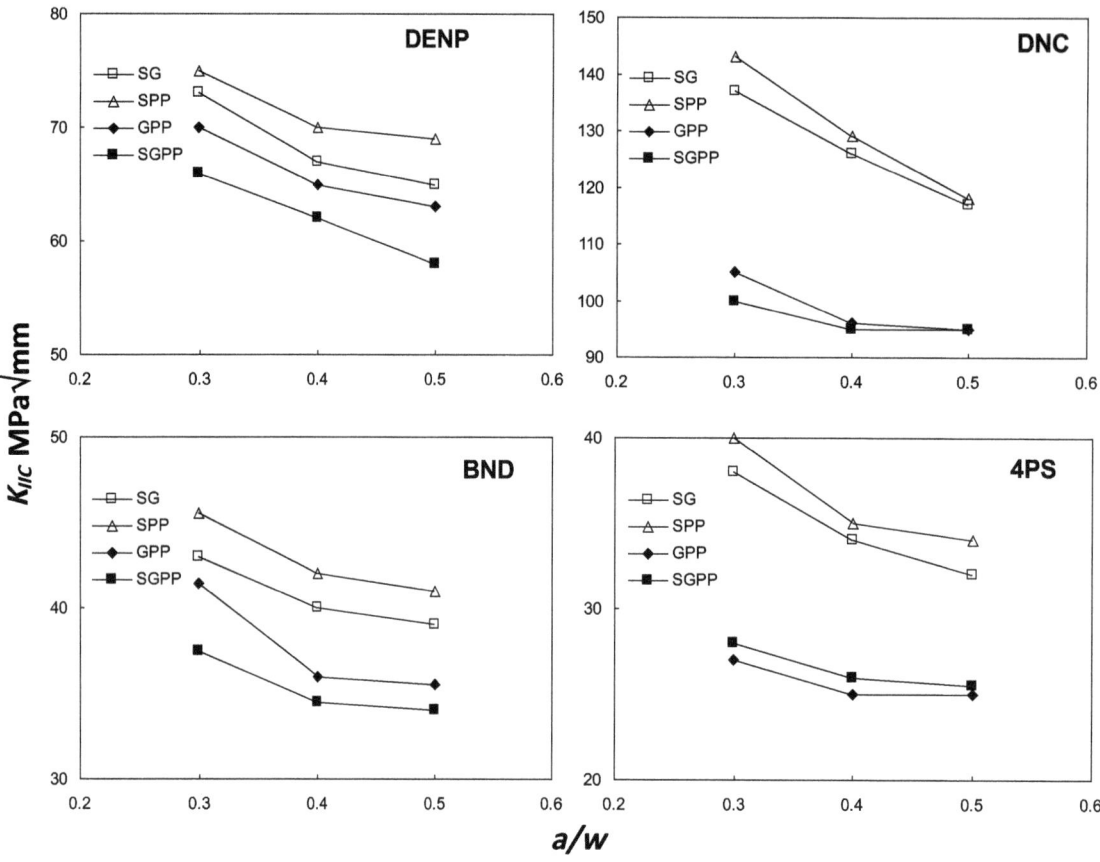

Fig. 2 Notch depth to specimen width ratio (*a/w*) versus mode II fracture toughness k_{IIc} for various test setups.

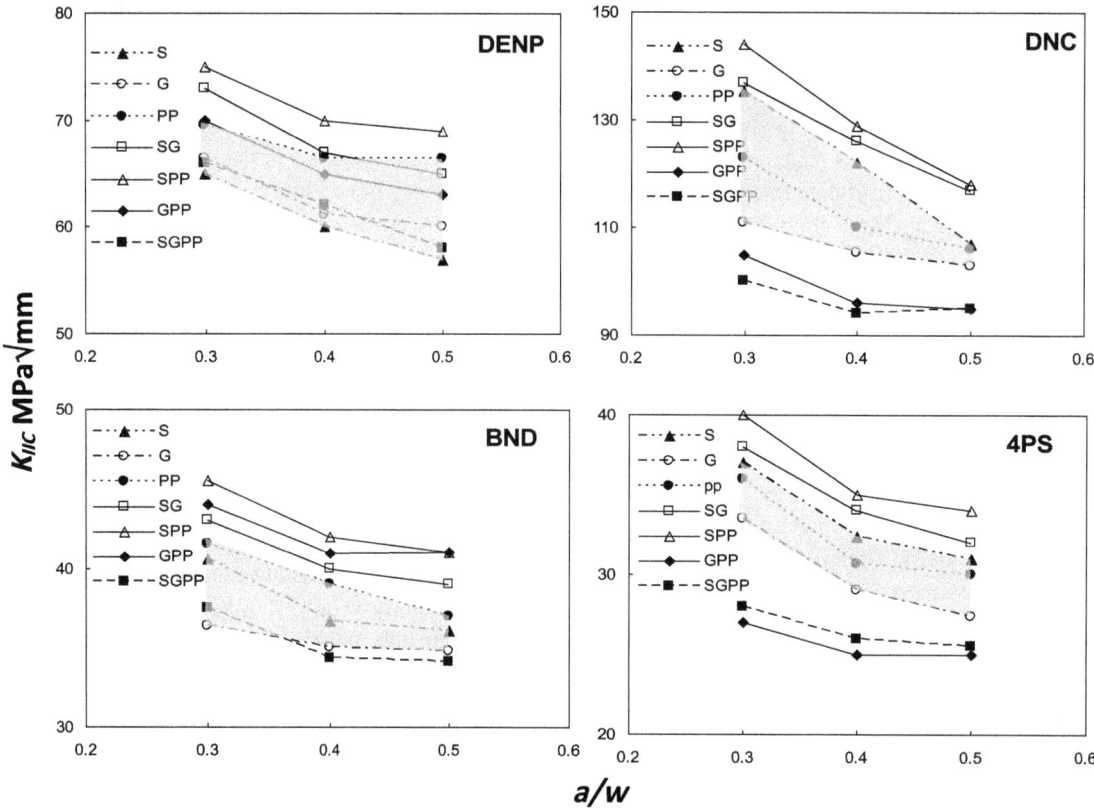

Fig. 3 Notch depth to specimen width ratio (*a/w*) versus mode II fracture toughness k_{IIc} for single and hybrid fiber reinforced concrete.

Table 6 Mode II fracture toughness K_{IIc} for hybrid and single FRC.

	(a/w)	Current study				Sherbini (2014)		
		S/PP	S/G	G/PP	S/G/PP	S	G	PP
Mode II fracture toughness (Mean ± SD) K_{IIc} MPa$\sqrt{}$mm	DENP 0.3	75 ± 2.91	73 ± 2.10	70 ± 3.48	66 ± 1.54	65	66.4	69.6
	0.4	70 ± 2.00	67 ± 3.11	65 ± 1.72	62 ± 2.65	60	61.1	66.4
	0.5	69 ± 3.21	65 ± 1.58	63 ± 2.02	58 ± 1.19	56.9	60.1	66.4
	DNC 0.3	143 ± 4.72	137 ± 4.45	105 ± 4.79	100 ± 4.89	135	111	123
	0.4	129 ± 5.16	126 ± 4.63	96 ± 3.78	95 ± 2.45	122	105	110
	0.5	118 ± 5.88	117 ± 5.50	95 ± 3.30	95 ± 4.57	107	103	106
	BND 0.3	45.5 ± 1.68	43 ± 1.02	44 ± 2.02	37.5 ± 0.78	40.6	36.4	41.5
	0.4	42 ± 1.69	40 ± 1.57	41 ± 1.14	34.5 ± 0.99	36.7	35	39
	0.5	41 ± 1.89	39 ± 1.13	41 ± 1.73	34 ± 1.57	36.1	34.8	37
	4PS 0.3	40 ± 1.97	38 ± 1.66	27 ± 0.84	28 ± 0.58	37	33.5	36
	0.4	35 ± 1.62	34 ± 1.48	25 ± 0.72	26 ± 1.13	32.5	29.2	30.7
	0.5	34 ± 1.02	32 ± 0.99	25 ± 0.72	25.5 ± 0.76	31	27.4	30

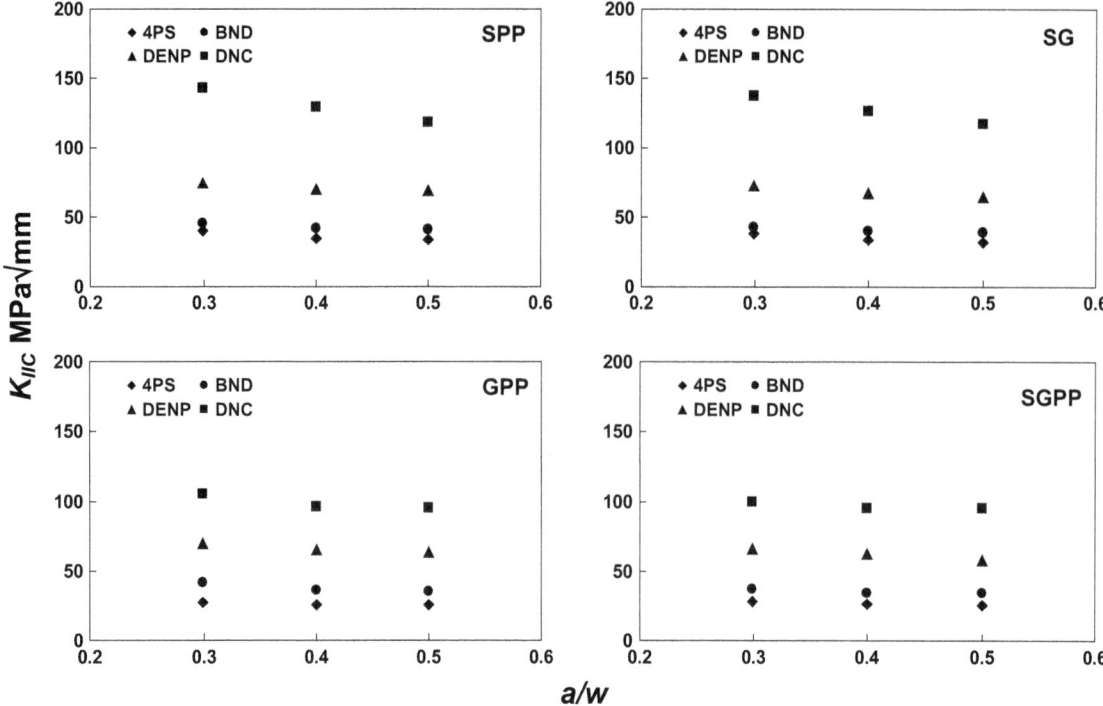

Fig. 4 Notch depth to specimen width ratio (a/w) versus mode II fracture toughness k_{IIc} for different hybridization patterns.

with the NMAS. Here, d_{max} is calculated by incorporating the strength of the material, f, instead of the critical applied stress along with the fracture toughness of the material, K_C, hence, d_{max} instead of the notch depth is as follows:

$$K_C = Y\sigma_{cr}\sqrt{\pi a} = Yf\sqrt{\pi d_{max}} \qquad (2)$$

Applying this concept in the present case, i.e. $K_C = K_{IIC}$ and $f = \tau_f$, hence

$$d_{max} = \frac{1}{\pi}\left(\frac{K_{IIC}}{\tau_f}\right) \qquad (3)$$

3. Results and Discussion

A comparison between the values of K_{IIc} of concrete according to four different test techniques is reported to make further assessment of the resulting data. Highlighting the effects of controlling the matrix fracture toughness, the fiber/matrix interface, and the matrix flaw size on the composite behavior might also enhance the production of engineered cementitious composites (ECC) (Li 2012).

Figure 2 shows the relation between a/w and mode II fracture toughness K_{IIc} for different test setups and hybridization patterns. For all hybrid patterns of FRC, a significant discrepancy of mode II fracture toughness K_{IIc} values (13–43 %) is clearly observed with small values of a/w, i.e. a/w = 0.3. while it ranges from (19–24 %) with high values of a/w, i.e. a/w = 0.5. This is evidence that mode II fracture toughness K_{IIc} is affected by a non material characteristic parameter (a/w), indicating that mode II fracture toughness K_{IIc} in hybrid fiber reinforced concrete can not be

assumed as a material property. Mode II fracture toughness K_{IIc} is inversely proportional to a/w for all concretes and test configurations, which strengthens the previous argument. This argument is in good agreement with the reported works by Swartz et al. (1988) and Reinhardt et al. (1997). By increasing a/w. The mode II fracture toughness K_{IIc} decreasing rate reduces. This behavior may be due to that, by increasing a/w both length and severity of crack increase, while the defense zone represented in the crack forehead ligament decreases.

Hybrid FRC containing steel fibers in combination of either glass or pp showed higher values of mode II fracture toughness k_{IIc} than all other hybrid patterns, i.e. synergistic effect. The increment percentages ranges from (11.5–16 %) for DENP test (20–30 %) for DNC test (17–18 %) for BND test, and (25–30 %) for 4PS test. Steel fiber represents the fiber type with higher strength and stiffness, while either glass or pp fiber represents the relatively flexible type. In accordance with the 1st synergic mechanism (Hybrids based on fiber constitutive response), this hybridization pattern leads to improved toughness and strain capacity in the post-crack zone. For different test geometries the (S/PP) results are higher than those of all other hybridization patterns. The 2nd synergic mechanism (Hybrids based on fiber dimensions) explains that phenomenon, the pp fiber represents the smaller type that bridges micro-cracks and therefore controls their growth and delays coalescence leading to a higher tensile strength of the composite. Steel fiber represents the larger type that is intended to arrest the propagation of macro-cracks and therefore results in a substantial improvement in the fracture toughness of the composite. In this specific hybridization pattern (S/PP), a dual synergic

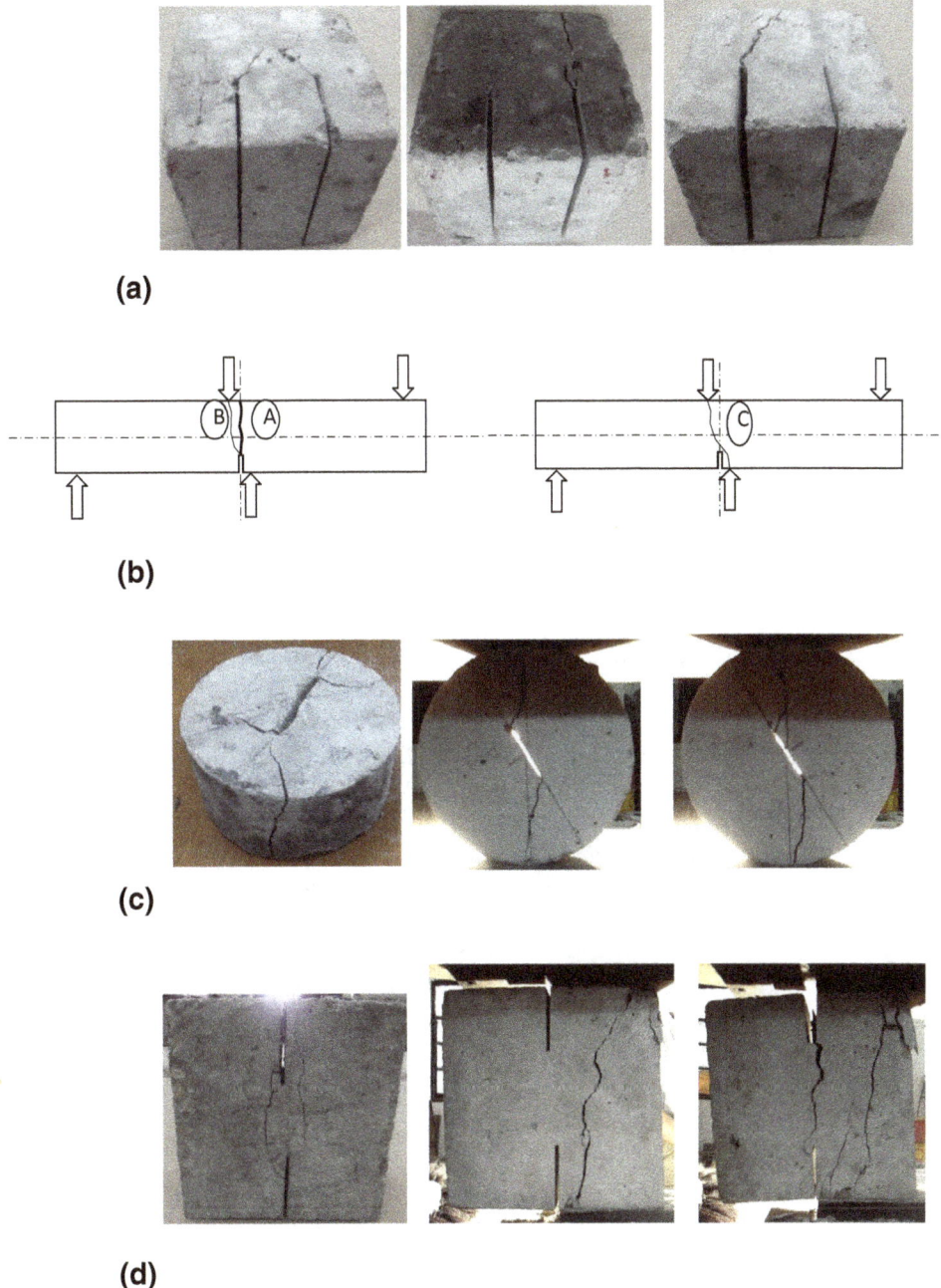

Fig. 5 Crack patterns in different mode II specimens under various test setups. **a** DNC specimens. **b** 4PS specimens. **c** BND specimens. **d** ENP specimens.

mechanism takes place resulting supreme values of mode II fracture toughness k_{IIc}. For different test setups, the (G/PP) showed lower mode II fracture toughness k_{IIc}. Both glass and pp are deficient in the required stiffness to provide reasonable crack propagation arrest. The S/G/PP hybrid FRC showed lower values of mode II fracture toughness k_{IIc} for most geometries due to the reduction of V_f of steel fiber from 1 to 0.5 %. For all hybrid FRCs except S/G/PP, the ration of V_f of steel fiber to V_f of other fibers is 2.0, while, that ration is 1.0 in S/G/PP hybrid FRC. For both DNC and 4PS test sctups, a wide gap bctwccn (S/PP & S/G) on one hand and (G/PP & S/G/PP) on the other hand is observed, which support the idea of dual synergic mechanism when adding steel fibers to the hybrid. In the other test setups DENP & BND, that gap exists but with narrower range.

In comparison with reported test results of single FRC by Sherbini (2014), the mode II fracture toughness k_{IIc} values fitted in those gaps as shown in Fig. 3 for almost all geometries. The difference between compressive strengths for different hybridization patterns reported in the current study is minimal (less than 6 %), and does not reflect the wide discrepancy in the values of mode II fracture toughness K_{IIc} (up to 43 %). In comparison with the previous study of single FRC reported by Sherbini (2014), the S/PP hybrid FRC pattern shows supreme behavior of both mechanical properties (compressive and tensile strengths) and mode II fracture toughness K_{IIc}.

Table 6 represents the mean values and standard deviations of mode II fracture toughness K_{IIc} reported in the current study in comparison to the results of single FRC

Table 7 Calculated values of d_{max}/NMAS.

		(a/w)	Current study			
			S/PP	S/G	G/PP	S/G/PP
d_{max}/NMAS	DENP	0.3	0.65	0.65	0.65	0.54
		0.4	0.57	0.55	0.56	0.48
		0.5	0.55	0.52	0.53	0.42
	DNC	0.3	2.36	2.3	1.46	1.24
		0.4	1.92	1.95	1.22	1.12
		0.5	1.61	1.68	1.2	1.12
	BND	0.3	0.24	0.23	0.26	0.17
		0.4	0.2	0.2	0.22	0.15
		0.5	0.19	0.19	0.22	0.14
	4PS	0.3	0.18	0.18	0.1	0.1
		0.4	0.14	0.14	0.08	0.08
		0.5	0.13	0.13	0.08	0.08

previously found by Sherbini (2014) for the same concrete matrix. The sensitivity effect of the adopted test is almost the same in either single or hybrid FRC. Figure 4 shows that, the mode II fracture toughness K_{IIc} measured from DNC test are the highest values due to the crack propagation miss alignment opposing sliding of crack surfaces. Figure 5a shows the different crack patterns of DNC specimens. However, the mode II fracture toughness K_{IIc} values measured from 4PS test are the lowest as shown in Fig. 4. It is obvious that the common drawback of the 4PS mode II testing method is that, in the direction perpendicular to crack plane a tensile stress cannot be avoided, especially for crack pattern (c), as shown in Fig. 5b. Similarly in BND test as shown in Fig. 5c, the tensile stress component certainly causes a mode I stress intensity. For the materials with low tensile strength like concrete a small mode I stress intensity could result in tensile failure prior to shear failure in those specimen geometries employed in mode II tests. On the other hand, DENP specimens suffer from indirect tensile cracks near the main shear crack, as shown in Fig. 5d.

Maximum size of undamaged defect (d_{max}) is defined as, the maximum defect size that does not affect the material prosperities, i.e. the damage size beyond which, the material properties decreases. The values of d_{max} should be normalized to an internal structure parameter of concrete such as the NMAS. The values of d_{max}/NMAS should not equal unity, i.e. d_{max} must be less than NMAS. On the other hand, d_{max} should not be of trivial value less than air voids in concrete. In the case of mode I fracture toughness, the value of d_{max}/NMAS was reported to be 0.7 by Sallam et al. (2014). On the present study, to check the reliability of the present results, d_{max}/NMAS are calculated and tabulated in Table 7. It is clear that, the values of d_{max}/NMAS in DNC test are greater than unity which is unacceptable. On the other hand, the values of d_{max}/NMAS in 4PS test are very low, ranged from 0.08 to 0.18. This may be attributed to the existence of

tensile stress at the tip of mode II crack as mentioned above. d_{max}/NMAS ranged from 0.42 to 0.65, and from 0.14 to 0.26 in DENP and BND test setups respectively, which represent acceptable values. Although DENP suffered from tensile cracks near the main shear crack, but still the most reliable test setup according to the non damage defect concept, i.e. The closer value to that obtained for Mode I (0.7) reported by Sallam et al. (2014).

4. Conclusions

The results of the present experimental work support the following conclusions:

1. Hybridization of fiber relatively increased compressive strength (2–13 %), and tensile strength (up to 14.8 %) in comparison with single fiber addition to concrete at the same fiber volume fraction.
2. Mode II fracture toughness of concrete K_{IIc} decreased with the increment of a/w ratio for all hybridization patterns and test setups (5–17.5 %).
3. Hybrid FRC containing steel fibers in combination of either glass or pp showed higher values of mode II fracture toughness k_{IIc} than all other hybrid patterns. The increment percentages ranges from (11.5–16 %) for DENP test (20–30 %) for DNC test (17–18 %) for BND test, and (25–30 %) for 4PS test.
4. Due to dual synergic mechanism, the (S/PP) mode II fracture toughness K_{IIc} results are the highest among all other hybridization patterns.
5. Mode II fracture toughness K_{IIc} of hybrid fiber reinforced concrete is found to be sensitive to a/w, geometry of test specimen, and loading condition. i.e., mode II fracture toughness K_{IIc} of hybrid fiber reinforced concrete could not be assumed as a real material property.

6. According to the non damage defect concept, DENP test setup is the most reliable test to measure pure mode II of concrete.

Acknowledgments

H. E. M. Sallam is on sabbatical leave from Materials Engineering Department, Zagazig University, Zagazig, 44519, Egypt.

References

ACI Commitee 544. (2011). *Fiber reinforced concrete*. Detroit, MI: American Concrete Institute.

Al Hazmi, H. S. J., Al Hazmi, W. H., Shubaili, M. A., & Sallam, H. E. M. (2012). Fracture energy of hybrid polypropylene–steel fiber high strength concrete. *HPSM, High Performance Structure and Materials, VI*, 309–318.

Banthia, N., & Gupta, R. (2004). Hybrid fiber reinforced concrete: Fiber synergy in high strength matrices. *RILEM, Materials and Structures, 37*(274), 707–716.

Banthia, N., Moncef, A., Chokri, K., & Sheng, J. (1995). Uniaxial tensile response of microfiber reinforced cement composites. *Journal of Materials and Structures, RILEM, 28*(183), 507–517.

Banthia, N., & Sheng, J. (1991). Micro reinforced cementitious materials, Materials Research Society Symposia Proceedings, Materials Research Society (Vol 211, pp. 25–32), Pittsburgh, PA.

Banthia, N., & Soleimani, S. M. (2005). Flexural response of hybrid fiber reinforced cementitious composites. *ACI Materials Journal, 102*(6), 382–389.

Bentur, A., & Mindess, S. (1990). *Fiber reinforced cementitious composites*. London, UK: Elsevier Applied Science.

Boulekbache, B., Hamrat, M., Chemrouk, M., & Amziane, S. (2012). Influence of yield stress and compressive strength on direct shear behaviour of steel fibre-reinforced concrete. *Construction and Building Materials, 27*, 6–14.

Feldman, D., & Zheng, Z. (1993). Synthetic fibers for fiber concrete composites, *Materials Research Society Symposia Proceedings, Materials Research Society* (Vol 305, pp. 123–128), Pittsburgh, PA.

Iosipescu, N. (1967). New accurate procedure for single shear testing of metals. *Journal of Materials, 2*(3), 537–566.

Irobe, M., & Pen, S.-Y. (1992). Mixed-mode and mode II fracture of concrete. In Z. P. Bazant (Ed.), *Fracture mechanics of concrete structures* (pp. 719–726). New York, NY: Elsevier Applied Science.

Kamlos, K., Babal, B., & Nurnbergerova, T. (1995). Hybrid fiber reinforced concrete under repeated loading. *Nuclear Engineering and Design, 156*(1–2), 195–200.

Kim, N. W., Saeki, N., & Horiguchi, T. (1999). Crack and strength properties of hybrid fiber reinforced concrete at early ages. *Transactions of the Japan Concrete Institute, 21*, 241–246.

Larson, E. S., & Krenchel, H. (1991). Durability of FRC-materials. *Materials Research Society Symposia Proceedings, Materials Research Society, Pittsburgh, PA 211*, 119–124.

Lawler, J. S., Zampini, D., & Shah, S. P. (2002). Permeability of cracked hybrid fiber reinforced mortar under load. *ACI Materials Journal, 99*(4), 379–385.

Lee, J., & Lopez, M. M. (2014). An experimental study on fracture energy of plain concrete. *International Journal of Concrete Structures and Materials, 8*(2), 129–139.

Li, V. C. (2012). Tailoring ECC for special attributes: A review. *International Journal of Concrete Structures and Materials, 6*(3), 135–144.

Mobasher, B., & Li, C. Y. (1996). Mechanical properties of hybrid cement-based composites. *ACI Materials Journal, 93*(3), 284–292.

Prokopski, G. (1991). Influence of water-cement ratio on microcracking of ordinary concrete. *Journal of Materials Science, 26*, 6352–6356.

Qian, C., & Stroeven, P. (2000). Fracture properties of concrete reinforced with steel–polypropylene hybrid fibers. *Cement & Concrete Composites, 22*(5), 343–351.

Rao, A. G., & Rao, A. S. (2009). Toughness indices of steel fiber reinforced concrete under mode II loading. *Materials and Structures, 42*, 1173–1184.

Reinhardt, H. W., Josko, O., Shilang, X., & Abebe, D. (1997). Shear of structural concrete members and pure mode II testing. *Advanced Cement Based Materials, 5*, 75–85.

Sallam, H. E. M. (2003). Fracture energy of fiber reinforced concrete. *Al-Azhar University Engineering Journal, 6*, 555–564.

Sallam, H. E. M., & Mubaraki, M. (2015). *Evaluation of the fracture energy methods used in fiber reinforced concrete pavements by the maximum undamaged defect size concept 94th Annual Meeting of TRB*, Washington, DC.

Sallam, H. E. M., Mubaraki, M., & Yusoff, N I Md. (2014). Application of the maximum undamaged defect size (d_{max}) concept in fiber-reinforced concrete pavements. *Arabian Journal for Science and Engineering, 39*(12), 8499–8506.

Shah, S. P. (1991). Do fibers increase the tensile strength of cement-based matrices. *ACI Materials Journal, 88*(6), 595–602.

Sherbini, A. S. (2014). Mode II fracture toughness estimates for fiber reinforced concretes using a variety of testing geometries. *Engineering Research Journal, 37*(2), 239–246.

Swartz, S. E., Lu, L. W., Tang, L. D., & Refai, T. M. E. (1988). Mode II fracture-parameter estimates for concrete from beam specimens. *Experimental Mechanics, 28*, 146–153.

Walton, P. L., & Majumdar, A. J. (1975). Cement-based composites with mixtures of different types of fibers. *Composites, 6*, 209–216.

Watkins, J. (1983). Fracture toughness test for soil–cement samples in mode II. *International Journal of Fracture, 23*, RI35–RI38.

Xu, G., Magnani, S., & Hannant, D. J. (1998). Durability of hybrid polypropylene–glass fiber cement corrugated sheets. *Cement & Concrete Composites, 20*(1), 79–84.

10

Creep Mechanisms of Calcium–Silicate–Hydrate: An Overview of Recent Advances and Challenges

Hailong Ye*

Abstract: A critical review on existing creep theories in calcium–silicate–hydrate (C–S–H) is presented with an emphasis on several fundamental questions (e.g. the roles of water, relative humidity, temperature, atomic ordering of C–S–H). A consensus on the rearrangement of nanostructures of C–S–H as a main consequence of creep, has almost been achieved. However, main disagreement still exists on two basic aspects regarding creep mechanisms: (1) at which site the creep occurs, like at interlayer, intergranular, or regions where C–S–H has a relatively higher solubility; (2) how the structural rearrangement evolutes, like in a manner of interlayer sliding, intra-transfer of water at various scales, recrystallization of gelled-like particles, or dissolution–diffusion–reprecipitation at inter-particle boundary. The further understanding of creep behavior of C–S–H relies heavily on the appropriate characterization of its nanostructure.

Keywords: calcium–silicate–hydrate, nanostructure, creep mechanism, relative humidity, shrinkage.

1. Introduction

Creep is a material property commonly defined as a time-dependent deformation under sustaining loads. Creep is essentially important to cementitious materials since excess deformation can negatively impact the long-term service-ability and sustainability of concrete structures (Neville 1981; Bažant 2001; Mindess et al. 2003; Singh et al. 2013; Li 2012; Ye et al. 2015). The creep behavior of cementitious materials was reported to be internally associated with the viscous nature of its main hydration product, i.e. calcium–silicate–hydrate (C–S–H), proportion of other secondary hydrated phases (e.g. portlandite), as well as externally affected by atmospheric conditions, e.g. relative humidity (RH) and temperature (Neville 1981; Bažant 2001; Powers 1968; Wittmann 2008; Alizadeh et al. 2010; Bu et al. 2015; Nguyen et al. 2013). The so-called basic creep and drying creep in cementitious materials are termed as the additional acquired deformation under constant water content (i.e. no moisture exchange with ambient and specimens, but not necessarily in saturated state) and decreasing water content (i.e. drying condition), respectively (Wittmann and Roelfstra 1980). Regardless of what type of creep, its behavior is drastically related to the interior water status in pore structure, which is further influenced by ambient RH (Ali and

Kesler 1964; Bažant et al. 1997). For instance, cementitious material can reduce up to 80 % deformation after a sever drying pre-treatment (e.g. heating to 105 °C) (Mindess et al. 2003; Glucklich and Ishai 1962), whilst its deformation can be more than three times higher when it is exposed to a simultaneous loading and drying conditions (Mindess et al. 2003; Pickett 1942). Nevertheless, different experimental observations lead researchers to propose different creep mechanisms to explain their data. Therefore, no one creep mechanism has been universally accepted despite the extensive investigation was conducted over decades.

One of the major debates among various theories (will be elaborated in Sect. 2) lies on the role of water involving in creep, as some researchers argued that water is inessential to creep (Feldman 1972) while others believed that water is important (Powers 1968; Alizadeh et al. 2010). Additionally, for the creep mechanisms that are relevant to water, discrepancy also exists regarding the function of water at various scales in C–S–H (Powers 1968; Alizadeh et al. 2010; Glucklich and Ishai 1962; Ruetz 1968). Roughly, water in C–S–H is classified into three types: capillary water, adsorbed water and interlayer water (Powers and Brownyard 1946). However, the classification of water at various dimensions also varies slight among different nanostructures of C–S–H (Powers and Brownyard 1946; Aligizaki 2006). In addition, the role of water is reported to be different for different mechanisms, and even same type of water is reported to have different functions in different mechanisms. The first subject of this paper is to review the role of water proposed in various modern creep theories and potentially provide some insights on further research.

On the other hand, it is essentially necessary to introduce shrinkage since creep (especially drying creep) and shrinkage has some sorts of intimate relation (Wittmann 2008;

Department of Civil and Environmental Engineering,
The Pennsylvania State University, State College,
PA 16801, USA.
*Corresponding Author; E-mail: huy131@psu.edu

Wittmann and Roelfstra 1980). Drying shrinkage is defined as the deformation of a specimen during drying without external loading. Both creep and shrinkage are believed to be affected by the water content in material interior, as well as external conditions. However, the mechanism between these two may not be exactly identical. It is difficult or even impossible to distinguish drying shrinkage and creep components when a material is under a circumstance of simultaneous loading and decreasing RH (i.e. drying creep) (Bažant et al. 1997). First of all, drying shrinkage (especially for irreversible component) also exhibits a time-dependent characteristic, which is intrinsically correlated to reorganization of C–S–H induced by internally drying-induced stress (e.g. capillary stress, desorption-induced solid surface tension increase, and disjoining pressure) rather than externally-applied stress (Jennings 2000; Ye et al. 2014). All of these stresses may be considerably different in their locations and magnitudes. For instance, capillary pressure exerts stress on solid skeleton by putting adjacent particle closer, external load exerts stress on the bulk materials, while desorption-induced stress exerts primarily on the solid surface (Kovler and Zhutovsky 2006; Beltzung and Wittmann 2005). The second part of this paper is to understand how the drying and thermal condition can potentially influence the creep performance.

Finally, this paper briefly discusses how the chemical composition, pore solution, and atomic ordering (presence of defects) of C–S–H can affect creep. In addition, the potential application of generalized creep equation on C–S–H research is also discussed.

2. Nanostructure of Calcium Silicate Hydrate

The nanostructure of C–S–H, which is believed to be strongly related to its creep performance, is still mysterious and controversial. Therefore, this paper briefly reviews some popular models for nanostructure of C–S–H before going deep to its creep mechanism (in Sect. 3). A more comprehensive review on nanostructure of C–S–H can be found in the literature (Papatzani et al. 2015).

2.1 Powers and Brownyard Model (First Colloidal Model) (Powers and Brownyard 1946)

The first nanostructure model (denoted as P–B model) that published by Powers and Brownyard, basically announces that C–S–H is a colloidal material with more or less crystalline phases. In P–B model, water is classified into three types: capillary water, physically adsorbed water (gel water) and interlayer water. As illustrated in Fig. 1a, capillary water refers to the water that is unreacted by cement hydration, primary remaining in the space between partial-/fully-hydrated particles; gel water refers to the physically adsorbed evaporable water in gel pores of C–S–H; interlayer water refers to the chemically bonded non-evaporable water incorporated in the solids of C–S–H (loss of interlayer water causes dehydration of C–S–H) (Mindess et al. 2003; Powers 1958). It should be noted that the above classification is somewhat arbitrary since differentiating capillary water and gel water is not easy in realistic circumstance. As further noted by Powers, the C–S–H consists mostly of fibrous particles with straight edges, where bundles of such fibers seem to form a cross-linked network, containing some more or less amorphous interstitial material (Powers 1958).

2.2 Feldman and Sereda Mode (Layered Structure Model) (Feldman and Sereda 1968)

Feldman and Sereda modified the P–B model by noticing that the C–S–H particle structurally resembles to a layered tobermorite-like crystal (denoted as F–S model) (see Fig. 1b). The significant improvement of F–S model is that it incorporates the structural function of the interlayer water in C–S–H 'particle', and emphasizes the importance of interlayer water in the mechanical and physical properties of hardened cement. As introduced later, this model argues the interlayer of C–S–H is crucial to creep of C–S–H.

2.3 Jennings and His Co-workers Model (Colloidal model) (Jennings 2000, 2008; Allen et al. 2007)

The colloidal model (denoted as CM-I and II) proposed by Jennings and his co-workers is basically a comprehensive combination of a layer model and a colloid model (see Fig. 1c, d). The basic unit in the CM model is a grain-like particle with a thickness of few layers stacked together. These particles can cluster together to form the globules (basic building block) of about 5.6 nm diameter. The packing density of this globule is reported to be important to the creep of C–S–H. The nano-granular nature of C–S–H is experimental verified by atomic force microscopy (AFM) (Nonat 2004) and nano-identification (Vandamme and Ulm 2009; Jones and Grasley 2011), as well as theoretically proven by molecular dynamics simulation (Pellenq et al. 2008). However, different experiments or simulations give different particle sizes of C–S–H (Papatzani et al. 2015).

3. Role of Water on Creep Mechanisms

3.1 The Role of Capillary Water

The theory is in nature analogous to the concept of consolidation in soil mechanics, in which reduction in volume takes place by expulsion of water under long-term static loads. It roughly treats C–S–H as a composite filled with free liquid (i.e. not physically/chemically bonded) and solid (i.e. C–S–H solid sheets with bonded water). The external load is distributed between the bulk liquid and the solid phases, where compressed water diffuses from high to low pressure areas and, consequently, a gradual transfer of the load from the water to the solid phase takes place. Hence, the stress in the solid gradually increases causing, in turn, a gradual volumetric reduction, i.e. creep. That is, creep may be regarded as a delayed elastic deformation. Accordingly, a lower creep is to be expected in concrete with a higher modulus of elasticity and lower moisture content, which is

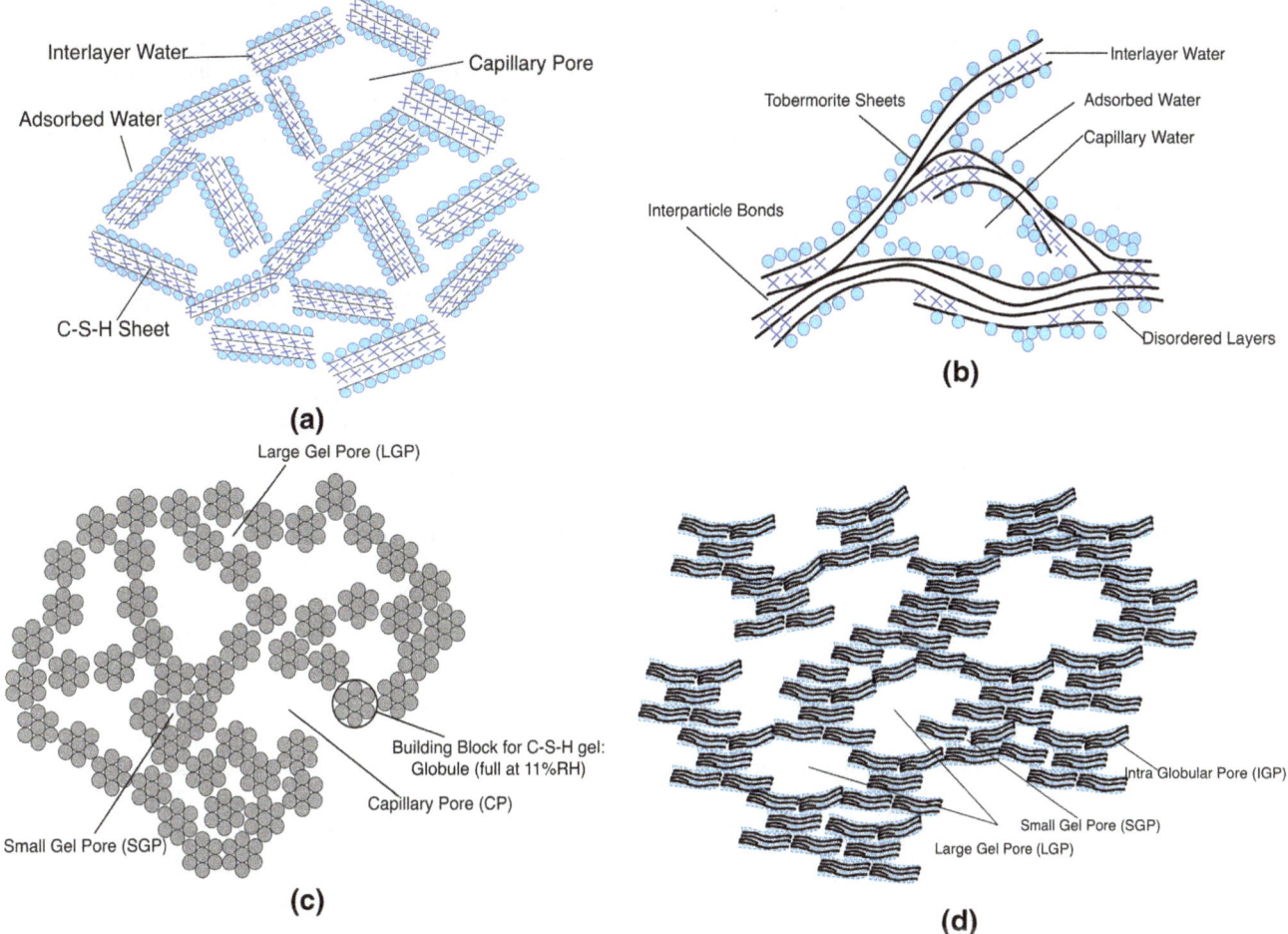

Fig. 1 Nanostructures of C–S–H (**a**) Powers and Brownyard model [adopted from Powers and Brownyard (1946)], **b** Feldman and Sereda model [adopted from Feldman and Sereda (1968)], **c** CM-I model [adopted from Jennings (2000)], and **d** CM-II model [adopted from Jennings (2008)]

however proven to be wrong by numerous experiments (Mindess et al. 2003).

The capillary water is not commonly regarded as the main status of water that contributes to creep deformation. The role of capillary water is inessential to creep mainly due to two main drawbacks: one is that it cannot account for the irreversible creep, which is believed to be in a considerable proportion of total deformation (Mindess et al. 2003). Another one is that re-immersing creeped sample into organic liquid, which can only penetrate into capillary pores but not gel and interlayer spaces, shows no volumetric change (Maekawa et al. 2003).

3.2 The Role of Adsorbed Water (Gel Water)
3.2.1 Seepage Theory

Based on the P–B model, Powers proposed the seepage theory, which ascribes creep due to the change in the gel water content as a manner of time-dependent seepage (Powers 1968). The process of transferring water from load-bearing areas into non-load-bearing results in the time-dependent volumetric reduction. Particularly, the external applied load violates the pre-existed balance between attractive and disjoining forces, which is gradually alleviated by the moisture diffusion among different phases. The

particular type of water which is easiest to be squeezed out and responsible for main volumetric reduction is believed to be the load-bearing water or water in the hindered adsorption region. As the hindered adsorption water decrease, the pre-existed disjoining pressure also decreases. It should be noted that the withdrawal of disjoining pressure is also commonly regarded as a main mechanism for drying shrinkage (Beltzung and Wittmann 2005). But different than external applied loading, the driving force for the water movement is attributed to decreasing RH for shrinkage. The seepage theory can account for several experimental evidence, including that oven-dried sample experiences very low creep rate (Glucklich and Ishai 1962), and drying creep is significantly higher than either dry or wet samples due to triggered water movement (Ali and Kesler 1964).

Some researchers argue that creep can occur in sealed or immersed specimens where water loss will be inhibited, and creep can even occurs for complete dried samples, both of which seems to question the seepage theory (Ali and Kesler 1964; Glucklich and Ishai 1962). It is important to note that creep occurs in sealed or immersed conditions does not necessarily mean that there is on moisture transfer and movement in material interior. It is likely the water is redistributed in various types of pores, and does not apparently exhibit any loss of moisture. The water in seepage theory is simply described as hindered adsorption water and commonly believed to be adsorbed water instead of capillary water.

In seepage theory, the irreversible creep is attributed to the formation of new bonds between surfaces of C–S–H when they are pressed together for the first time. Wittmann (1973) found that the creep rate only increased significantly when dried samples were re-exposed to a RH above approximately 45 %. These results cannot be accounted for by the seepage theory.

3.2.2 Microprestress-Solidification Theory

Extending the seepage theory, a microprestress-solidification theory was proposed by Bažant and his co-workers to predict creep using thermodynamic approach (Bažant 1972; Jirásek and Havlásek 2014). The microprestress is generated by the disjoining pressure of the hindered adsorbed water in the micropores and by very large and highly localized volume changes caused by hydration or drying. However, Bažant maintains that the same equations are valid whether the evaporable water is capillary, interlayer, or adsorbed. It is, however, necessary to define the state of the water associated with creep in order to understand and perhaps control the mechanism of creep.

3.2.3 Viscous Shear Theory

The viscous shear theory as proposed by Ruetz (1968) suggests that creep occurs through slip between C–S–H particles in a shear process. The sliding process takes place also in adsorbed water, in which water acts as a lubricant. Different to seepage theory, viscous shear theory considers the rearrangement of overall C–S–H particles under shear force. Although the overall tendency of C–S–H particles is to approach and contributes to volumetric reduction, the local movement of each C–S–H particle may be even more complicated rather than merely being close as disjoining pressure decreases. Since the rearrangement of C–S–H particles finally presents as a volumetric reduction, it is reasonable to assume that there should have some sorts of squeezed and redistributed water as well. The difference from seepage theory is that viscous shear theory emphasizes the role of gel water as lubricator to promote C–S–H slip rather than merely being squeezed out.

3.2.4 Thermal Activation Energy Theory

Wittmann (1973) proposed a thermal activation theory, which considers the absorbed water plays an indirect role on weakening interparticle bonds (Klug and Wittmann 1974). This mechanism is actually similar to viscous shear theory since weakening interaction between C–S–H particles is also the function of water lubricant. This theory also suggests that after the strength of these forces is reduced, the particles slide apart with respect to each other and creep is therefore increased. Furthermore, thermal activation theory hypothesizes that time-dependent strains are the result of thermally activated processes described as the rate-determining and structural-deforming process theory. The creep occurs

towards a lower energy state when additional external energy is provided to the materials. The argument is intrinsically similar to the rearrangement and redistribution of C–S–H particles which evolve towards lower energy. In addition, a concept of creep centers is introduced in this theory to further distinguish the particle region with slip occurs between adjacent particles of C–S–H. Klug and Wittmann (1974) concluded that individual solid particles are responsible for creep, and that the movement of single molecules of water is negligible. Admittedly, the energy-relaxed or arranged and organized structures are mostly irreversible.

3.2.5 Rearrangement of Globules

Based on the CM-I/II model, Jennings and his co-workers ascribe creep as a rearrangement of the globules under stress, resulting a reduction in gel porosity and enhancement of bonds which is directly correlated to the degree of silicate polymerization (Thomas and Jennings 2006). However, as noticed in CM-II model, although Jennings and his co-workers do agree that the C–S–H particle is comprised of a bunch of tobermorite-like layered sheets, how these layered sheets enter C–S–H particle rearrangement is still unknown. In other words, the structural role and physical description of adsorbed and interlayer water are not addressed in this theory. Recently, based on nano-indentation techniques, Vandamme and Ulm (2009) confirmed that creep is due to the rearrangement and tighter packing of C–S–H agglomerate around limit packing densities. As well, the structural role of water on creep is not elaborated in this argument.

3.2.6 Dissolution–Diffusion–Reprecipitation Theory

A recent study indicates that the nanoparticle rearrangement could be a result of dissolution–diffusion–reprecipitation process in which the grain boundary dissolves at the high stress fields and then transport and precipitate at lower stress fields (i.e. regions where solubility is lower), contributing to an overall lower energy (Pachon-Rodriguez et al. 2014). A computational model has also shown that loading can affect the dissolution–formation process of load-bearing phases at early age, which contributes to a reasonable viscous characteristic (Li et al. 2015). Therefore, it would be appealing to observe the change of compositions in hydrated phases and pore solution due to loading, since other theories (e.g. denser packing of globules) do not necessarily require an evolution of phase assemblages and composition.

3.3 The Role of Interlayer Water
3.3.1 Crystallization/Aging Theory

Feldman and Sereda stated that adsorbed water playing a relatively minor role, whilst interlayer water in C–S–H is responsible for the major portion of creep (Feldman 1972). The interlayer theory hypothesizes that creep of cement paste is a manifestation of the gradual crystallization or aging of a poorly crystallized layered silicate material, accelerated by drying or stress. During creep, compression of C–S–H sheets causes the interactions between adjacent sheets and formation of new interlayer space.

3.3.2 Sliding/Translation Theory

Based on beam-bending technique (Vichit-Vadakan and Scherer 2001), Alizadeh et al. (2010) suggested that the viscoelastic behavior of C–S–H is attributed to the sliding and translation of the C–S–H sheets under stress, which actually supports the interlayer theory proposed by Feldman and Sereda. The stress relaxation during beam-bending test is comprised of hydrodynamic relaxation (i.e. flow of liquid in the porous body) and viscous relaxation of the solid network (Vichit-Vadakan and Scherer 2001). Their study indicated that the water seepage accounts for minor proportion of total creep, but the interlayer water severs as a lubricator contributing to the layered structure sliding.

3.4 Consensus and Discrepancy

Although the nanostructure models of C–S–H proposed by Feldman and Sereda (layered structure) and Jennings (colloid model) are intrinsically different, the consequence of creep has almost been reached a consensus. They all attribute a featured consequence of creep as a rearrangement of nanostructures. Figure 2 illustrates the possible manners of nanostructure evolution based on different nanostructure

models of C–S–H (see Fig. 1). Accordingly, some new features should be observed in C–S–H after structural reorganization, such as the closure of small pores, increased bonding between C–S–H chains, elongated mean chain length, formation of interlayer regions.

However, main disagreement still exists on several fundamental aspects regarding creep mechanisms:

1. At which site the creep occurs. It can be in the interlayer, intergranular (particle) or regions where the solubility of C–S–H is relatively higher (due to a locally stress concentration).
2. How the nanoscale rearrangement evolutes. In the mechanism based on Feldman and Sereda, the sliding or collapse of interlayer has been explicitly indicated, although there is no direct evidence regarding the inter-crystalline slip. However, the exact manners of how rearrangement occurs were not been extensively provided by previous researchers. It is probably be one of the interlayer sliding or dissolution–diffusion–precipitation at intergranular boundary, or intra-transfer of water, or stress and drying-induced re-crystallization of gelled-like particles.

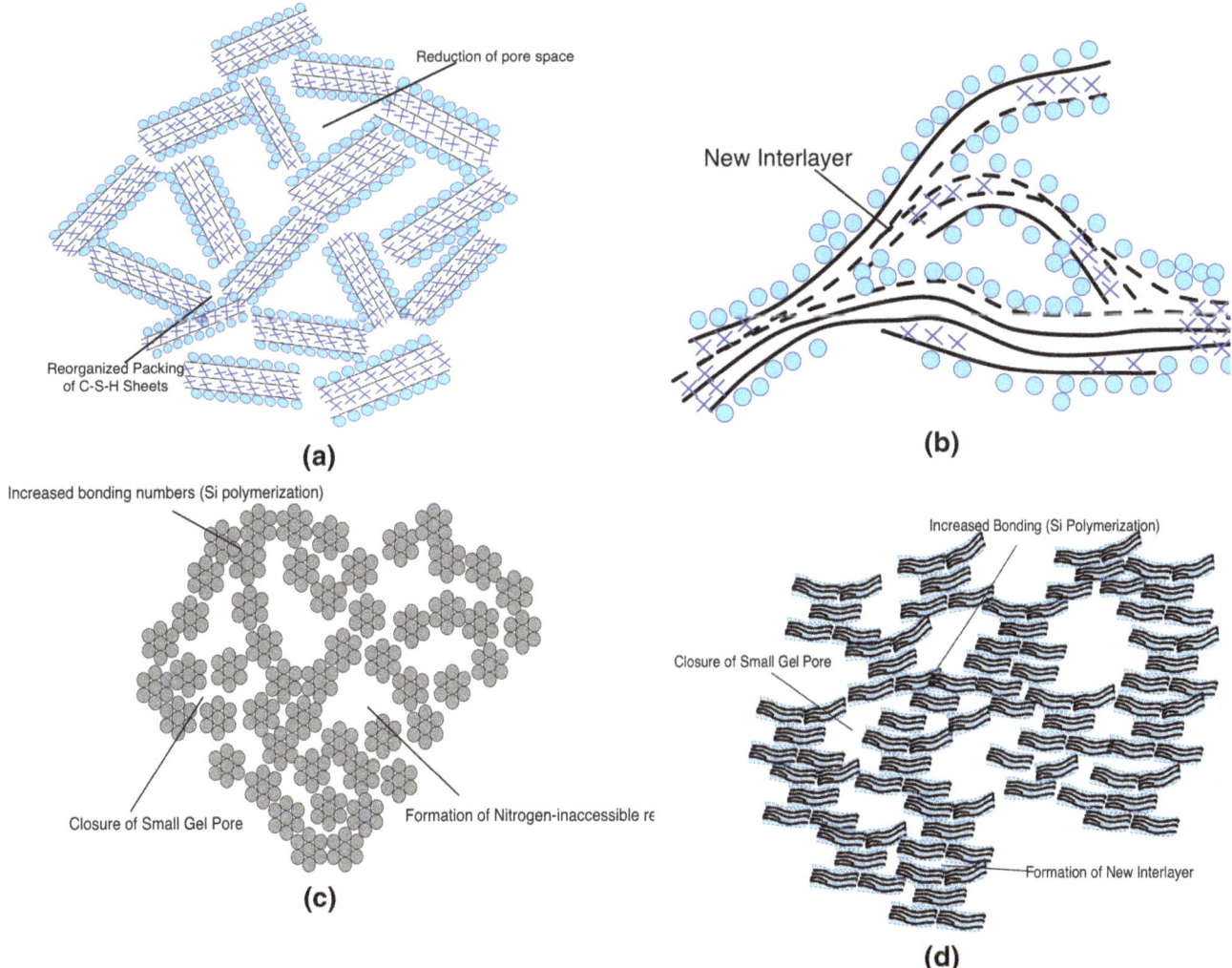

Fig. 2 Reorganized nanostructure of C–S–H (**a**) Powers and Brownyard model [adopted from Powers and Brownyard (1946)], **b** Feldman and Sereda model [adopted from Feldman and Sereda (1968)], **c** CM-I model [adopted from Jennings (2000)], and **d** CM-II model [adopted from Jennings (2008)]

3. Regardless of at which site creep occurs, reorganization process involves some sorts of break and reform bonds, which contributes to the irreversible portion. Therefore, it is hard to differentiate the rearrangement manners by simply observing the nanostructure changes after creep using nuclear magnetic resonance spectroscopy. More attention needs to be paid on stress-induced changes of phase assemblages, chemical composition, pore solution species, and morphology of C–S–H.

4. Influence of Drying Condition

Drying creep (also known as Pickett effect) (Wittmann and Roelfstra (1980); Pickett (1942)) is the excess of creep at drying over the sum of shrinkage and basic creep. It indicates that there is a strong correlation between creep and shrinkage, and no-linear relation between deformation and loading for drying creep. However, the exact mechanism of Pickett effect is still mysterious although several dubious hypotheses were proposed. For example, Ali and Kesler (1964) suggested drying creep is a stress-modified shrinkage, while Ruetz (1968) suggested that shearing action takes place under loading, accelerated by a concurrent moisture movement.

Thermodynamically, ambient RH plays an important role on creep by influencing several aspects: First of all, it controls the total amount of water in the C–S–H and hydrostatic capillary pressure due to drying. The capillary pressure is determined as the difference between the gas pressure above the meniscus and pressure inside the liquid (i.e. $p_c = p_g - p_l$). The well-known Kelvin-Laplace equation is commonly used to establish the mechanical and thermodynamic equilibrium between liquid and gas (Chen et al. 2013; Cohan 1938; Radlinska et al. 2008):

$$p_c = -\frac{2\gamma \cdot \cos\theta}{r_c} = \rho_l \frac{RT}{M} \ln RH \qquad (1)$$

where p_c is the capillary pressure, p_g is the gas pressure, p_l is the liquid pressure, ρ_l is density of liquid, M is molar mass of liquid, R is the universal gas constant, T is temperature in Kelvin, r_c is capillary radius at the position of meniscus, also named Kelvin radius; γ is surface tension between pore water and vapor, θ is the contact angle denoting the hydrophilicity of pore wall. Since the liquid pressure excess significantly that of atmospheric pressure during drying (i.e. $\|p_l\| \gg \|p_g\|$), the gas pressure is eliminated and $p_l \approx p_c$. As a consequence of Eq. (1), at a given RH, all pores whose radii are smaller than Kelvin radius are completely filled with water, whereas larger pores have dried with layers of adsorbed water (Maekawa et al. 2003). In addition, the capillary pressure is primarily controlled by the ambient RH, irrespective of the pore size (Radlinska et al. 2008).

Secondly, the RH controls the diffusion rate of interior moisture moving to the exposed boundary. For instance, at high RH, the gradient of chemical potential of vapor is less and diffusion occurs slowly. Under that circumstance, the slow water movement may contribute to the long-term

rearrangement of C–S–H nanoparticles or sheets. As evidenced by a recently study (Vlahinić et al. 2012), transient creep occur as the RH changes, mainly originating from the water movement due to the triggered chemical potential gradient. However, as mentioned before that capillary pressure decreases as RH increases, therefore there may exist a tradeoff for the influence of RH on the creep behaviors. In addition, RH affects the viscoelastic behavior of the C–S–H due to the modification of its bulk properties (Maruyama et al. 2014).

Additionally, the initial formed gel-like colloid particle in pore solution are held by a equilibrium of various forcers, like crystal-forming tendency, surface tension, solid to liquid attraction and electrical attraction and repulse. The drying of C–S–H destroys the initial quasi-balance forces in wet condition, and promotes the semi-amorphous phases to re-crystallize. Admittedly, crystallization result in a volumetric reduction as the structure becomes more stable and periodic.

Another important aspect is that the stress distribution of C–S–H under simultaneous external loads and drying is extremely complicated, and has not been addressed yet. It may help to understand the creep mechanism during drying if one could unveil the constitutive relation of C–S–H in nano-scale.

5. Effects of Temperature

The creep performance of cementitious material is also dependent on the ambient temperature and thermal history (Bažant 1983). It was reported that short-term creep increases approximately linearly with temperature up to 80 °C, where it is about three times value at ambient temperature (Mindess et al. 2003). Regardless of the creep mechanisms involved, temperature can alter several factors those are important to creep rate. First, temperature affects the rates of dissolution, diffusion, and reaction process during any chemical action in C–S–H structural reorganization. Furthermore, temperature can slightly alter the physical properties of pore solution (e.g. viscosity) and statured vapor pressure, both of which influence the water status in concrete (Chaube et al. 1993). Another important aspect is that creep is an activation energy-associated process, which is considerably controlled by temperature as well (Green 1998). However, a pre-heating treatment can reduce the long-term creep deformation for C–S–H (Mindess et al. 2003), which may be attributed to a modification in the pore structure, (packing) density, and crystallinity of C–S–H (Thomas and Jennings 2006).

6. Atomic Ordering of C–S–H

Considering the rearrangement of nanostructure under external or internal stress, the dislocation site (i.e. defects) in C–S–H is likely to be a major factor. Generally, a crystalline phase is more resistant to plastic deformation than its glassy state with a similar chemical composition due to the extensive presence of defects (Green 1998). A molecular dynamic simulation of crystalline and glassy C–S–H under shear stress indicates that glassy C–S–H has a lower shear strength and larger irreversible deformation (Manzano et al. 2013). This would be a reason why the creep of semi-amorphous C–S–H prepared by synthesis is higher than that of tobermorite minerals as experimentally measured (Nguyen et al. 2014). However, it is still unknown how to accurately access the atomic ordering in C–S–H due to its small dimensions, although some advanced nanoscale characterization techniques have shed some lights on it (Nonat 2004; Lothenbach and Nonat 2015; Chae et al. 2013). In addition, the chemical composition and pore solution species are both likely to affect the structure of C–S–H (Lothenbach and Nonat 2015). Therefore, it is not surprising that the Ca/Si ratio of C–S–H, additive of admixtures (e.g. slag, silica fume, colloid silica) can affect the creep performance (Alizadeh et al. 2010; Li and Yao 2001; Singh et al. 2015), probably due to a modification in the degrees of micro-defects and polymerization (Nguyen et al. 2013). On the other hands, a large incorporation of alumina and alkalis into C–S–H has shown to drastically change its structure and behaviors (Lodeiro et al. 2010). This becomes increasingly crucial to innovative alternative binders like alkali-activated slag, which has shown larger creep deformation than ordinary Portland cement (Ye et al. 2014; Häkkinen 1986).

7. Application of the Generalized Creep Equation

It is still unknown whether there is more than one mechanism for creep, or whether the creep mechanism varies depending on the experienced stress levels and ambient conditions (e.g. RH and temperature). Nevertheless, the primary issue regarding the creep mechanism is the criteria to differentiate various mechanisms at least qualitatively.

In the case of polycrystals (e.g. metals, minerals), the generalized creep equation has been widely implemented to differentiate the creep mechanisms, as shown below (Green 1998):

$$\dot{\varepsilon} = \frac{AD\mu b}{kT} \left(\frac{b}{d}\right)^m \left(\frac{\sigma}{\mu}\right)^n \qquad (2)$$

where $\dot{\varepsilon}$ is the creep rate at secondary steady period; A is a dimensionless constant; D is the diffusion coefficient associated with the creep process; μ is the shear modulus; b is the Burgers vector; d is the grain size, m is the grain size exponent and n is the stress exponent. The various creep mechanisms give rise to different m and n values (as shown in Table 1), which can be obtained experimentally by varying the grain size of materials and applied stress.

Upon application on C–S–H, the primary challenge is how to change the grain size of C–S–H nanoparticles, since varying the stress level is much easier. However, the determination of the particle size of C–S–H and how the interlayer structure incorporated into the colloid structure needs to be pre-understood. According to the CM-II model, the interlayer is perfectly incorporated and stacked together to form a grain called globules, while in Feldman and Sereda model, the separation between particle boundary and interlayer is not clear. By admitting the CM-II model, another important aspect before applying the generalized creep equation is how to vary the particle size experimentally. Physically, the application of stress and high temperature evaluation could possibly reduce the grain size for some

Table 1 Creep equation exponents and diffusion paths for various creep mechanisms [adopted from Green (1998)]

Creep mechanism	m	n	Diffusion path
Dislocation creep mechanisms			
Dislocation glide climb (climb controlled)	0	4–5	Lattice
Dislocation glide climb, glide controlled	0	3	Lattice
Dissolution of dislocation loops	0	4	Lattice
Dislocation climb without glide	0	3	Lattice
Dislocation climb by pipe diffusion	0	5	Dislocation core
Diffusional creep mechanisms			
Vacancy flow through grains	2	1	Lattice
Vacancy flow along boundaries	3	1	Grain boundary
Interface reaction control	1	2	Lattice/grain boundary
Grain boundary sliding mechanisms			
Sliding with liquid	3	1	Liquid
Sliding without liquid (diffusion control)	2–3	1	Lattice/grain boundary

materials, like metals (Green 1998). However, CM-II model emphasizes that both of them basically increase the packing density of grains rather than reduce the size. Nevertheless, it is still mysterious whether the grain boundary involved in creep behavior is identical to that described by CM-II.

8. Summaries

Creep of cementitious materials is a complicated phenomenon, which is influenced by the loading magnitude/history, temperature, relative humidity, thermal and drying histories, as well as chemical composition and structure of C–S–H itself. These mechanical- thermal- physical- chemical interactions are simultaneously presented and non-linearly coupled. Understanding the creep mechanism requires a proper pre-examination of the nanostructure of C–S–H. The exact creep mechanism should be reflected by a concurrent evolution of nanostructure, chemical composition of hydrated phases, and pore solution composition during creep. Although a combination of various advanced technique for nanostructure characterization may shed some lights on the creep mechanism, there still exist several disagreements among the results obtained by different techniques. As elaborated in the text, two fundamental questions regarding creep are still unclarified yet. One is 'where does the creep take place', and another is 'how does creep occur'. Answering these questions will provide the theoretical backgrounds on how to mitigate creep, quantitatively predict creep deformation, and enhance the volumetric stability of cementitious materials.

Acknowledgments

The author would like to thank the anonymous referees for improving the quality of this manuscript.

References

Ali, I., & Kesler, C. E. (1964). *Mechanisms of creep in concrete.* Champaign, IL: University of Illinois.

Aligizaki, K. K. (2006). *Pore structure of cement-based materials: Testing, interpretation and requirements.* Boca Raton, FL: CRC Press.

Alizadeh, R., Beaudoin, J. J., & Raki, L. (2010). Viscoelastic nature of calcium silicate hydrate. *Cement & Concrete Composites, 32*(5), 369–376.

Allen, A. J., Thomas, J. J., & Jennings, H. M. (2007). Composition and density of nanoscale calcium–silicate–hydrate in cement. *Nature Materials, 6*(4), 311–316.

Bažant, Z. (1972). Thermodynamics of interacting continua with surfaces and creep analysis of concrete structures. *Nuclear Engineering and Design, 20*(2), 477–505.

Bažant, Z. P. (1983). Mathematical model for creep and thermal shrinkage of concrete at high temperature. *Nuclear Engineering and Design, 76*(2), 183–191.

Bažant, Z. P. (2001). Prediction of concrete creep and shrinkage: Past, present and future. *Nuclear Engineering and Design, 203*(1), 27–38.

Bažant, Z. P., Hauggaard, A. B., Baweja, S., & Ulm, F.-J. (1997). Microprestress-solidification theory for concrete creep. I: Aging and drying effects. *Journal of Engineering Mechanics, 123*(11), 1188–1194.

Beltzung, F., & Wittmann, F. H. (2005). Role of disjoining pressure in cement based materials. *Cement and Concrete Research, 35*(12), 2364–2370.

Bu, Y., Saldana, C., Handwerker, C., & Weiss, J. (2015). *The role of calcium hydroxide in the elastic and viscoelastic response of cementitious materials: A Nanoindentation and SEM-EDS Study* (pp. 25–34). Nanotechnology in Construction: Springer.

Chae, S. R., Moon, J., Yoon, S., Bae, S., Levitz, P., Winarski, R., et al. (2013). Advanced nanoscale characterization of cement based materials using X-ray synchrotron radiation: A review. *International Journal of Concrete Structures and Materials, 7*(2), 95–110.

Chaube, R., Shimomura, T., & Maekawa, K. (1993). Multi-phase water movement in concrete as a multi-component system. In *RILEM proceedings* (p. 139). Chapman & Hall.

Chen, H., Wyrzykowski, M., Scrivener, K., & Lura, P. (2013). Prediction of self-desiccation in low water-to-cement ratio pastes based on pore structure evolution. *Cement and Concrete Research, 49*, 38–47.

Cohan, L. H. (1938). Sorption hysteresis and the vapor pressure of concave surfaces. *Journal of the American Chemical Society, 60*(2), 433–435.

Feldman, R. F. (1972). Mechanism of creep of hydrated Portland cement paste. *Cement and Concrete Research, 2*(5), 521–540.

Feldman, R. F., & Sereda, P. J. (1968). A model for hydrated Portland cement paste as deduced from sorption-length change and mechanical properties. *Matériaux et Construction, 1*(6), 509–520.

Glucklich, J., & Ishai, O. (1962). Creep mechanism in cement mortar. *ACI Journal Proceedings, 59*(7), 923–948.

Green, D. J. (1998). *An introduction to the mechanical properties of ceramics.* Cambridge, UK: Cambridge University Press.

Häkkinen, T. (1986). Properties of alkali-activated slag concrete. Valtion teknillinen tutkimuskeskus, Betoni-ja silikaattitekniikan laboratorio.

Jennings, H. M. (2000). A model for the microstructure of calcium silicate hydrate in cement paste. *Cement and Concrete Research, 30*(1), 101–116.

Jennings, H. M. (2008). Refinements to colloid model of CSH in cement: CM-II. *Cement and Concrete Research, 38*(3), 275–289.

Jirásek, M., & Havlásek, P. (2014). Microprestress–solidification theory of concrete creep: Reformulation and improvement. *Cement and Concrete Research, 60*, 51–62.

Jones, C. A., & Grasley, Z. C. (2011). Short-term creep of cement paste during nanoindentation. *Cement & Concrete Composites, 33*(1), 12–18.

Klug, P., & Wittmann, F. (1974). Activation energy and activation volume of creep of hardened cement paste. *Materials Science and Engineering, 15*(1), 63–66.

Kovler, K., & Zhutovsky, S. (2006). Overview and future trends of shrinkage research. *Materials and Structures, 39*(9), 827–847.

Li, V. C. (2012). Tailoring ECC for special attributes: A review. *International Journal of Concrete Structures and Materials, 6*(3), 135–144.

Li, X., Grasley, Z., Garboczi, E., & Bullard, J. (2015). Modeling the apparent and intrinsic viscoelastic relaxation of hydrating cement paste. *Cement & Concrete Composites, 55*, 322–330.

Li, J., & Yao, Y. (2001). A study on creep and drying shrinkage of high performance concrete. *Cement and Concrete Research, 31*(8), 1203–1206.

Lodeiro, I. G., Fernández-Jimenez, A., Palomo, A., & Macphee, D. (2010). Effect on fresh CSH gels of the simultaneous addition of alkali and aluminium. *Cement and Concrete Research, 40*(1), 27–32.

Lothenbach, B., & Nonat, A. (2015). Calcium silicate hydrates: Solid and liquid phase composition. *Cement and Concrete Research, 78*, 57–70.

Maekawa, K., Ishida, T., & Kishi, T. (2003). Multi-scale modeling of concrete performance. *Journal of Advanced Concrete Technology, 1*(2), 91–126.

Manzano, H., Masoero, E., Lopez-Arbeloa, I., & Jennings, H. M. (2013). Shear deformations in calcium silicate hydrates. *Soft Matter, 9*(30), 7333–7341.

Maruyama, I., Nishioka, Y., Igarashi, G., & Matsui, K. (2014). Microstructural and bulk property changes in hardened cement paste during the first drying process. *Cement and Concrete Research, 58*, 20–34.

Mindess, S., Young, J. F., & Darwin, D. (2003). *Concrete* (2nd ed.). Upper Saddle River: Pearson Education, Inc.

Neville, A. (1981). *Properties of concrete* (3rd ed.). London: Pitman Publishing Ltd.

Nguyen, D.-T., Alizadeh, R., Beaudoin, J. J., Pourbeik, P., & Raki, L. (2014). Microindentation creep of monophasic calcium–silicate–hydrates. *Cement & Concrete Composites, 48*, 118–126.

Nguyen, D.-T., Alizadeh, R., Beaudoin, J., & Raki, L. (2013). Microindentation creep of secondary hydrated cement phases and C–S–H. *Materials and Structures, 46*(9), 1519–1525.

Nonat, A. (2004). The structure and stoichiometry of CSH. *Cement and Concrete Research, 34*(9), 1521–1528.

Pachon-Rodriguez, E. A., Guillon, E., Houvenaghel, G., & Colombani, J. (2014). Wet creep of hardened hydraulic cements—Example of gypsum plaster and implication for hydrated Portland cement. *Cement and Concrete Research, 63*, 67–74.

Papatzani, S., Paine, K., & Calabria-Holley, J. (2015). A comprehensive review of the models on the nanostructure of calcium silicate hydrates. *Construction and Building Materials, 74*, 219–234.

Pellenq, R.-M., Lequeux, N., & Van Damme, H. (2008). Engineering the bonding scheme in C–S–H: The iono-covalent framework. *Cement and Concrete Research, 38*(2), 159–174.

Pickett, G. (1942). The effect of Chang in moisturecontent on the crepe of concrete under a sustained load. *ACI Journal Proceedings, 38*, 333–356.

Powers, T. C. (1958). Structure and physical properties of hardened Portland cement paste. *Journal of the American Ceramic Society, 41*(1), 1–6.

Powers, T. (1968). The thermodynamics of volume change and creep. *Matériaux et Construction, 1*(6), 487–507.

Powers, T. C., & Brownyard, T. L. (1946). Studies of the physical properties of hardened Portland cement paste. *ACI Journal Proceedings, 43*(9), 249–336.

Radlinska, A., Rajabipour, F., Bucher, B., Henkensiefken, R., Sant, G., & Weiss, J. (2008). Shrinkage mitigation strategies in cementitious systems: A closer look at differences in sealed and unsealed behavior. *Transportation Research Record: Journal of the Transportation Research Board., 2070*(1), 59–67.

Ruetz, W. (1968). A hypothesis for the creep of hardened cement paste and the influence of simultaneous shrinkage. In *Proceedings of the structure of concrete and its behavior under load* (pp. 365–387).

Singh, L. P., Goel, A., Bhattachharyya, S. K., Ahalawat, S., Sharma, U., & Mishra, G. (2015). Effect of morphology and dispersibility of silica nanoparticles on the mechanical behaviour of cement mortar. *International Journal of Concrete Structures and Materials, 9*(2), 1–11.

Singh, B. P., Yazdani, N., & Ramirez, G. (2013). Effect of a time dependent concrete modulus of elasticity on prestress losses in bridge girders. *International Journal of Concrete Structures and Materials, 7*(3), 183–191.

Thomas, J. J., & Jennings, H. M. (2006). A colloidal interpretation of chemical aging of the CSH gel and its effects on the properties of cement paste. *Cement and Concrete Research, 36*(1), 30–38.

Vandamme, M., & Ulm, F.-J. (2009). Nanogranular origin of concrete creep. *Proceedings of the National Academy of Sciences, 106*(26), 10552–10557.

Vichit-Vadakan, W., & Scherer, G. (2001). Beam-bending method for permeability and creep characterization of cement paste and mortar. In *Proceedings of the 6th international conference on creep, shrinkage and durability mechanics of concrete and other quasi-brittle materials* (pp. 27–32). Cambridge, MA: Elsevier.

Vlahinić, I., Thomas, J. J., Jennings, H. M., & Andrade, J. E. (2012). Transient creep effects and the lubricating power of water in materials ranging from paper to concrete and Kevlar. *Journal of the Mechanics and Physics of Solids, 60*(7), 1350–1362.

Wittmann, F. (1973). Interaction of hardened cement paste and water. *Journal of the American Ceramic Society, 56*(8), 409–415.

Wittmann, F. (2008). Heresies on shrinkage and creep mechanisms. In *Proceedings of the 8th international conference on creep, shrinkage and durability mechanics of concrete and concrete structures (CONCREEP 8)* (pp. 3–9).

Wittmann, F., & Roelfstra, P. (1980). Total deformation of loaded drying concrete. *Cement and Concrete Research, 10*(5), 601–610.

Ye, H., Cartwright, C., Rajabipour, F., & Radlińska, A. (2014). Effect of drying rate on shrinkage of alkali-activated slag cements. In *4th international conference on the durability of concrete structure (ICDCS 2014)* (pp. 254–261). Purdue University.

Ye, H., Fu, C., Jin, N., & Jin, X. (2015). Influence of flexural loading on chloride ingress in concrete subjected to cyclic drying-wetting condition. *Computers and Concrete, 15*(2), 183–198.

11

Shear Strength of Prestressed Steel Fiber Concrete I-Beams

Padmanabha Rao Tadepalli[1],*, Hemant B. Dhonde[2], Y. L. Mo[3], and Thomas T. C. Hsu[3]

Abstract: Six full-scale prestressed concrete (PC) I-beams with steel fibers were tested to failure in this work. Beams were cast without any traditional transverse steel reinforcement. The main objective of the study was to determine the effects of two variables—the shear-span-to-depth ratio and steel fiber dosage, on the web-shear and flexural-shear modes of beam failure. The beams were subjected to concentrated vertical loads up to their maximum shear or moment capacity using four hydraulic actuators in load and displacement control mode. During the load tests, vertical deflections and displacements at several critical points on the web in the end zone of the beams were measured. From the load tests, it was observed that the shear capacities of the beams increased significantly due to the addition of steel fibers in concrete. Complete replacement of traditional shear reinforcement with steel fibers also increased the ductility and energy dissipation capacity of the PC I-beams.

Keywords: shear, steel fibers, prestress concrete, full-scale beams.

1. Introduction

When a concrete element is subjected to shear stress it causes principal diagonal tensile and compressive stresses in the element. Concrete starts cracking when the applied principal tensile stress exceeds the tensile strength of concrete. This cracking causes softening in the other principal direction and reduces the compressive strength of concrete. When the applied principal compressive stress exceeds the softened compressive strength, crushing of concrete occurs. This phenomenon is known as shear failure. This failure could be very brittle since tensile strength of concrete is much less than its compressive strength. Therefore, to enhance the behavior of concrete subjected to shear forces, one of the methods is to improve its tensile strength by adding steel fibers.

Steel Fiber Reinforced Concrete (SFRC) is conventional concrete reinforced with discrete fibers of a short length and small diameter. It has been extensively used by many researchers over the past two decades to improve the post cracking behavior of concrete. SFRC is important in seismic, impact and blast resistant structures due to its improved properties over conventional concrete. Full or partial

replacement of mild steel reinforcement with steel fibers can also save considerable labor costs and time for construction. Although SFRC is mainly used along with mild steel bars as transverse reinforcement, studies have shown (Dhonde 2006) that presence of only steel fibers could enhance the shear behavior of concrete even in the absence of mild steel.

Dhonde (2006) found that most of the fiber reinforced beams have performed better in controlling shear crack widths than the beam with mild steel shear reinforcement. It is noted that the 4.2 % of shear steel, has its crack width greater than that of beam with 0.83 fiber factor that is the volume percentage multiplied by the aspect ratio of steel fibers. Study clearly indicated that the replacement of stirrups by steel fibers plays an important role in the crack control of the beams. It was found that the steel fiber reinforced beam had higher shear strength and greater ductility than the control beam.

In fully prestressed beams with a fiber volume fraction of 1.5 %, Padmarajaiah and Ramaswamy (2001) found an increase of shear strength up to 20 % at the first crack as well as at the peak. They also found that the fiber inclusion alters the a/d ratio that divides the flexure and shear critical failure mode. Thomas and Ramaswamy (2006) observed similar results in increment of shear strength when fibers were added to concrete. Additionally, they found that high strength concrete benefits more from fibers. That may be attributed to better bond characteristics between fibers and high strength concrete.

Meda et al. (2005) conducted a series of experiments on prestressed SFRC beams and concluded that the beams reinforced only with steel fibers show a similar, or even better, post cracking behavior than the beams with minimum amount of transverse reinforcement. The study also showed that the addition of steel fibers to replace conventional

[1]American Global Maritime, Houston, TX 77079, USA.
*Corresponding Author; E-mail: tvvssprao@gmail.com
[2]Civil Engineering Department, Vishwakarma Institute of Information Technology, Pune 411048, India.
[3]Department of Civil and Environmental Engineering, University of Houston, Houston, TX 77204, USA.

transverse reinforcement improved the shear strength significantly. Steel fibers were also found to reduce the width of shear cracks more effectively than conventional steel reinforcement, thus improving durability of concrete.

In another study (Cho and Kim 2003) conducted on SFRC beams, it was found that fibrous concrete beams eventually collapsed from the severely localized deformations at one or two major cracks regardless of the failure mode.

Tan et al. (1995) showed through his experiments that inclusion of steel fibers enhanced the ultimate shear strength of partially prestressed concrete (PC) beams. The load at which the first shear crack appeared increased with an increase in steel fiber content. The study also found that stirrups may be replaced by an equivalent amount of steel fibers without significantly affecting the behavior and strength in shear of partially prestressed concrete beams.

Langsford et al. (2007) tested 13 fully prestressed steel fiber reinforced concrete beams without stirrups under shear loading. The shear span to depth ratio was varied from 1.5 to 2 and volume fraction of hooked end steel fibers from 0.5 to 1.2 %. They found out that addition of 0.5 % volume fraction of steel fibers increased the shear carrying capacity by 30 and 25 % at an a/d ratio of 2.0 and 1.5, respectively. The increase is 50 and 33 % with addition of 1.2 % volume fraction of steel fibers for both cases, respectively.

Narayanan and Darwish (1987) carried out 36 shear tests on simply supported rectangular prestressed concrete beams, containing steel fibers as web reinforcement. It was found that the failure modes for beams with prestress and without prestress are similar. Ultimate shear strengths increased up to 95 % when steel fibers were added to concrete.

Abdul-Wahab and Al-Kadhimi (2000) found similar results as mentioned above even with unbounded tendons. They also observed that increase in shear strength is more significant at lower a/d ratios. Junior and De Hanai (1999) found steel fibers are more effective when used in combination with traditional stirrups.

With regard to the behavior of steel fiber concrete under shear, when a concrete element is subjected to pure shear stress, it imposes tension and compression on the element in principal directions as shown in Fig. 1. In the case of normal concrete, the cracking due to tension in one direction causes the concrete to soften in the orthogonal direction. When steel fibers are added to concrete they contribute to shear strength in three ways.

i.   By improving the post cracking tensile behavior of concrete, which in turn reduces the cracking and hence the softening of the compression direction.

ii. Addition of steel fibers also improves the compressive behavior by confining the lateral strain. Many researchers found that addition of steel fibers improves the compressive strength of concrete up to 15 %.

iii. The compression in the normal direction acts as a confinement for the steel fibers as shown in Fig. 1, improving their bond with concrete and hence the tensile behavior of SFRC.

2. Research Significance

Addition of deformed steel fibers improves the shear behavior of prestress concrete. However, limited studies are available on large or full-scale prestressed concrete specimens with high strength concrete. This research focuses on full-scale testing of fully prestressed SFRC beams made with high strength concrete to reduce or completely eliminate transverse steel reinforcement and aiming towards the development of rational and simple shear design provision for SFRC structures. Furthermore, the replacement of stirrups with steel fibers will potentially reduce the labor and construction costs.

3. Experimental Program

The test specimens consisted of Texas Department of Transportation (TxDOT, Texas-USA) Type-A beams (prestressed I-Beams). Six 7.6 m (25 ft) long beams (R1–R6) were fabricated with Prestressed Steel Fiber Concrete (PSFC) to study the behavior of the beams in web-shear and flexure-shear mode of failure under monotonic loading. Steel fibers, which have double hooked ends and are collated, were chosen to produce the PSFC beams. The beam cross section is shown in Fig. 2. The primary testing variables investigated were the amount of steel fiber (fiber factor) and the mode of shear failure (i.e., shear span-to-effective depth ratio, a/d). No traditional transverse rebars (stirrups) were used in any of the beams; the shear reinforcement consisted solely of steel fibers. Beams R1 through R4 were designed to fail in web-shear with a/d ratio of 1.6, while Beams R5 and R6 were designed to fail in flexure-shear with a/d ratio of 4.2.

Table 1 summarizes the test variables for Beams R1–R6. Beam R1 with a fiber factor of 0.4 was designed to fail in web-shear. Beams R2, R3 and R4 were made using fiber factor of 0.55, 0.83 and 1.23, respectively and were also designed to fail in web-shear. Beams R5 and R6 with a fiber factor of 0.4 and 1.23, respectively were designed to fail in flexural-shear.

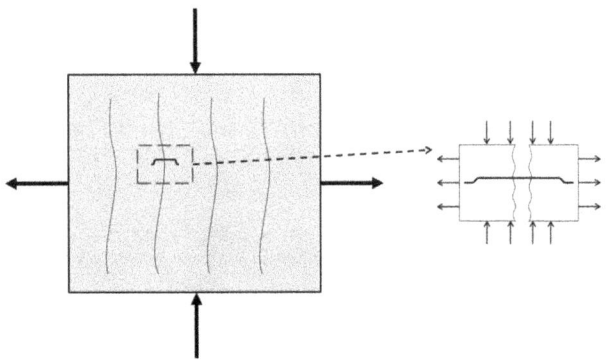

Fig. 1 Fiber concrete under principal tension and compression.

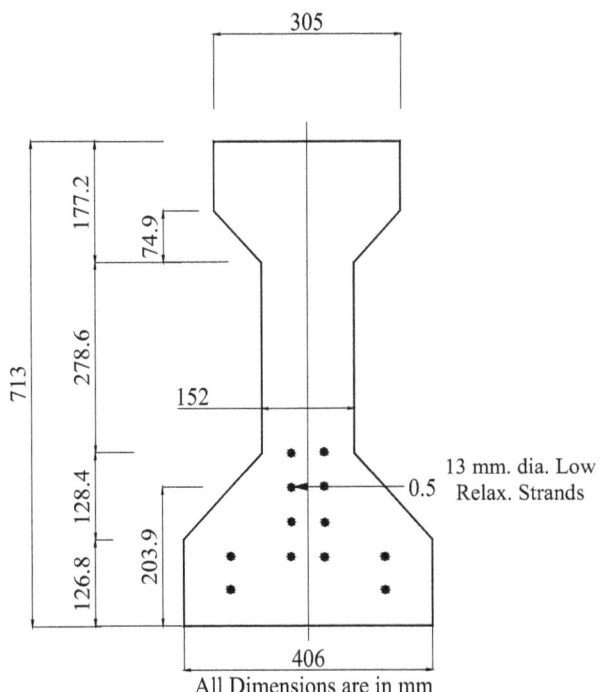

Fig. 2 Cross section of PSFC I-Beam. *Note* 1″ = 25.4 mm.

3.1 Details of PSFC I-Beams

The cross-section of the TxDOT Type A beam is shown in Fig. 2. The total height of the beam was 713 mm (28 in.) and the widths of the top and bottom flange were 305 mm (12 in.) and 406 mm (16 in.), respectively. The width of the web was 152 mm (6 in.). The prestressing tendons in all beams were straight. The location of prestressing tendons is also shown in Fig. 2. Twelve 13 mm (0.5-inch) diameter, 7-wire, low-relaxation strands were used as prestressing steel to resist flexure. The prestressing strands had an ultimate tensile strength of 1863 MPa (270 ksi). The total length of the beams tested was 7.62 m (25 ft) while the test span-length was 7.32 m (24 ft). The W, V, R and Y rebars (Fig. 3) were installed to resist the end zone bearing, spalling and bursting stresses. The sizes of the rebars are as follows: # 4 for R bars and V rebars, # 5 for W rebars, and # 6 for Y rebars. A view of the beam prestressing bed just before installing the formwork is also shown in Fig. 4.

3.2 Materials and Mix Design

Two types of steel fibers were used to cast the PSFC I-Beams. The type of steel fibers, selection of optimum and practical fiber dosage, and suitable fiber factors to cast the beams were selected based on previously carried out experiments at the University of Houston (Tadepalli et al. 2013). The steel fibers were 'trough' shaped with hook at both ends and were collated together with water soluble glue. The long fiber (LF) is shown in Fig. 5a and the short fiber (SF) is shown in Fig. 5b. The long fibers had a length of 60 mm, a diameter of 0.75 mm (aspect ratio of 80) and had a tensile strength of 1040 MPa. The short fibers were 30 mm long and 0.55 mm in diameter (aspect ratio of 55) and had a tensile strength of 1100 MPa.

Table 2 gives the details of the steel fibers used in this experimental study. The steel fibers were relatively stiff and glued into bundles, i.e., collated. The glue dissolved in the water during mixing, thus dispersing the fibers in the mix as shown in Fig. 6.

Table 3 shows the details of different constituent materials of concrete used to cast the PSFC I-Beams. Locally available materials were used to prepare the high strength fibrous concrete mixes.

3.2.1 Cement

High early strength cement was used in all the mixes, since it was necessary to develop high release strength at an early age in the PSFC I-Beams. Portland cement (Type-III) conforming to ASTM-C150, and fly ash (Class-F) conforming to ASTM-C618, were the only powder materials used. Fly ash was added to the mix to enhance workability, curtail rise in temperature and reduce cost.

3.2.2 Coarse and Fine Aggregates

The mixes utilized uniformly-graded, rounded, river-bed, coarse aggregates of 3/4 inch nominal size (AASHTO-T27, 1996) and well-graded, river-bed sand (AASHTO-M43, 1998).

3.2.3 Admixtures

A Polycarboxylate-based High Range Water Reducing (HRWR) agent conforming to ASTM C 494-1999, Class-F was used to achieve workable concrete mixes. A retarder

Table 1 Test variables of PSFC I-Beam.

Beam ID	Designed mode of failure	Concrete compressive strength (MPa) ksi	Volume of steel fiber reinforcement V_f	Fiber factor $[(L_f/D_f)V_f]/100$
R1	Web shear	86.9 (12.6)	0.5 % LF	0.40
R2	Web shear	90.4 (13.1)	1 % SF	0.55
R3	Web shear	82.1 (11.9)	1.5 % SF	0.825
R4	Web shear	73.1 (10.6)	1.5 % SF + 0.5 % LF	0.825 + 0.40 = 1.225
R5	Flexural shear	84.2 (12.2)	0.5 % LF	0.40
R6	Flexural shear	88.3 (12.8)	1.5 % SF + 0.5 % LF	0.825 + 0.40 = 1.225

LF long fibers, L_f/D_f 80, *SF* short fibers, L_f/D_f 55, L_f length of steel fiber, D_f diameter of steel fiber.

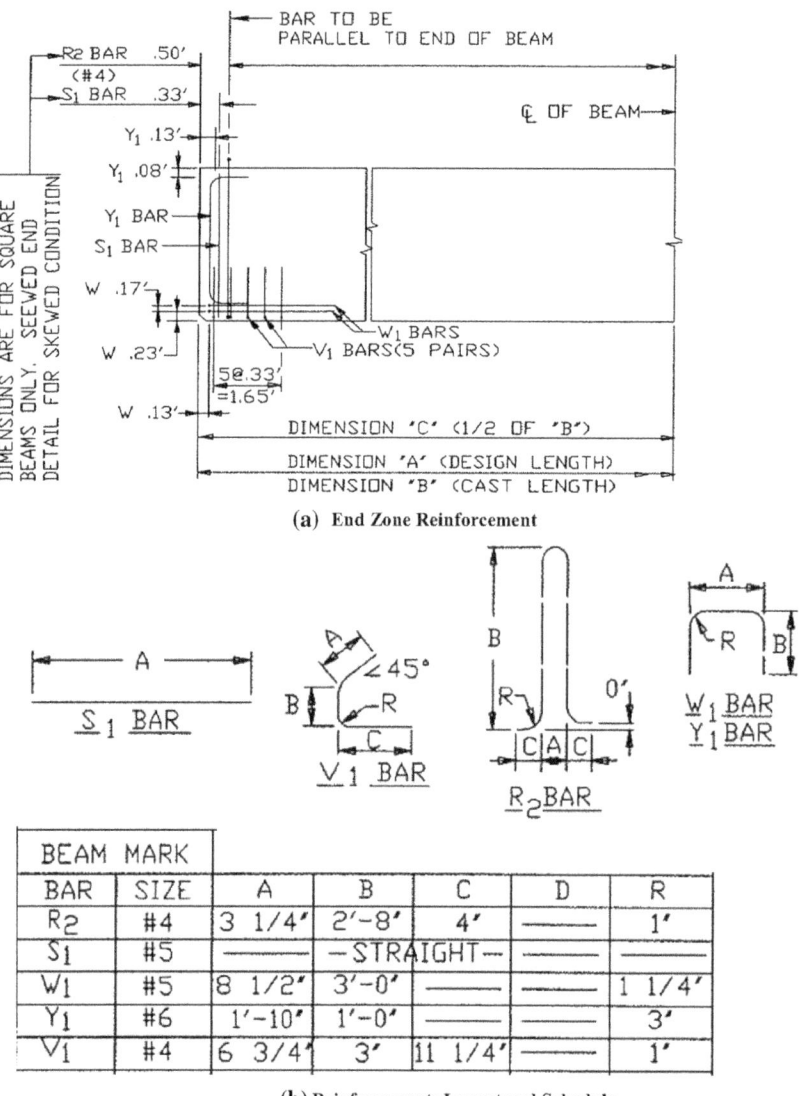

(a) End Zone Reinforcement

BEAM MARK						
BAR	SIZE	A	B	C	D	R
R2	#4	3 1/4'	2'-8'	4'	———	1'
S1	#5	———	—STRAIGHT—		———	———
W1	#5	8 1/2'	3'-0'	———	———	1 1/4'
Y1	#6	1'-10'	1'-0'	———	———	3'
V1	#4	6 3/4'	3'	11 1/4'	———	1'

(b) Reinforcement: Layout and Schedule

Fig. 3 Details of end zone reinforcement in PSFC I-Beams. *Note* $1'' = 25.4$ mm.

Fig. 4 Picture of end zone reinforcement.

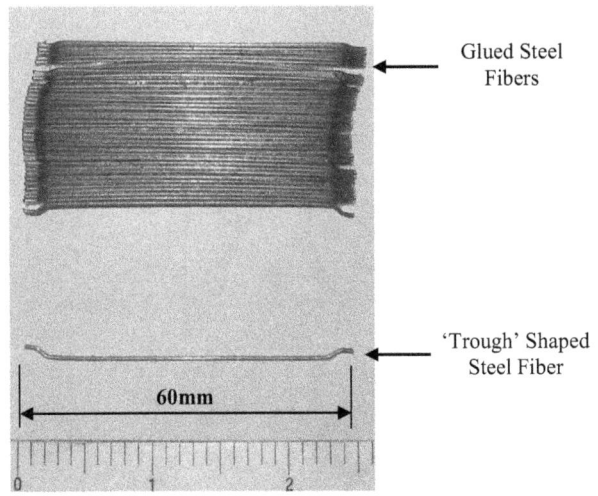

(a) Hooked Steel Fiber -Long

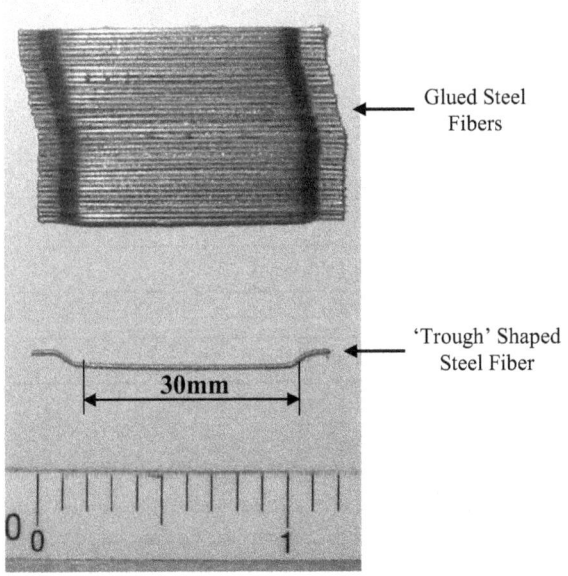

(b) Hooked Steel Fiber -Short

Fig. 5 Steel fibers used in PSFC I-Beams.

conforming to ASTM-C494/C494 M, Class-B was added to the mixes as required to delay the initial *setting* of the mix. Concrete mix design used to cast each of the PSFC beam is given in Table 4. The amount of fibers used in a concrete mix can also be reported as its fiber-factor, which is the product of the aspect ratio of the fibers and the volume of fibers in the mix, i.e., $(L_f \mathrm{ha}/D_f)V_f$.

3.3 Fabrication of PSFC I-Beams

All steel fiber concrete mixes were mixed in a 4.6 m^3 (6 yd^3) drum mixer at the Texas Concrete Company's (Victoria, Texas) precast plant. 1.5 m^3 (2 yd^3) of concrete was mixed for each beam. The six PSFC I-Beams were cast in two groups on two different days. Beams R2, R3 and R6 were first cast concurrently in a long-line prestressing bed using Type-A steel formwork. The strands were prestressed by hydraulic jacks against the prestressing bed ends. The second group of three Beams R1, R4 and R5 were cast 1 week after the first group. Concrete for both the groups was prepared in the plant's mixer, transported to the casting location (prestressing bed), and placed into the formwork using a mobile hopper. During concrete placement, spud vibrators were used for compacting the fibrous concrete.

Casting and compaction of PSFC I-Beams was relatively fast and easy in comparison with the conventional I-beams, even when the mix used large dosage of steel fibers. This was because transverse reinforcement in the beams was totally absent, causing no hindrance to the compaction of the fiber reinforced mix. Thus, fiber reinforced concrete was found to be relatively easy to compact in the absence of any traditional reinforcement. Just after mixing the steel fiber concrete (i.e., before casting the beams), slump tests were carried out for all the mixes.

It has been found out that the true workability of SFC cannot be ascertained by the slump tests as the fibers significantly affect the rheology of fresh concrete. At lower fiber factors concrete was more workable than at higher fiber factors based on the laboratory and field casting experiences. During casting of full-scale beam specimens it was interesting to observe that the compaction energy required for SFC was almost same as for non-fibrous concrete. This was evidently due to the absence of the traditional rebars resulting in lower hindrance to compact fresh concrete in the beams. A satisfactory level of workability and finish was achieved for all the fibrous mixes used in casting the beams.

Curing of the PSFC I-Beams was carried out until a minimum concrete compressive strength of 28 MPa (4000 psi) was obtained in the beams, sufficient for release of prestress. One day after casting, the prestressing strands were slowly released and the beams were de-molded.

3.4 Test Setup

The PSFC I-Beams were placed in a vertical loading system at the University of Houston and were subjected to vertical load up to their maximum shear capacity, until failure. The testing system was a specially built steel loading

Table 2 Properties of steel fiber used in PSFC I-Beams.

Fiber type		Length (mm) L_f	Diameter (mm) D_f	Aspect ratio L_f/D_f	Tensile strength (MPa)
Hooked end, collated	Long fiber (LF)	60	0.75	80	1040
	Short fiber (SF)	30	0.55	55	1100

Steel Fibers Mixed in Fresh Concrete

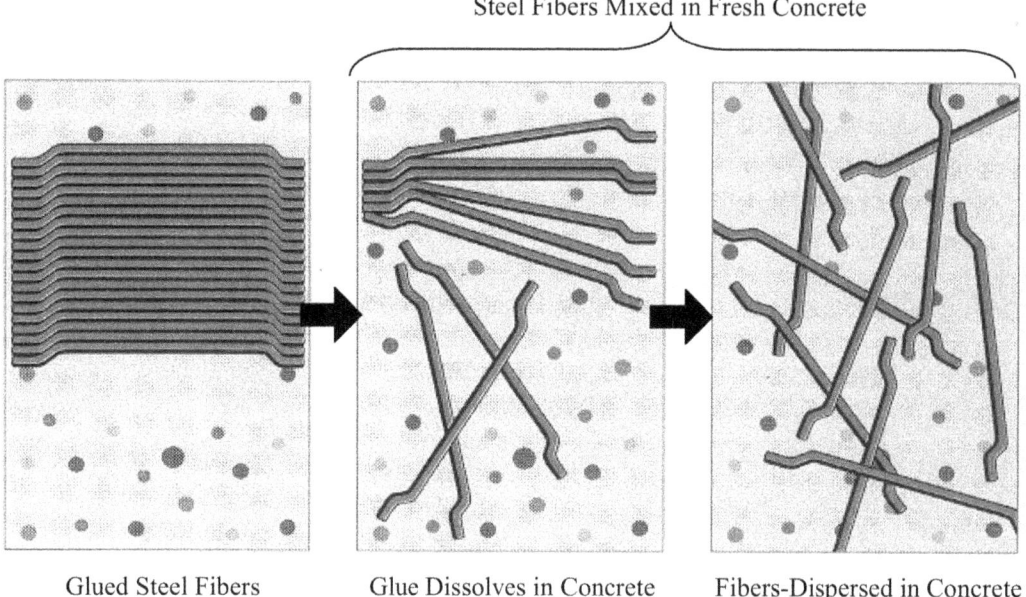

Glued Steel Fibers   Glue Dissolves in Concrete   Fibers-Dispersed in Concrete

Fig. 6 Dispersion of glued (collated) steel fibers in concrete.

Table 3 Materials used in steel fiber concrete.

Material	Type
Cement	ASTM C150 Type-III
Fly ash	ASTM C618 Class F
Coarse aggregate	AASHTO T27
Fine aggregate	AASHTO M43

frame with four actuators as depicted in Fig. 7. Two of the four actuators (namely actuator B and actuator C) attached to the steel frame were used to apply vertical loads on the beams. Each of the actuators had a maximum load capacity of 1420 kN (320 kips). Details regarding the design, layout and capabilities of the loading system can be found in Laskar (2009).

Load application points and support locations for PSFC I-Beams are shown in Fig. 8. Support bearings beneath the beams were located 6 in. from each beam end. The applied loads from actuators B and C were 0.92 m (3 ft) away from each of the supports for Beams R1, R2, R3 and R4, and at 2.44 m (8 ft) from each of the supports for Beams R5 and R6. Actuator loads were applied on the beam via a steel roller and bearing plate assembly. This assembly consisted of two steel rollers of 51 mm (2 in. diameter and 305 mm (12 in.) long, sandwiched between two steel bearing plates 152 mm wide × 305 mm long × 51 mm thick (6 in. × 12 in. × 2 in.). This ensured a uniform and friction-less load transfer from the actuators to the top surface of the beam.

A freely movable roller assembly (roller-support) and a fixed roller assembly (hinged-support) were provided at the North and South beam ends, respectively. This enabled free

rotation and longitudinal movement of the simply supported beam during test. All the steel bearing plates and rollers were heat-treated to maximum hardness in order to minimize local deformations. Lead sheets were also used between the load bearing plates and beam surface to help distribute the load evenly.

Beam displacements and concrete strains at important locations on the beam were measured continuously throughout the load test using Linear Variable Differential Transformer Linear Variable Differential Transformer s (LVDTs). A group of seven LVDTs was used at either end and on each side of the beam to measure smeared (average) concrete strains within the beam-web. The LVDTs were arranged in a rosette form as shown in Fig. 9. Each rosette consisted of two vertical, three horizontal, and two diagonal LVDTs. The rosettes were mounted on the beam adjacent to the loading points where the web-shear or flexure-shear failure was anticipated (Figs. 8a, b, 9).

A total of six LVDTs were used to continuously monitor and measure the vertical deflections of the beam. LVDTs were placed under each beam support (North and South ends) on either sides of the beam (West and East). Two pairs of LVDTs were positioned under the beam at each of the two loading points. These LVDTs were used to measure the

Table 4 Concrete mix design for PSFC I-Beams.

Component (kg/m³)	R1 and R5	R2	R3	R4 and R6
Cement	364.0	364.0	364.0	364.0
Fly ash	121.5	121.5	121.5	121.5
Cementitious material	485.6	485.6	485.6	485.6
Water	146.3	146.3	146.3	146.3
Water/cement ratio (w/c)	0.4	0.4	0.4	0.4
Water/cementitious ratio	0.3	0.3	0.3	0.3
Coarse aggregate (CA)	1125.1	1125.1	1125.1	1125.1
Fine aggregate (FA)	596.5	596.5	596.5	596.5
CA/FA ratio	1.88	1.88	1.88	1.88
HRWR/superplastisizer (gm/100 kg)	688	688	688	688
Fibers	39.5 LF	79 SF	118.5 SF	119 SF + 39.5 LF
Retarder (gm/100 kg)	250	250	250	250

1 kg/m³ = 1.686 lb/yd³.
LF long fibers, *SF* short fibers.

Fig. 7 Test set-up at University of Houston.

deflections of the beam. An additional set of LVDTs was used to monitor potential lateral displacements of the beam.

Two 2225 kN (500 kips) capacity load cells were used to monitor support reactions at each beam-end. Two load cells, attached to the loading actuators (B and C), were used to measure the applied load on top of the beam. During a test, force equilibrium between the applied loads (actuators B and C) and the measured reactions (load cells) was always verified.

Non-stop measurement of all the experimental data (beam deflections, strains, loads, and support reactions) from the above sensors were continuously monitored and stored by data acquisition system, during a load test. Shear cracks,

which formed on the beam web during a load test, were regularly marked on a grid, as shown in Fig. 9. The crack widths were measured using a hand-held microscope having a 0.025 mm (0.001 in.) measuring precision.

The two hydraulic actuators were precisely controlled in force or displacement modes by a servo-controlled system. Actuators B and C were initially used to apply shear force on the beam in force control mode at a rate of 22 kN/min (5 kips/min). During a test, the shear load–displacement curve for a beam was continuously monitored visually on a display screen. When the slope of this load–displacement curve started to decreasing (flatten-out), the control mode of the actuators was switched to displacement control with a rate of

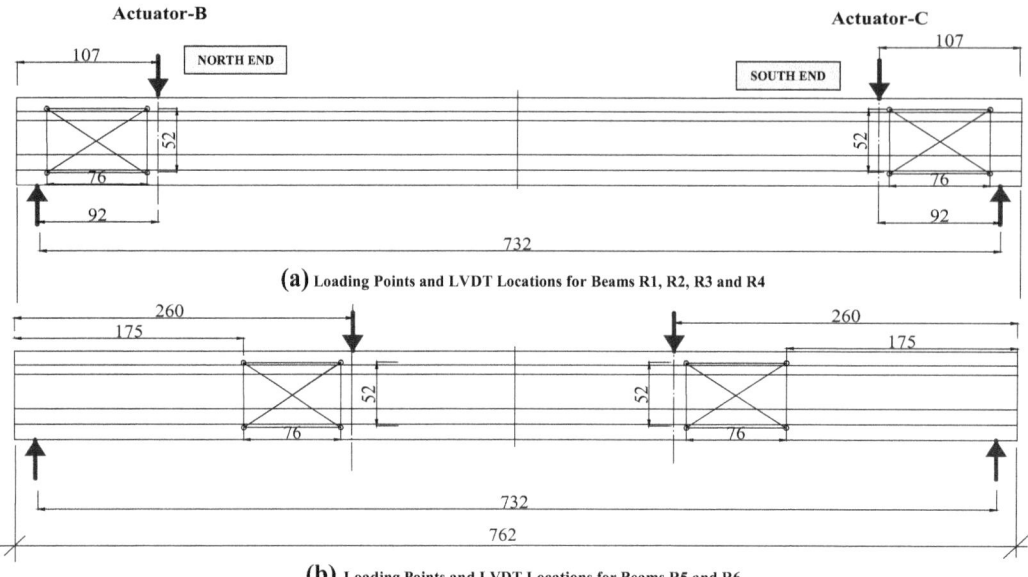

(a) Loading Points and LVDT Locations for Beams R1, R2, R3 and R4

(b) Loading Points and LVDT Locations for Beams R5 and R6

Fig. 8 Loading and support locations in PSFC I-Beams. *Note* All dimensions in cm: $1'' = 25.4$ mm.

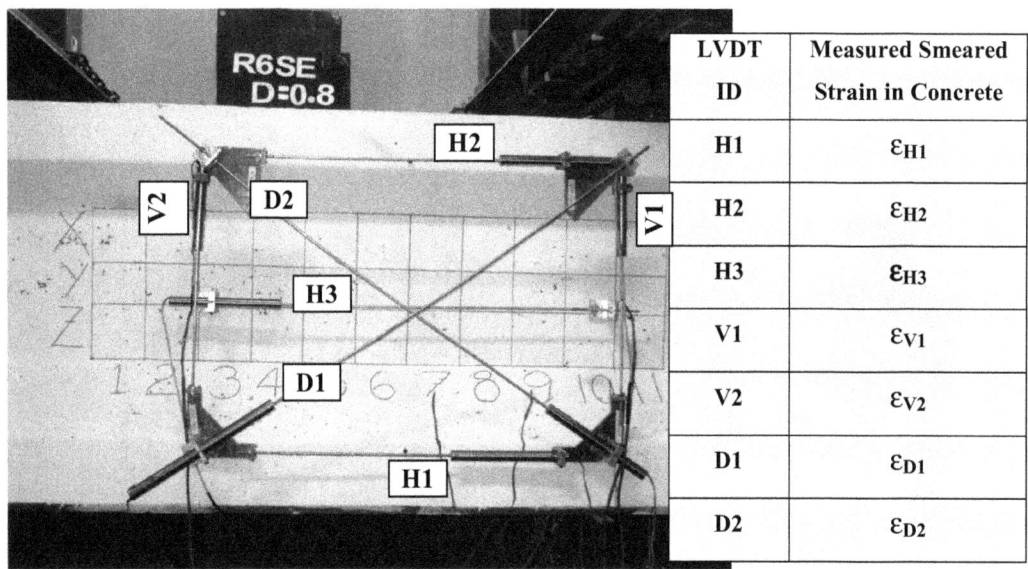

Fig. 9 Typical LVDT rosette used to measure smeared/average concrete strains in PSFC beams.

5 mm/h (0.2 in./h). This displacement control mode was maintained until the failure occurred at either end of the beam. The displacement control feature was essential in capturing the ductility/brittleness behavior of the beam as it failed in shear.

4. Experimental Results

Table 5 shows the experimental ultimate strengths at failure for the six beams tested (R1–R6). During the test, although application of load and support arrangements were symmetric for all the beams; only in the case of Beam R2 web-shear failures occurred simultaneous at both the ends. In all the other beams, the weaker end failed first. Even though Beam R3 ultimately failed in flexure, the shear load at failure at both the

ends was close to the web-shear capacity, as indicated by the spalling of concrete struts in the web region of this beam.

While testing Beam R4 it was found that the shear capacity was surprisingly increased beyond the anticipated value due to the use of higher fiber-factor. Hence, the beam would have failed in flexure instead of the desired web-shear failure mode. Therefore, Beam R4 was reinforced with FRP sheets (installed on the beam soffit at the bottom flange) to increase its flexural capacity. The beam was then tested using a shorter span of 4.26 m (14 ft), which failed in web-shear. This test is denoted as 'R4-Short' hereafter in the discussion.

Beams R5 and R6 failed near a region adjacent to the loading point (i.e., at one third span of the beam) in flexural-shear and flexure failure mode, respectively. As a result, both these beams did not have a sufficiently long undamaged

Table 5 Experimental ultimate strengths at failure for PSFC I-Beams.

Beam ID and failed end	Steel fiber by volume (%)	Fiber factor	Concrete compressive strength (MPa)	Failure mode	Ultimate shear capacity (kN)	Ultimate moment capacity (kN/m)	Max. shear at ultimate moment (kN)	Max. moment at ultimate shear (kN/m)
R1-North	0.5 % LF	0.40	86.9	Web-shear	1175	–	–	1078
R2-North	1 % SF	0.55	90.4	Web-shear	1250	–	–	1146
R2-South	1 % SF	0.55	90.4	Web-shear	1313	–	–	1205
R3	1.5 % SF	0.825	82.1	Flexure/web-shear	–	1191	1299	–
R4-North (short beam)	1.5 % SF + 0.5 % LF	1.225	73.1	Web-shear	1540	–	–	–
R5-South	0.5 % LF	0.40	84.2	Flexural-shear	472	–	–	1153
R6	1.5 % SF + 0.5 % LF	1.225	88.3	Flexure	–	1243	507	–

1 MPa = 145 psi, 1 kN = 0.225 kip, 1 kN/m = 0.735 kip-ft.

LF long fibers, *SF* short fibers.

length for another re-test in flexure-shear mode. Hence each of these two beams could provide only one failure capacity. Beam R5 failed on the South side without any prior warning. The sudden brittle failure of beams subjected to flexure-shear was explained by Kani (1964). When the strength of concrete "teeth" formed between the flexural cracks is smaller than the remaining arch, the beam fails suddenly as soon as the strength of teeth is compromised. Specimen R6 apparently failed in flexure mode instead of the targeted flexure-shear mode. Beam R6 demonstrated much higher web-shear capacity than expected, owing to the use of higher fiber-factor.

The comparison of shear strength of PSFC I-Beams tested in this work (Table 5) shows that shear capacity of beams can be significantly increased due to the addition of steel fibers in concrete. The beam test results reveal a good co-relation between the fiber-factor and shear strength. The general trend detected was that with an increasing fiber-factor, shear strength also increased. The failure of beam R5 suggested that a fiber-factor of more than 0.4 (0.5 % by volume of LF) may be necessary to serve as minimum shear reinforcement. ACI 318 code recommends a minimum of 0.75 % by volume of steel fibers as minimum shear reinforcement.

The ACI provisions are basically formulated on experimental studies of non-PC beams and majority of them had a cylinder compressive strength less than 42 MPa (6000 psi). Nevertheless, in a prestressed concrete beam, the beneficial effect from prestressing forces could further relax the minimum required fiber volume fraction. Further, concrete compressive strengths much higher than 42 MPa (6000 psi) are commonly used in PC for higher structural performance. The achievement of high shear strength in the PSFC beams in this study suggests a minimum amount fiber volume less than 0.75 % is feasible for replacing traditional shear reinforcement in PC members. Therefore, the current ACI 318

requirement may hamper the use of SFC in structures with PC members made of high strength concrete. However, further research would be required to establish the minimum fiber-factor required for PSFC members to avoid brittle failure.

The crack pattern and photograph at failure of all the PSFC I-Beams are shown in Fig. 10. The web-shear failures in beams R1–R4 were noticeably along a single shear crack which formed between the support and loading points at failure. Studying the failure photographs closely, it can be observed that the damage to the beams with web-shear failure mode (R1–R4) was less pronounced in comparison to the damage in beams with a destructive flexure-shear mode of failure (R5 and R6).

From the shape of the load–deflection curves of the PSFC I-Beams, shown in Fig. 11, it can be seen that the beams which failed in web-shear mode (R1–R4) demonstrated higher shear capacities compared to the beams that failed in flexural-shear mode (R5 and R6). It is therefore evident that the shear span-to-effective depth ratio (a/d) has a significant effect on the web-shear and flexure-shear strengths of PSFC I-Beams. Laskar (2009) reported similar results for traditional TxDOT PC I-Beams. The PSFC I-Beams that failed in flexural-shear or flexure mode displayed higher ductility than the beams which failed in web-shear mode.

The advantageous effect of steel fibers on shear strength of PSFC I-Beams can be observed by examining Fig. 11. The values of shear force plotted in this figure were obtained from the load cells under the beam's end-supports and were also verified by the load equilibrium computations. The net deflection was obtained from the difference in readings of LVDT placed under the beam at the particular actuator location and the readings of LVDT placed at the corresponding support. Hence, the beam total deflection values were subtracted by the support settlement and then used to plot the load–deflection curves (Fig. 11).

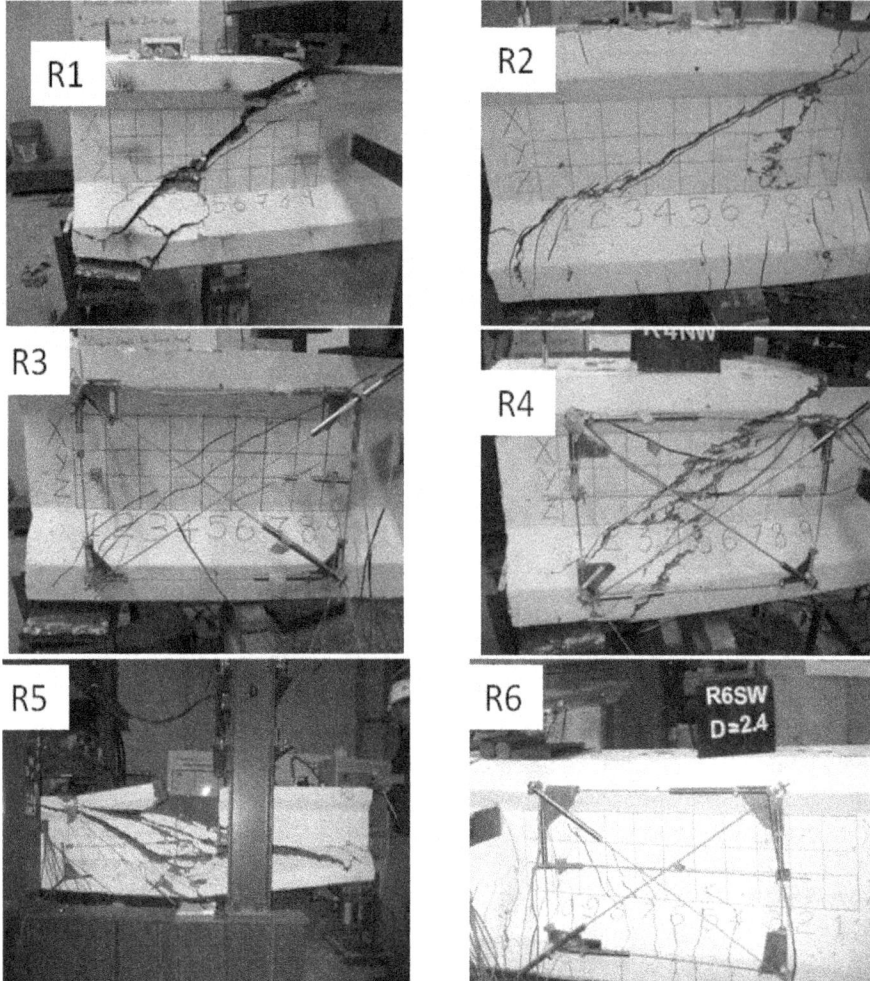

Fig. 10 PSFC I-Beams at failure.

Since the compressive strength of concrete for various I-Beams tested were different, the beam's ultimate shear capacity was normalized with the corresponding compressive strength of concrete to better compare all beam results. Normalized shear was calculated as:

$$\text{Normalized Shear Force of PSFC I-Beam} = \frac{Shear\,Capacity}{bd\sqrt{f_c}}$$

$$(1)$$

where experimental shear capacity is in N. f_c is in MPa., and b and d are in mm.

The normalized shear force versus net deflection curves for PSFC I-Beam are shown in Fig. 12. It can be clearly seen that the shear behavior of beams improves with increasing fiber-factor. The ductility in beams also increased with an increase in the fiber factor. This performance shows that the complete replacement of traditional transverse steel by steel fibers is very effective in resisting the shear force.

To understand the true effectiveness of steel fibers as shear reinforcement, the results of PSFC I-Beams are compared with the results of conventional beams (LB2 and LB4) having mild steel as shear reinforcement, tested by Laskar (2009). Laskar's beams had the same compressive strength of concrete, a/d ratios, test span and total prestressing force

as the PSFC I-Beams. I-Beam LB2 had a transverse steel ratio of 1 % by volume of concrete and failed in web-shear mode, while LB4 had a transverse steel ratio of 0.17 % by volume of concrete and failed in flexure-shear mode. The comparisons of web-shear and flexural-shear failures for fibrous and non-fibrous PC beams are shown in Figs. 13 and 14, respectively.

Figure 13 shows that the PSFC I-Beams demonstrated superior shear performance when compared with the traditional PC I-Beam. Not only the shear strengths, but also the ductility and stiffness were greater in all the PSFC I-Beams in comparison with the PC I-Beam. In case of web-shear failure, the increase in shear strengths of PSFC I-Beams over the PC I-Beam due to addition of steel fibers ranged from 15 to 50 % corresponding to a fiber factor of 0.40–1.225, respectively. Hence, based on the limited work, the authors suggest an optum fiber factor of 0.40 for minimum shear strength and a fiber-factor of 1.225 for maximum shear strength based on workability and constructibility requirements in PC beams made with high strength concrete.

Figure 14 shows that the PSFC I-Beam also demonstrated superior flexure-shear performance when compared with the traditional PC I-Beams. Not only the flexure-shear strengths, but also the ductility was greater in all the PSFC I-Beams in comparison with the PC I-Beams. The increase in flexure-

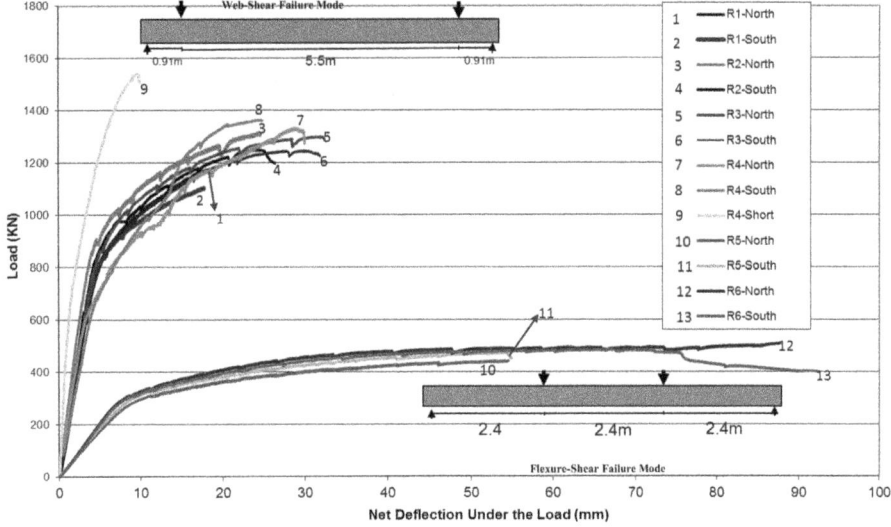

Fig. 11 Shear force versus net deflection curves for PSFC I-Beams. *Note* 1″ = 25.4 mm, 1 KN = 0.225 kip.

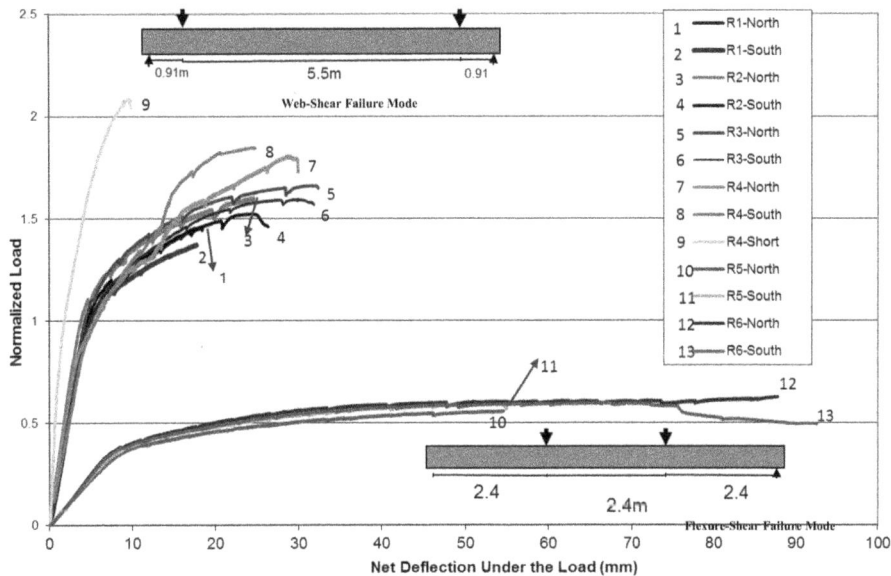

Fig. 12 Normalized shear force versus net deflection curves for PSFC I-Beams. *Note* 1″ = 25.4 mm, 1 KN = 0.225 kip.

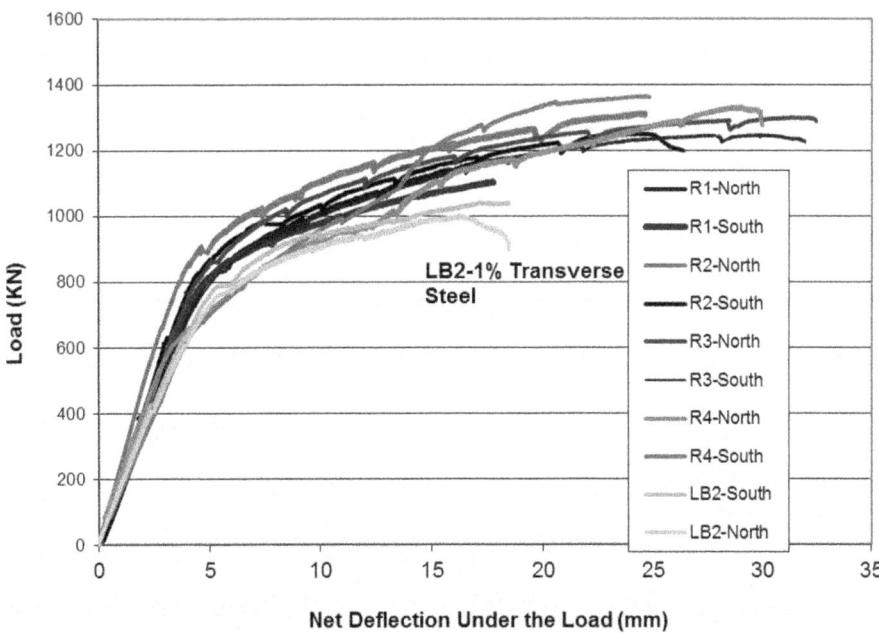

Fig. 13 Comparison of PSFC and PC I-Beams in web-shear failure mode. *Note* 1″ = 25.4 mm, 1 KN = 0.225 kip.

(Note: 1"=25.4mm, 1KN=0.225kip)

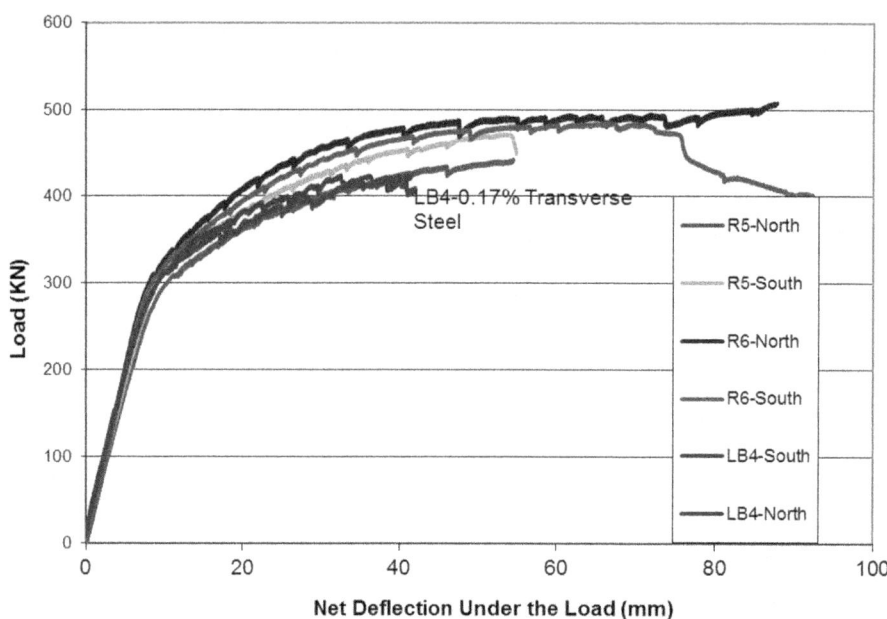

Fig. 14 Comparison of PSFC and PC I-Beams in flexure-shear failure mode. *Note* 1″ = 25.4 mm, 1 KN = 0.225 kip.

Table 6 Ultimate strains in specimens R1–R6 measured by LVDTs.

Beam ID -End side	Strains ($\times 10^{-6}$)						
	ε_{V1}	ε_{V2}	ε_{H1}	ε_{H2}	ε_{H3}	ε_{D1}	ε_{D2}
R1-North E	−1980	−367	−1430	42.6	−2097	1890	−4000
W	−2300	−28	−981	97.2	−2052	1950	−2890
R1-South E	−1390	−1100	−821	234	−1296	−2660	1120
W	−1960	−1150	−807	109	−1256	−2260	1180
R2-North E	−1270	−655	−2920	140	−2388	2490	−5080
W	−2090	−484	−2510	−324	−2741	1870	−4920
R2-South E	−3320	−1190	−817	−169	−3066	−4680	2330
W	−3490	−2140	−867	−35.9	−2975	−4540	2360
R3-North E	−1290	−845	−1550	459	−1319	1870	−3360
W	−738	21	−1380	−694	−1797	1210	−3010
R3-South E	−2030	−444	−2120	275	−1504	−2910	875
W	−1900	20.9	−3500	116	−1684	−4890	2010
R4-South E	−2930	−1000	−1720	−109	−2900	3040	−5490
W	−1140	−1130	−3740	−850	−3326	6770	−7100
R5-North E	−3560	−673	−2900	771	−1590	−47.8	−3140
W	−3140	−1140	−3030	562	−1610	248	−3370
R5-South E	−578	−6110	−2660	712	−1260	−3700	560
W	−122	−5610	−2920	519	−1400	−2940	−42.5
R6-North E	144	−1210	−5020	687	−1208	−18.2	−10,600
W	−99.6	−436	−2610	874	−1192	428	−1870
R6-South E	−811	−6.62	−2395	572	−653	−1684	−9864
W	−183	−142	−1814	275	−527	−890	87

E East side, *W* West side.

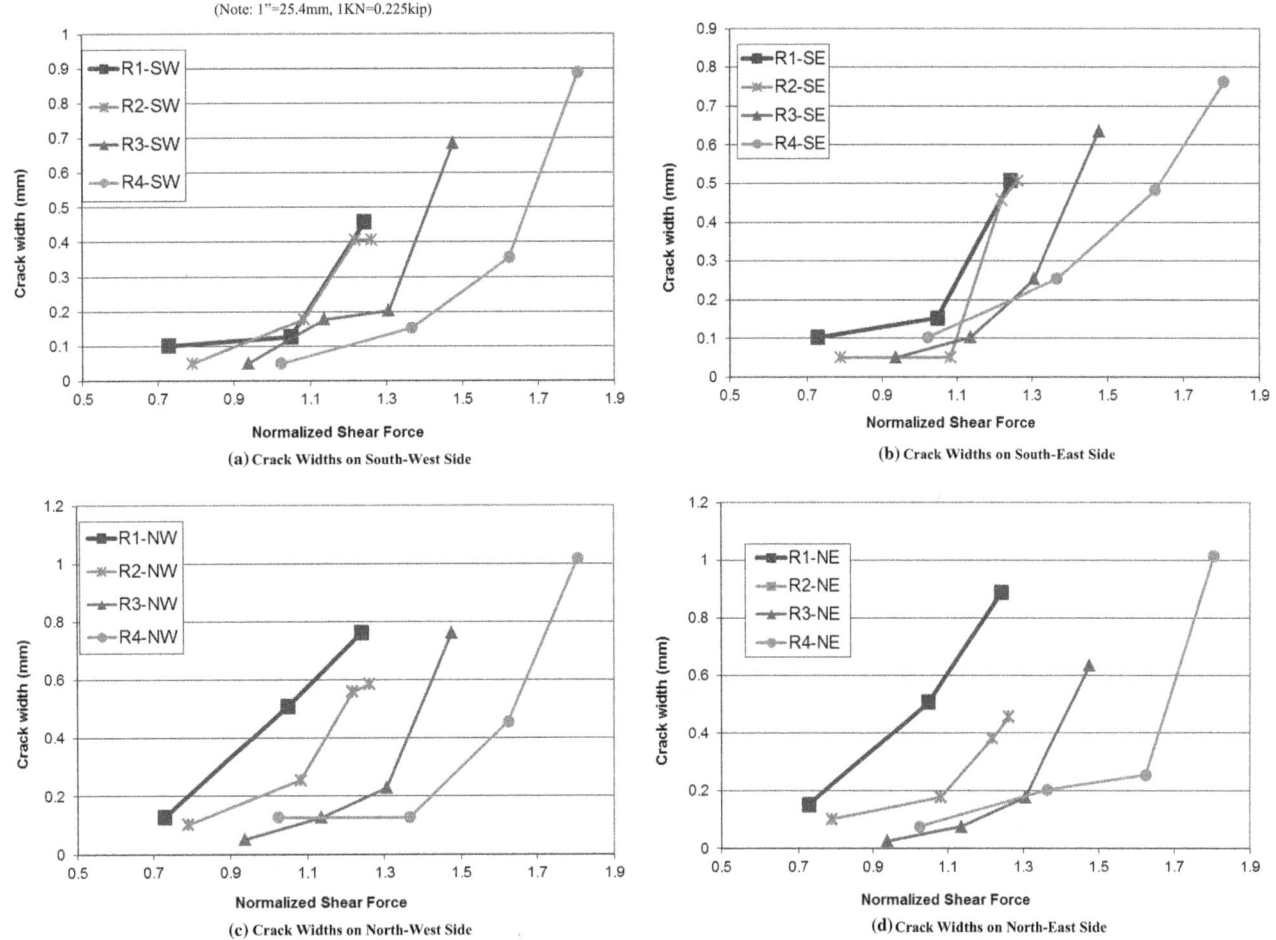

Fig. 15 Shear crack widths versus normalized shear force in beams R1–R4 (North). *Note* $1'' = 25.4$ mm, 1 KN = 0.225 kip.

shear strengths of PSFC I-Beams over the PC I-Beam due to addition of steel fibers ranged from 15 to 24 % corresponding to a fiber factor of 0.40–1.225, respectively. It can be clearly observed from the Figs. 13 and 14 that the web-shear behavior is affected more than the flexure-shear behavior of PC beams owing to the addition of steel-fibers.

Table 6 gives the ultimate strains in Beams R1–R6 measured by LVDTs at failure. The LVDTs were located adjacent to the loading point as indicated in Fig. 8. A set of six LVDTs is shown in Fig. 9. Each set had two vertical, two horizontal and two diagonal LVDTs. Out of the two vertical LVDTs, the one that was situated closer to the load was named V2 (strain of ε_{V2}) while the other was named V1 (strain of ε_{V1}). The horizontal LVDT situated on the top flange was named H1 (strain of ε_{H1}) while the one on the bottom flange was named H2 (strain of ε_{H2}). The diagonal LVDT that was connected to the top flange near the load point was named D1 (strain of ε_{D1}) and was subjected to compressive strains during the loading of the beams. Diagonal LVDT D2 (strain of ε_{D2}) was connected to the lower flange near the load point and subjected to tensile stresses during loading of the beams. The tensile and compressive strains measured are more than what had been observed in conventional PC I-beam (Laskar 2009).

5. Shear Crack Widths and Crack Patterns

As mentioned earlier, shear cracks were continuously tracked and measured during the load tests of the beams. A grid was marked on the beam-web at both the beam-ends to facilitate easy identification and location of the shear cracks. Hand-held microscopes were utilized to precisely measure the shear crack width with an accuracy of 0.025 mm (0.001 in.). Figure 15a–d shows the plot of the normalized shear force and corresponding shear crack width in Beams R1–R4 (having web-shear mode of failure) measured on four different sides of the beams, during the test. The represented shear crack widths for a given beam were the maximum crack widths recorded along the most dominating shear crack in a beam during the test.

The onset of shear crack formation in all the beams initiated at the mid height of the beam web and was oriented along a line joining the loading and support points. Shear cracks of this nature are referred to as "diagonal tension cracks", because the general direction of principal tension is perpendicular to this crack. The ligament of concert formed between adjacent diagonal tension cracks is referred to as a concrete compression strut. In the conventionally reinforced PC beams, the applied shear force is resisted by tension in transverse rebars and compression in the concrete strut

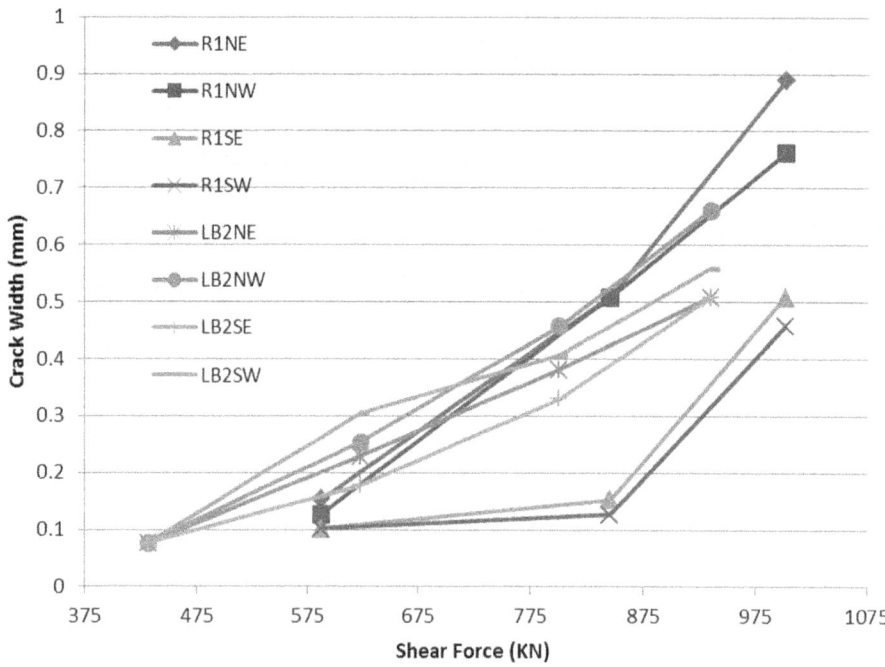

Fig. 16 Shear crack widths versus shear force in beams R1 and LB2. *Note* 1″ = 25.4 mm, 1 KN = 0.225 kip.

(Schlaich et al. 1987). In the case of PSFC girders, diagonal tension is resisted solely by the steel fibers. In the test beams, the initial diagonal tension crack did not generally progress to form the failure surface, but as the load increased, other cracks appeared and further developed into a failure surface with a single dominant failure shear crack (see Fig. 10).

Steel fibers were clearly observed to restrict the width of the shear cracks, as shown in Fig. 15. Generally, it was observed that as the fiber-factor increased the shear crack width for a given load decreased. Also, the load at which first visible shear crack appeared increased as the fiber-factor increased. This can be attributed to the fact that with the use of higher fiber-factor, more steel fibers are available in bridging and intersecting the shear crack. The stresses across the shear crack will therefore be shared by a larger number of steel fibers, thereby reducing the tensile strain across the crack. As the strains across the crack and in the steel fibers are reduced, the crack widths will be less.

To better understand the effectiveness of steel fibers in controlling the shear crack widths in PC beams, Fig. 16 is plotted depicting the crack widths of fibrous (Beam R1) and non-fibrous (Beam LB2) beams. It can be seen from Fig. 16 that the onset of shear cracking for beams with steel fibers occurred at a higher normalized shear force than those without steel fibers. This indicates that the addition of steel fibers in beams is helpful in preventing the development and growth of initial shear cracks. This property of steel fibers can be helpful particularly at service load level in PC highway-bridge beams.

The above discussion signifies that the replacement of traditional transverse rebars with steel fibers enhances the shear crack resistance in PC beams. The test results demonstrated that steel fibers more effectively delayed the opening of cracks

beyond the service load level in the PSFC I-Beams in comparison with the traditionally reinforced PC beams.

6. Conclusions

The shear behavior of PSFC beams was critically examined by full-scale tests on six TxDOT Type-A I-beams with web-shear or flexural-shear failure modes. From the experimental results of six PSFC I-beams, it was observed that with increase in the amount of steel fibers (fiber factor) the shear capacity of the beam was increased.

The amount of steel fibers had an effect on first crack load and crack widths in shear span of the beam. As the amount of steel fibers increased, the first crack load increased and crack widths at a given load were decreased. The optimum fiber-factor is dependent on many parameters such as the desired minimum and maximum shear capacities, the required degree of workability to cast full-scale beams, compressive strength of concrete and mode of failure based on a/d-ratio. Based on the results of this limited study, the authors suggest an optum fiber factor of 0.40 for minimum shear strength and a fiber-factor of 1.225 for maximum shear strength based on workability and constructibility requirements in PC beams made with high strength concrete.

From the above observations it can be concluded that steel fibers were found very effective in resisting the shear loads and arresting the shear cracks. In order to replace shear reinforcement with steel fibers in PC beams, however, further research would be required on the serviceability and strength of PSFC members considering a reliable margin of safety.

Acknowledgments

This research was funded by the Texas Department of Transportation. The researchers would like to thank the Texas Concrete Company, Victoria, Texas, for continued cooperation during this project. The researchers are grateful to Bekaert Corporation (USA) for supplying the steel fibers for this research.

References

Abdul-Wahab, H. M. S., & Al-Kadhimi, S. G. (2000). Effect of SFRC on shear strength of prestressed concrete beams. *Magazine of Concrete Research, 52*(1), 43–51.

Cho, S. H., & Kim, Y. I. (2003). Effects of steel fibers on short beams loaded in shear. *ACI Structural Journal, 100*(6), 765–774.

Dhonde, H. (2006). Steel fibers and self-consolidating concrete in prestressed concrete beams. PhD. Dissertation, Department of Civil and Environmental Engineering, University of Houston, TX.

Junior, S. F., & De Hanai, J. B. (1999). Prestressed fiber reinforced concrete beams with reduced ratios of shear reinforcement. *Cement and Concrete Composites, 21*(3), 213–221.

Kani, G. N. J. (1964). Riddle of shear failure and its solution. *American Concrete Institute Journal, 61*(4), 441–467.

Langsford, R. P., Lloyd, N., & Sarker, P. K. (2007). Shear strength of steel fibre reinforced prestressed concrete beam. In *Proceedings of the 4th International Structural Engineering and Construction Conference (ISEC-4)—Innovations in Structural Engineering and Construction* (pp. 441–446). Melbourne, Australia: Taylor & Francis/Balkema.

Laskar, A. (2009). Shear behavior and design of prestressed concrete members. PhD Dissertation, Department of Civil and Environmental Engineering, University of Houston, TX.

Meda, A., Minelli, F., Plizzari, G. A., & Riva, P. (2005). Shear behaviour of steel fibre reinforced concrete beams. *Materials and Structures, 38*(3), 343–351.

Narayanan, R., & Darwish, I. Y. S. (1987). Shear in prestressed concrete beams containing steel fibers. *The International Journal of Cement Composites and Lightweight Concrete, 9*(2), 81–90.

Padmarajaiah, S. K., & Ramaswamy, A. (2001). Behavior of fiber-reinforced prestressed and reinforced high-strength concrete beams subjected to shear. *ACI Structural Journal, 98*(5), 752–761.

Schlaich, J., Schafer, K., & Jennewein, M. (1987). Toward a consistent design of structural concrete. *PCI Journal, 32*(3), 74–150.

Tadepalli, P. R., Mo, Y. L., & Hsu, T. C. (2013). Mechanical properties of steel fibre concrete. *Magazine of Concrete Research, 65*(8), 462–474.

Tan, K. H., Paramasivam, P., & Murugappan, K. (1995). Steel fibers as shear reinforcement in partially prestressed beams. *ACI Structural Journal, 92*(6), 643–652.

Thomas, J., & Ramaswamy, A. (2006). Shear strength of prestressed concrete T-beams with steel fibers over partial/full depth. *ACI Structural Journal, 103*(3), 427–435.

12

Image Analysis and DC Conductivity Measurement for the Evaluation of Carbon Nanotube Distribution in Cement Matrix

I. W. Nam, and H. K. Lee*

Abstract: The present work proposes a new image analysis method for the evaluation of the multi-walled carbon nanotube (MWNT) distribution in a cement matrix. In this method, white cement was used instead of ordinary Portland cement with MWNT in an effort to differentiate MWNT from the cement matrix. In addition, MWNT-embedded cement composites were fabricated under different flows of fresh composite mixtures, incorporating a constant MWNT content (0.6 wt%) to verify correlation between the MWNT distribution and flow. The image analysis demonstrated that the MWNT distribution was significantly enhanced in the composites fabricated under a low flow condition, and DC conductivity results revealed the dramatic increase in the conductivity of the composites fabricated under the same condition, which supported the image analysis results. The composites were also prepared under the low flow condition (114 mm < flow < 126 mm), incorporating various MWNT contents. The image analysis of the composites revealed an increase in the planar occupation ratio of MWNT, and DC conductivity results exhibited dramatic increase in the conductivity (percolation phenomena) as the MWNT content increased. The image analysis and DC conductivity results indicated that fabrication of the composites under the low flow condition was an effective way to enhance the MWNT distribution.

Keywords: cement composites, carbon nanotubes, distribution evaluation, image analysis, electrical conductivity, fluidity.

1. Introduction

The electrical and mechanical properties of cement composites with conductive fibers are significantly influenced by the fiber distribution (Sorensen et al. 2014; Liu et al. 2011; Lee et al. 2009). The electrical properties can be enhanced by the formation of conductive fiber networks. The mechanical properties can also be improved by the effect of fibers bridging micro-cracks, whereas they deteriorated in the base of a non-uniform distribution of fibers (Sorensen et al. 2014; Liu et al. 2011). Accordingly, evaluation of the fiber distribution is important in understanding its influence on the physical properties of the composite materials and making full use of fibers (Liu et al. 2011).

Evaluation methods for the fiber distribution can be classified into two categories—image analyses and electrical property analyses. An image analysis entails acquisition of images via a microscope (optical microscope, scanning electron microscope (SEM), etc.) or a transmission X-ray and image processing and analysis. Lee et al. (2009) and Kang et al. (2011) obtained cross sectional images of composite materials via an

optical microscope (Lee et al. 2009; Kang et al. 2011). Lee et al. (2002) and Liu et al. (2011) obtained images via digital cameras (Lee et al. 2002; Liu et al. 2011). Redon et al. (1999) used X-ray photography and Fan et al. (2000) used a SEM to capture fiber distribution images (Redon et al. 1999; Fan et al. 2000). On the other hand, the fiber distribution state can also be evaluated by the DC conductivity or AC impedance. Li et al. (2007) analyzed MWNT distribution states using the DC conductivity and Ozyurt et al. (2006) evaluated the degree of carbon fiber clumping and fiber orientation by AC impedance spectroscopy (Li et al. 2007; Ozyurt et al. 2006). In the case of the electrical property analysis method, it is only applicable to conductive fiber-incorporated composite materials.

The image analysis is a more direct and convincing approach in that actual fiber distribution states in the composite materials can be evaluated. The electrical property analysis method is an indirect approach for evaluation of the fiber distribution, but sometimes it is affected by other parameters (matrix type, moisture content in matrix, etc.), possibly leading to incorrect electrical property results. Accordingly, the image analysis plays an important role in the evaluation of fiber distributions and it is a critical technique that should be carried out along with an electrical property analysis (Li et al. 2007; Ozyurt et al. 2006).

Uncovered in 1991, carbon nanotube (CNT) has gained great attention from researchers attributable to its remarkable physical properties (Li et al. 2007; Kim et al. 2014). From 2010s, CNT has been used as a nano-scale fiber in cement-based composites in an effort to enhance physical properties

Department of Civil and Environmental Engineering,
Korea Advanced Institute of Science and Technology,
Daejeon 34141, South Korea.
*Corresponding Author; E-mail: haengki@kaist.ac.kr

of the composites (Kim et al. 2014). Various approaches have been explored to evaluate CNT distributions in the literature. SEM observation has been the most common approach to study the CNT distribution state, carried out in tandem with an electrical property analysis (Konsta-Gdoutos et al. 2010). However, the SEM observations may vary depending on observation spots in the same sample since it is mostly conducted with high magnification levels (up to 3000-fold). Accordingly, the CNT distribution should be observed and evaluated at low magnification levels and the evaluation results should be compared with electrical properties of the CNT-embedded composites.

In the present work, a novel image analysis method to evaluate the multi-walled carbon nanotube (MWNT) distribution in cement-based composites is proposed. In this approach, white cement was used instead of ordinary Portland cement (OPC) with MWNT in an effort to differentiate MWNT from the cement matrix. MWNT distribution images in the cement matrix materials were acquired by using an optical microscope in conjunction with image processing tools. The MWNT distribution was quantitatively assessed in terms of the planar occupation ratio of MWNT. In preparation of specimens, a novel method was adopted for dispersion of CNT in cement. This method was proposed on the basis of experimental attempts previously conducted by the authors. It was suggested that the distribution of CNT can be enhanced by means of lowering the fluidity of CNT/cement mixture at fresh state. It is notable that the suggested method does not require sonication technique, acid treatment, etc., which were widely demonstrated in the literature. To verify correlation between the MWNT distribution and flow of the mixtures, MWNT-embedded cement composites were fabricated under different flows, incorporating a constant MWNT content (0.6 wt%). The image analysis demonstrated that the MWNT distribution was significantly enhanced in the composites fabricated under a low flow condition, and DC conductivity measurement results revealed the dramatic increase in the conductivity of the composites fabricated under the low flow condition, which supported the image analysis results. The composites were also prepared under the low flow condition (114 mm < flow < 126 mm), incorporating various MWNT contents. The image analysis and DC conductivity results demonstrated remarkable enhancement in the planar occupation ratio of MWNT and the conductivity, respectively, which indicated that fabrication of the composites under the low flow condition was an effective way to enhance the MWNT distribution.

2. Materials

MWNT, Portland cement, nylon fiber, super-plasticizer, tap water, and silica fume (SF) were used in the present work. MWNT produced through the chemical vapor deposition (CVD) growth method, a proprietary product of Hyosung Inc. (M1111), was used (Nam et al. 2012). Their purity, diameter, and aspect ratio were 96.2 %, 12.29 ± 2.18 nm, and 930

(aspect ratio was approximate value), respectively. Type I ordinary Portland cement was used in the present work. SF, a proprietary product of Elkem Inc. (EMS-970 D), contained over 90 % silicon oxide (SiO) and 80 wt% of its primary particles have a diameter greater than 5 μm. White cement was used instead of OPC when image analysis samples were fabricated. Nylon fiber, a proprietary product of Nycon fibers Inc. (NYMAX), was used in an effort to prevent cracks that occur due to shrinkage while cement matrix materials cured. Their diameter and length were 23–36 μm and 3 mm, respectively. A poly-carboxylic acid-based super-plasticizer (SP), a proprietary product of BASF Pozzolith Ltd., (Rheobuild SP8HU), was utilized in an effort to improve the workability of the fresh cement matrix materials. The true specific gravity of MWNT, Portland cement, SP, tap water, and SF was 1.32, 3.15, 1.07, 1, and 2.1, respectively.

3. Methods

3.1 Image Analysis of MWNT Distribution in MWNT-Embedded Cement Composites

3.1.1 Specimen Preparation and Image Acquisition for Image Analysis

A novel image analysis approach was proposed in an effort to evaluate the MWNT distribution in cement matrix materials. In order to distinguish MWNT with black color from OPC, which also shows a fairly dark color, white cement was used in place of OPC when composite specimens were fabricated for the image analysis. In addition to using white cement instead of OPC, the incorporation of nylon fiber was omitted in the preparation of the samples for the image analysis since crack control was unnecessary. The constituent materials (white cement, MWNT, water, and SP) were weighed according to mixing ratios and placed together in a steel bowl.

Table 1 shows the mix proportions of composites fabricated with different W/C values, which will be dealt in Sect. 4.1.1. The mix proportions for composites fabricated under a controlled low flow condition (from 114 to 126 mm) with various MWNT contents are presented in Table 2, which will be dealt in Sect. 4.2.1. The weighed materials were thoroughly mixed for 20 min by an electric hand mixer. The electric hand mixer consisted of two curl shaped-beaters and their rotation rate was approximately 200 rpm. The fresh mixtures were poured into plastic molds that were designed in accordance with the sample size. The sample size was $10 \times 10 \times 80$ (mm^3) for the image analysis. After pouring the fresh mixtures in the molds, they were covered with a plastic plate (or wrapping film) in an effort to avoid moisture loss of the samples and cured for 24–48 h at room temperature (18–20 °C) under a humid condition (40–50 %), and then detached from the molds.

In an effort to produce a fracture surface, a cutting knife was used to make a cutting guide line with depth of approximately 1 mm at the center of the sample. The sample was split into two parts applying manual force by hand to the cutting guide. Each split part was sliced by means of a hand saw at 3 mm from the fracture surface. As a result, a

Table 1 Constituent materials and their mix proportions of the MWNT-embedded cement composites fabricated under different flow.

Denotations	Mix proportions (g)						MWNT (vol%)
	Water	Cement	SF	SP	Nylon fiber	MWNT	
W26–M06	26	100	20	1.6	0 (omitted)	0.6	0.66
W30–M06	30						0.62
W34–M06	34						0.59
W38–M06	38						0.56
W42–M06	42						0.53

Table 2 Constituent materials and their mix proportions of the composites fabricated under the low flow condition (114 mm < flow < 126 mm).

Denotations		Mix proportions (g)						Flow (mm)
Group	Type	Water	Cement	MWNT		SP	Nylon fiber	
				Wt% by cement	Vol%			
LF–M	LF–M0	25	100	0	0	0.6	0.2	160
	LF–M0.3	25		0.3	0.39	1.0		126
	LF–M0.6	25		0.6	0.76	3		125
	LF–M1.0	30		1.0	1.14	4.54		120
	LF–M1.5	36		1.5	1.56	4.4		114

specimen with size of $10 \times 10 \times 3$ mm^3 was prepared from the original sample.

A magnification to cover MWNT agglomerates whose diameter ranged from 3 to 200 µm was selected by a trial and error process and determined as 50-fold. An auto stage optical microscope (Olympus MX51) was used so that visible light could be used to distinguish MWNT (black) from the cement matrix (white). The microscope system illuminates the specimen with uniform brightness and transfers the observation image to image processing software, DotSlide. Since the optical focus at each local spot of the fractured specimen changes due to unevenness of the specimen surface, the microscope system captured observation images at every 3 µm along the out of plane direction from the deepest level to the highest level of the fractured surface. By using the image processing software, focused local spots in the captured images were collected and combined to form a single image that is in focus overall. Two images of each specimen were gained through the aforementioned procedure.

3.1.2 Image Processing and Analysis Procedures

In microscopic images, the black amorphous agglomerates indicated MWNTs. The contrast of white cement and MWNT was seen remarkably well by the naked eye. By image processing and analysis with the commercial software, the MWNT distribution can be quantitatively evaluated. Stepwise tasks were carried out in the following order (illustrations for each step can be found in Fig. 1) with the fractured specimens.

(1) Image acquisition by microscope in collaboration with a digital camera
(2) Capture areas that are optically well focused and assemble them so a well-focused image can be completed (conducted by DotSlide)
(3) Convert the image to a black and white image by thresholding
(4) Reverse the black and white image
(5) Convert the black and white image to a binary image
(6) Acquire data (area) of agglomerate regions

Steps (1) to (2) were conducted by the microscope system and the remaining steps were conducted using a MATLAB Image processing tool box. Automatic brightness control was applied in step (1) by the microscope system. The effect of change of brightness level on the image thresholding was considered a minor factor and hence was ignored in the present work. Otsu (1979)'s automatic thresholding method was used in the thresholding process of step (3) (Otsu 1979). The gray-level scale, which is determined between 0 and 1, was used with a 15 % reduction (gray-level scale × 0.85) since the brightness of MWNT was considered high when compared with the original image. After the reversion process (4), bright parts indicated MWNT. Accordingly, the planar occupation area of the bright parts expressed as 1 in the corresponding binary image, was summed. The planar occupation ratio of bright parts can be calculated by dividing the sum total of the bright parts by the total area of the corresponding image. The planar occupation ratio of MWNT

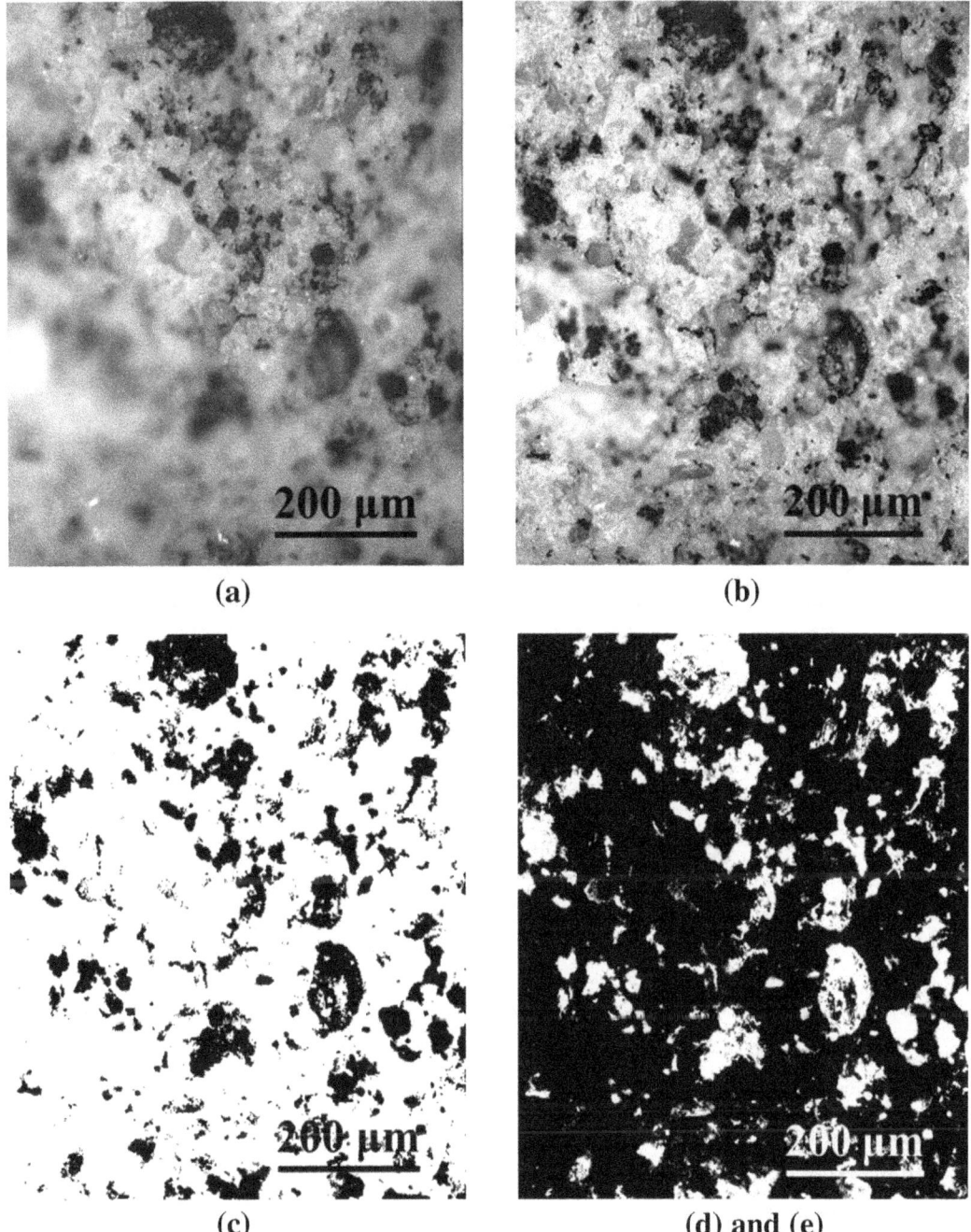

Fig. 1 Image processing procedures (image size: 295 × 355 pixel): **a** Image acquisition, **b** gathering of areas that are well optically focused and assemble them together, **c** conversion of the image to a black and white image by thresholding, **d** reversal of the image, and **e** conversion of the image to a binary image.

was calculated from each of two images and an average value was determined from the ratios.

3.2 DC Conductivity of the MWNT-Embedded Cement Composites

3.2.1 Specimen Preparation for DC Conductivity Measurement

The preparation procedures of samples for DC conductivity measurement followed the specimen preparation procedures for the image analysis in Sect. 3.1.1. However, OPC was used as cement material in this experiment and nylon fiber was used. The constituent materials of the composites were prepared in accordance with mix proportions. In particular, the

mix proportions for composites prepared with different flows and constant MWNT content are presented in Table 3, and DC conductivity of the composites will be dealt in Sect. 4.1.2. On the other hand, Tables 2 and 4 show the mix proportions of composites fabricated under the low flow condition and a high flow condition (over 250 mm), which will be dealt in Sect. 4.2.2 in terms of DC conductivity. The constituent materials were weighed according to the mix proportions and mixed by using the electric hand mixer. The resultant fresh mixture was decanted into plastic molds designed with the size $25 \times 25 \times 25\,(\text{mm}^3)$. After undergoing 1 day's curing under conditions explained in Sect. 3.1.1, the samples were separated from the molds. In an effort to gain mechanical strength

Table 3 The constituent materials and their weight content ratios of the composites fabricated under different flow in an effort to understand change of the MWNT distribution by examining the DC conductivity of the composites.

Denotations	Mix proportions (g)						
	Water	Cement	SF	SP	Nylon fiber	MWNT	Flow (mm)
M06–W26	26	100	20	1.6	0.2	0.6	102
M06–W30	30						127
M06–W34	34						151
M06–W38	38						211
M06–W40	40						250
M06–W42	42						>250
M06–SP0	40			0			130
M06–SP04				0.4			119
M06–SP16				1.6			250
M06–SP32				3.2			>250
M06–SP64				6.4			>250

Table 4 The constituent materials and their mix proportions of the composites fabricated under the high flow condition (flow > 250 mm).

Denotations		Mix proportions (g)					
Group	Type	Water	Cement	MWNT		SP	Nylon fiber
				Wt% by cement	Vol%		
HF-M	HF-M0	40	100	0	0	1.6	0.2
	HF-M0.3			0.3	0.31		
	HF-M0.6			0.6	0.62		
	HF-M1.0			1.0	1.02		

to prevent unexpected damages in the samples, they were submerged in water during 7 days.

3.2.2 Measurement Method

Li et al. (2007) and Xie et al. (1996) revealed that the CNT distribution in CNT-added composite materials can be assessed by examining the DC conductivity of the materials (Li et al. 2007; Xie et al. 1996). Once SEM or transmission electron microscopy (TEM) reveal that CNTs are homogeneously dispersed in the composite materials, DC conductivity evaluation can be used in an effort to identify the percolation network, which is a large cluster network spanning from one side to the opposite side without disconnection in the composite materials (Stauffer and Aharony 1994).

The measurement method of DC conductivity in the present work complied with a standard method presented in SEMI MF43 (2005). As electrodes for the measurement, a pair of copper plates having dimensions of $10 \times 35 \times 0.5$ (mm^3) was inserted in the center part of the sample with a spacing of 8 mm as shown in Fig. 1 of Nam et al. (2015). The copper electrodes were embedded when the composite mixtures were still fresh. As another pair of electrodes, both sides of the samples were coated with silver paste as shown

in Fig. 1 of Nam et al. (2015). A power supply equipment (Agilent E3642A) generated DC current, and it was passed through the silver paste electrodes. Due to the supply of current in resistors, which are the MWNT-embedded cement composites, potential difference in the composites was created, and it was measured by a digital multimeter (Agilent 34410A) connected with the copper plates. The supplied current was automatically controlled by the power supplier with a limitation up to 0.2 A and the resultant voltage was produced with a limitation up to 20 V. Figure 2 illustrates the measurement method for DC conductivity.

The resistance of the composites was calculated on the basis of Ohm's law. For the determination of DC conductivity, the calculated resistance value (R), cross sectional area of the electrode in contact with the composites (A), and the interval of the electrodes (L) were plugged into a following equation (Xie et al. 1996; Vance et al. 2014)

$$\sigma = \frac{1}{\rho} = \frac{L}{R \times A} = \frac{0.8\,\text{cm}}{R \times (2.5\,\text{cm} \times 1\,\text{cm})}\ (\text{S/cm}) \quad (1)$$

3.3 Flow of the Composite Mixtures

The flow of fresh mixture of the composites was tested on the basis of ASTM C1437 (ASTM International 2013). In

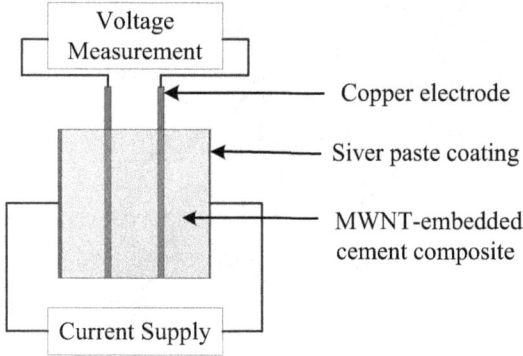

Fig. 2 A schematic illustration of the DC conductivity measurement for the composites.

addition, the flow was determined from an average of mixture's diameter after dropping the flow table.

4. Results and Discussion

4.1 MWNT Distribution in Composites Fabricated Under Different Flows

4.1.1 Image Analysis of MWNT Distribution State in the Composites Fabricated Under Different Flows

Through preliminary tests on MWNT-embedded cement composites, it is found that the flow of the fresh mixture may influence the MWNT distribution. Accordingly, MWNT-embedded cement composites fabricated with different W/C values were prepared and the MWNT distribution state of the composites was evaluated. The constituent materials are listed in Table 1. The mixing ratio of MWNT was 0.6 wt% in order to provide a sufficient amount of MWNT. 0.6 wt% of MWNT exceeded the percolation threshold as reported in the literature (Nam et al. 2015). Accordingly, the distribution of MWNT was expected to be pronouncedly visible. The volumetric fractions (vol%) of MWNT were also provided in the table. In determination procedures of volumetric fractions of MWNT, volume of each constituent material had to be calculated by using true specific gravity of the materials shown in the Sect. 2, then volume ratio of MWNT to total volume of the mixture was obtained. 20 % (by cement weight) SF was added in the mixtures under the consideration that SF can improve the MWNT distribution (Nam et al. 2012). The white cement was substituted for OPC and the incorporation of nylon fiber was omitted, as addressed in Sect. 3.1.1. The specimens were prepared and images of fractured surfaces of the specimens were obtained according to the method described in Sect. 3.1.2.

Figure 3 presents processed images that were obtained from the fabricated specimens. In the processed images, the total area of the MWNT agglomerates appeared to decrease with an increase of the flow (or W/C ratio). Moreover, MWNT clumps, which are excessively entangled-MWNT agglomerates, were generated as the flow increased. Based on these observations, it can be surmised that MWNT agglomerates disentangled as the flow (or W/C ratio) decreased.

In an effort to express the change of the MWNT distribution states in a quantitative manner, the proportion of MWNT agglomerates, indicated by white color, to the total area of the image, was calculated by the MATLAB image processing tool box. This proportion is designated as a q value. The q value refers to the planar occupation ratio of MWNT on a fractured surface of the composite.

Figure 4 shows an increase of the q value as the W/C ratio of the cement composites decreases. The q value doubled due to a 16 % reduction of the W/C ratio. The increase of the q value was attributed to disentanglement of the MWNT agglomerates. The disentanglement of the MWNT agglomerates can be explained by variation of the microstructure of the fresh mixture. Figure 5a shows the microstructure of the fresh mixture of the MWNT-embedded cement composite prepared under a high flow condition. The distance between unhydrated cements increased as the W/C ratio (or SP/C ratio, i.e. SP versus cement ratio) increased (Daimon and Roy 1979). When the interspace among the unhydrated cements increased, highly entangled MWNTs floated in the free water. This is ascribed to hydrophobic characteristics of MWNT. As a result, the images obtained from the W38–M06 and W42–M06 samples showed the highly entangled MWNTs, as presented in the Fig. 3. On the contrary, the microstructure of the fresh mixture of the composites can be changed if the flow of the mixture is decreased. The distance between unhydrated cements in the composites prepared under a low flow condition was narrowed, as shown in Fig. 5b. If the interspace is narrowed, the MWNT agglomerates can be disentangled during the mixing process. This is attributable to the characteristic that the size of the MWNT agglomerates can be restricted by the size of the interspace among the unhydrated cements. As a result, the images obtained from the W26–M06 and W30–M06 specimens in the Fig. 3 showed disentangled MWNT agglomerates.

The planar occupation ratio of MWNT, q, was increased twofold with decrease of W/C ratio from 42 to 26 % as shown in the image analysis result of Fig. 4. However, electrical conductivity of the composites is expected to increase more than hundreds of times as the W/C ratio decreases. This stems from that disconnected CNTs can be electrically connected attributable to the enhancement of MWNT distribution. Accordingly, it can be said that the twofold greater q value can lead to a dramatic increase in the electrical conductivity. This will be dealt in the electrical property characterization section.

4.1.2 DC Conductivity of the Composites Fabricated Under Different Flows

The image analysis of the MWNT distribution state of the MWNT-embedded cement composites fabricated under different flows (or W/C) studied in Sect. 4.1.1 showed that the MWNT distribution improved as the flow of the composite decreased. In the present section, the DC conductivity of the composites was examined in an effort to understand the change of the MWNT distribution in the composites and also to verify whether it supported the image analysis results. The constituent materials and respective weight content ratios of

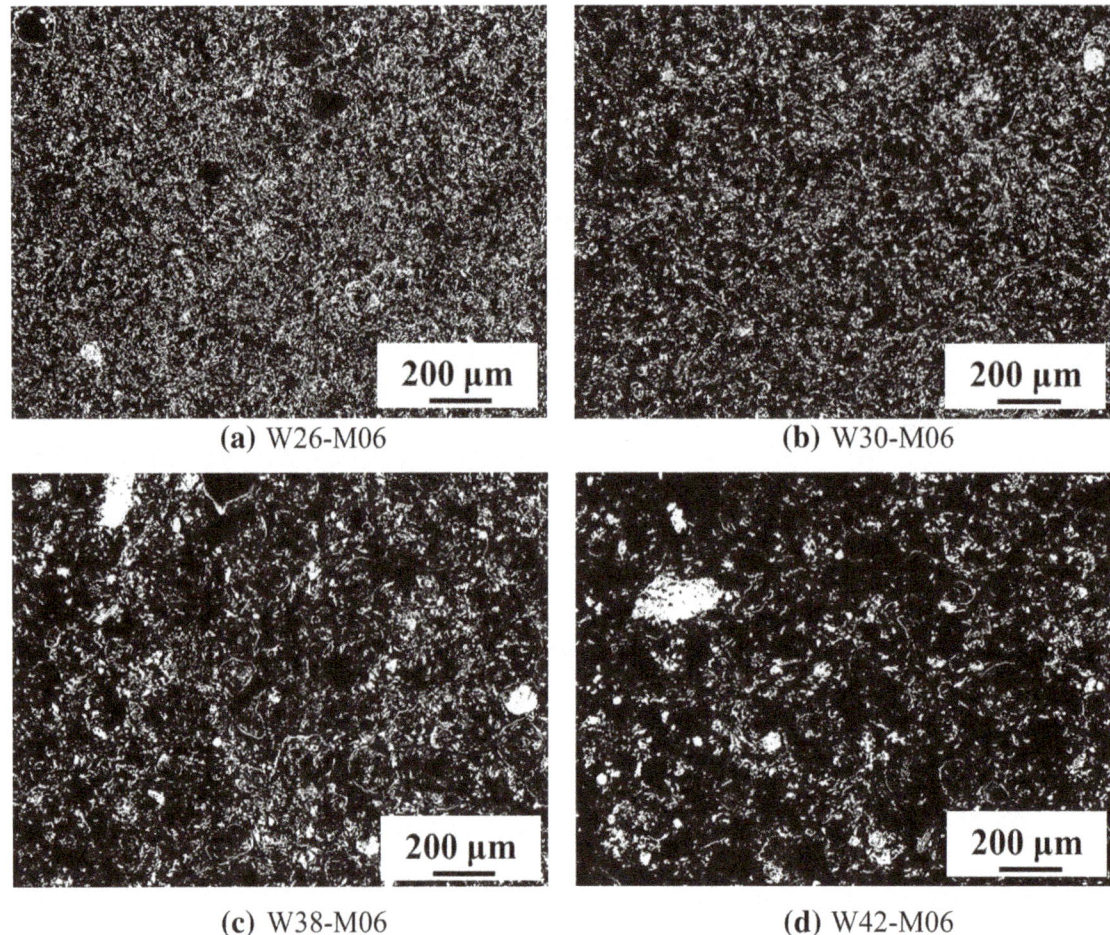

Fig. 3 The fractured surface of the MWNT-embedded cement composites fabricated under different flow (W/C ratio) after having the image thresholding, reversion, and binarization processes (image size: 570 × 426 pixel): **a** W26–M06, **b** W30–M06, **c** W38–M06, and **d** W42–M06.

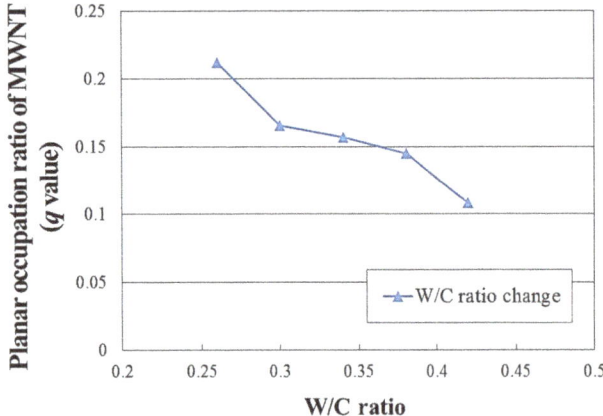

Fig. 4 The variation of the q value, which is the planar occupation of MWNT on the fractured surface of the composites, as a function of W/C ratio.

the composites are given in Table 3. To make specimens with different flows, one group of specimens was fabricated by varying the water content, as described in Sect. 4.1.1, and an additional group of specimens was fabricated by varying the SP content in the present section. The fabrication procedures of the specimens followed descriptions in

Sect. 3.2.1 and the measurement method was described in Sect. 3.2.2.

The variations in the DC conductivity of the specimens are plotted in Fig. 6a, b. The change of the DC conductivity of the specimen group prepared with different water content is shown in the Fig. 6a and the results of the specimen group prepared with different SP content are shown in Fig. 6b. An electrical percolation phenomenon can be observed in the electrical conductivity versus W/C ratio plot in the Fig. 6a. This phenomenon indicated that the water content or flow of the mixture is a crucial factor affecting the MWNT distribution of the composites, as found in the image analysis in Sect. 4.1.1. The dramatic increase of electrical conductivity with a decrease of the flow is attributed to disentanglement of the MWNT agglomerates, as shown in the image analysis. In the image analysis, the q value linearly increased with a decrease of the W/C ratio but the electrical conductivity exponentially increased with the decrease of the W/C ratio. This was due to the formation of the percolation network, which induced a remarkable increase of the DC conductivity of some orders. The Fig. 6b, which presents the electrical conductivity versus SP/C ratio plot, also shows the electrical percolation phenomenon. This indicates that the flow plays a decisive role in determining the MWNT distribution,

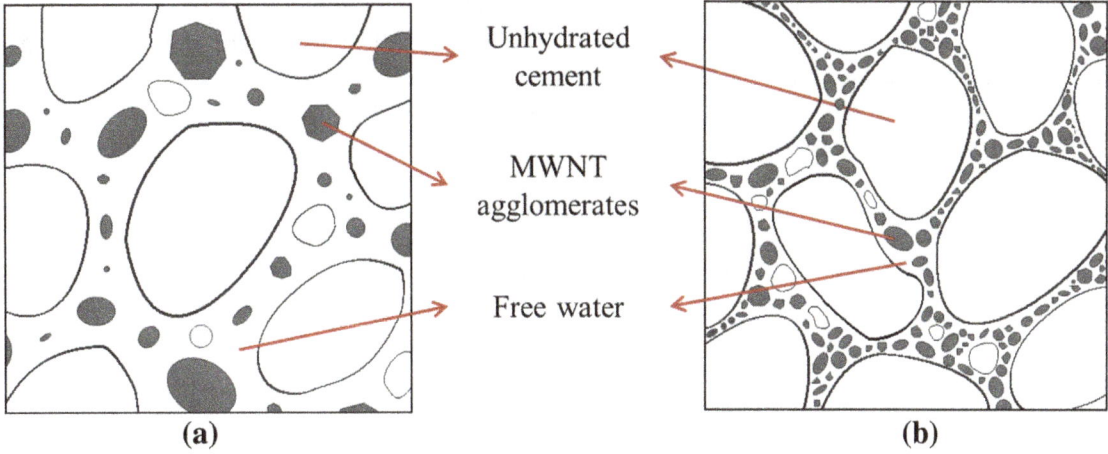

Fig. 5 Micro structure of fresh mixture of the MWNT-embedded cement composites prepared under the high flow condition (**a**) and the low flow condition (**b**).

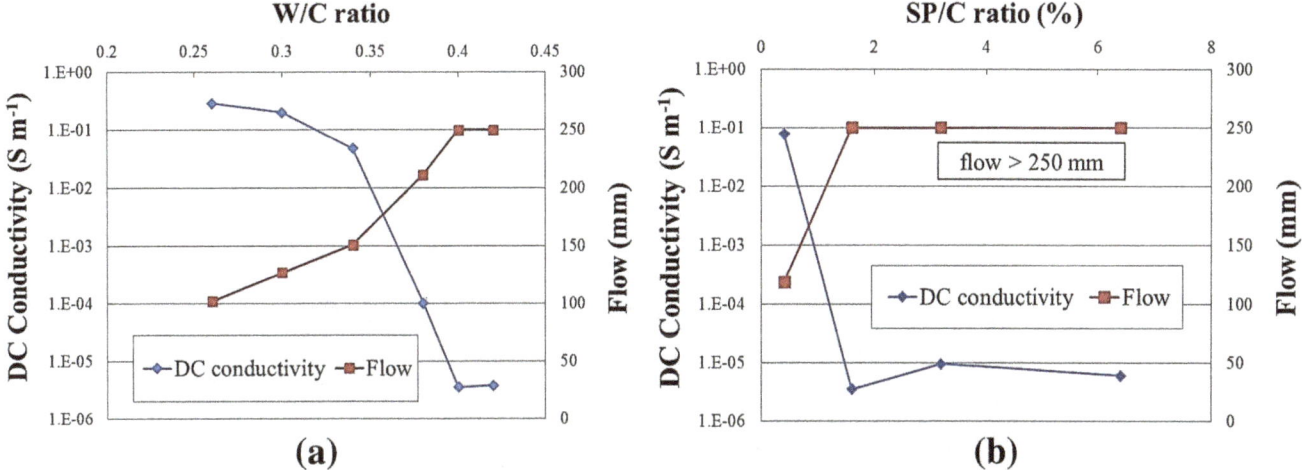

Fig. 6 The DC conductivity of the MWNT-embedded cement composites fabricated as a function of W/C ratio (**a**) and SP/C ratio (**b**).

because water content was constant in the specimen group with different SP/C ratio. It is consequently found that the flow of fresh mixture of the composites is an important factor that determines the MWNT distribution state, based on the image analysis and the electrical conductivity evaluation. The improvement of the MWNT distribution with a decrease of the flow is thought to be related to variation of the interspace among unhydrated cement, as discussed in Sect. 4.1.1.

It is worth noting that the electrical conductivity of the M06–W30 was greater than that of the M06–SP04 although the flow of the M06–W30 was higher than that of the M06–SP04. This indicates that control of water content can be a more effective way of enhancing the electrical conductivity of the composites.

Therefore, two categories of specimens were prepared in the subsequent experiments. One category included MWNT-embedded cement composites fabricated under a high flow condition (flow > 250 mm) and another category included composites fabricated under a low flow condition (114 mm < flow < 126 mm) by adjusting the W/C ratio and

the SP/C ratio. To understand the MWNT distribution states, the image analysis and DC conductivity measurement for the specimen groups were carried out.

4.2 Influence of the Flow on MWNT Distribution in the Composites Incorporating Various MWNT Contents

4.2.1 Image Analysis of the MWNT Distribution State in the Composites

It was found that the flow of fresh mixture was closely related with the MWNT distribution, as observed in Sect. 4.1.2. Accordingly, the MWNT distribution states in MWNT-embedded cement composites with various MWNT contents can be compared by setting the flow of mixtures to be consistently low (flow between 114 and 126 mm). Otherwise, the MWNT distribution is so poor that the reliability of the image analysis data obtained from the composites with various MWNT contents can diminish. To prepare composite mixtures with the low flow values, cement composites with various MWNT contents were fabricated by adjusting the content ratios of water and SP.

The water to cement ratio was controlled with reference to the MWNT content in each batch. This was based on a report that MWNT tends to absorb water within its hollow structure (Striolo et al. 2005). The SP content ratio was also adjusted in consideration of the amount of cement and MWNT, but the final SP content was determined by some trials of SP addition in the mixture and flow test. The constituent materials used in the present work are listed in Table 2 but nylon fiber was not used for the image analysis specimens. The mixing ratio of MWNT ranged from 0 to 1.5 %. Figure 7 presents processed images obtained from the fabricated composites. Images observed from LF–M0 showed some particles that were thought to be impurities of the white cement. However, their planar occupation area was so small that it was negligible in a comparison study of the q value.

It is generally agreed that an increase in the MWNT content is accompanied with an increase in the total area of MWNT if it is uniformly distributed throughout composites. Such phenomena was in close agreement with the test result provided in Fig. 7 where the total area of MWNT was observed to increase with the MWNT content. In addition, the MWNT clumps were not found even when the MWNT content was increased up to 1.5 %. Observation of the images thus indicated that MWNT was satisfactorily distributed throughout the composites in the LF–M group.

In an effort to express the change of MWNT distribution states in a quantitative manner, the q value was calculated for the LF–M group, as shown in Fig. 8. The q values of the group exhibited a steady increase as a function of the MWNT content in the composites. Accordingly, a dramatic increase in the electrical conductivity is expected as the MWNT content increases.

An image analysis of the MWNT distribution state in the composites fabricated under the high flow condition was not carried out because it was not possible to present reliable q values due to the presence of the MWNT clumps, as shown in the W42–M06 specimen of the Fig. 3.

4.2.2 DC Conductivity of the Composites Incorporating Various MWNT Contents

The change of the DC conductivity of composites fabricated under the two different flow condition, the low and high flow conditions, with various MWNT contents was investigated here. The constituent materials and respective weight content ratios of the composites fabricated under the low flow condition are given in the Table 2. The DC conductivity of the composites in the LF–M group is shown as a function of the MWNT content in Fig. 9. DC conductivity of 1.7 S m^{-1} was attained for specimen LF–M1.5. This value was greater than the electrical conductivity of 5 mm long

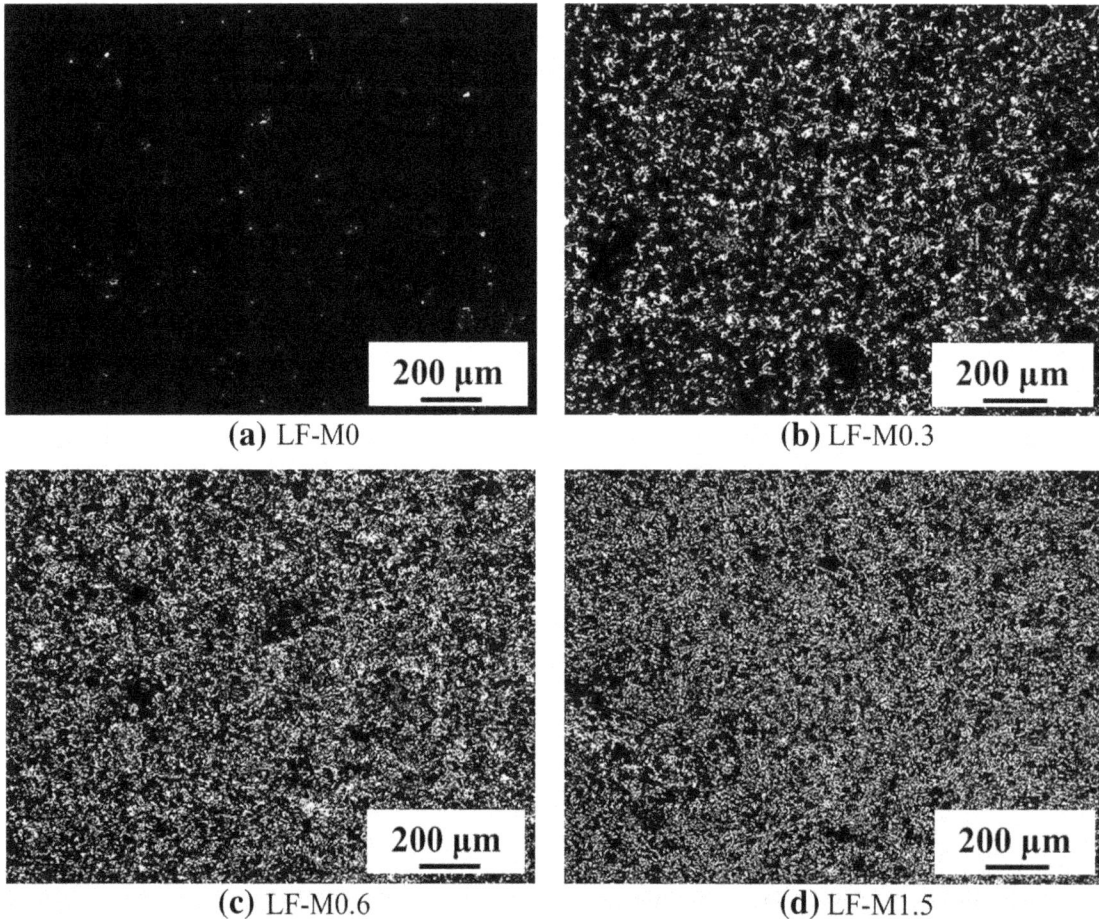

(a) LF-M0

(b) LF-M0.3

(c) LF-M0.6

(d) LF-M1.5

Fig. 7 The fractured surface of the cement composites of the LF–M group fabricated under the low flow condition after having image thresholding, reversion, and binarization processes (Image size: 570 × 426 pixel): a LF–M0, b LF–M0.3, c LF–M0.6, and d LF–M1.5.

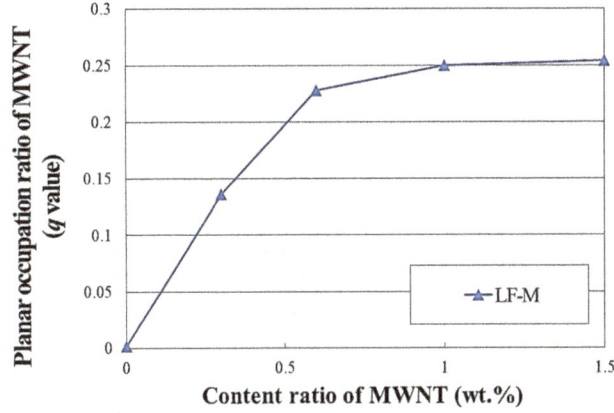

Fig. 8 A comparison of the *q* value, which is the proportion of MWNT agglomerates to total area of the image, of the specimens in the LF–M group.

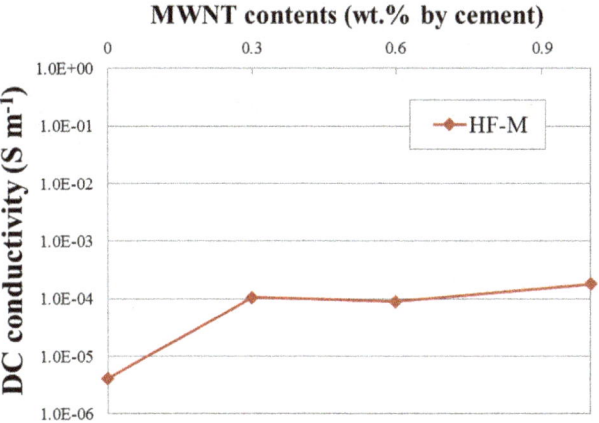

Fig. 10 The variations of the DC conductivity of the specimens fabricated under the high flow condition (flow > 250 mm) are plotted as a function of the MWNT content.

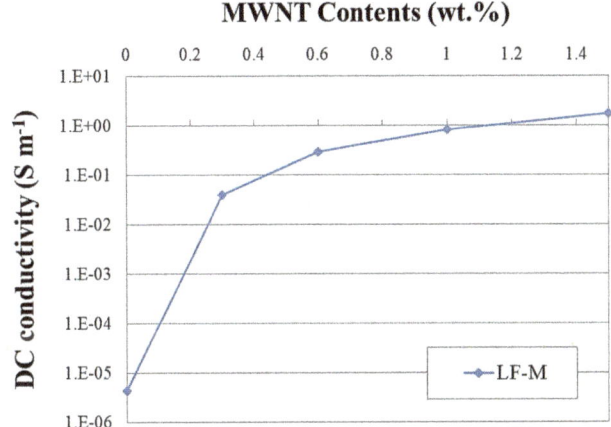

Fig. 9 The experimental conductivity values of the specimens fabricated under the low flow condition (114 mm < flow < 126 mm).

carbon fiber-incorporated cement composites and carbon black-filled cement composites (Wen and Chung 2007; Li et al. 2006). The high DC conductivity of the composites in the present section and the image analysis conducted in Sect. 4.2.1 indicated that MWNT was well distributed in the cement matrix.

The electrical percolation phenomenon of the LF–M group was found in the Fig. 9. The percolation threshold, which refers to the critical volume fraction of MWNT inducing remarkable change in the conductive phase, existed in a MWNT content range of 0–0.3 wt% for the LF–M group. Accordingly, the percolation threshold corresponded to 0.196 vol% if it is determined as the mean value of the percolation threshold range.

The percolation phenomena manifested in the MWNT-embedded cement composites fabricated under the low flow condition indicated that MWNT was well distributed throughout the composites and the MWNT distribution was consistent in the specimen group. The results supported the image analysis results obtained in Sect. 4.2.1. Based on the image analysis results and the electrical percolation

phenomena, an acceptable MWNT distribution was attained by reducing the flow of the composites. Consequently, it can be concluded that maintaining the low flow of fresh mixtures of the composites is crucial to improve the MWNT distribution.

The change of the DC conductivity of the composites fabricated under the high flow condition with various MWNT contents was also investigated. The constituent materials and respective weight content ratios of the composites are given in Table 4 (Nam et al. 2012). MWNT was incorporated in each composite type at 0, 0.3, 0.6, and 1.0 wt% by weight of cement. To fabricate specimens with high flow, the W/C ratio was fixed to 0.4 in the composites and the SP/C ratio was 0.016.

Figure 10 presents the change of the DC conductivity of the composites as a function of the MWNT content. The electrical percolation phenomenon was not found in Fig. 10. The DC conductivity of all the composites did not exceed 0.001 S m^{-1}, hence all the composites appeared to be under the percolation threshold. The under percolation behavior of the DC conductivity was attributed to a poor distribution of MWNT in the composites. In addition, the DC conductivity of the composites may be dependent on many factors such as the dry condition of the specimens, void ratio, contact resistance between electrodes and the composite specimens, etc. In conclusion, it is not recommended that MWNT-embedded cement composites are fabricated under a high flow condition unless appropriate dispersion methods were carried out.

5. Conclusion

In the present work, the MWNT distribution in MWNT-embedded cement composites was evaluated by a newly proposed image analysis method as well as electrical conductivity measurement. In addition, the influence of the fresh mixture's flow on the MWNT distribution in the composites was experimentally studied on the basis of the two evaluation methods. The conclusions derived from the present work can be summarized as follow.

(1) The MWNT distribution, which was evaluated by the planar occupation ratio of MWNT, q, was enhanced as the flow of fresh composite mixtures decreased.

(2) The DC conductivity of MWNT-embedded cement composites fabricated with different flows was examined and it was observed that the conductivity increased as the flow was decreased.

(3) The image analysis of the MWNT-embedded cement composites fabricated under the low flow condition (114 mm < flow < 126 mm) revealed that the q value linearly increased as the MWNT content increased. The linear relationship between the q value and the MWNT content demonstrated that fabrication of the composites under the low flow condition is an effective way to enhance the MWNT distribution.

(4) The DC conductivity of the MWNT-embedded cement composites fabricated under the low flow condition showed percolation phenomena, thus indicating that MWNT was well distributed in the composites, as evaluated in the image analysis.

The proposed image analysis procedures for evaluation of the MWNT distribution can be used as an evaluation method to quantify the distribution of carbon nano-materials in the cement matrix. The conductive MWNT-embedded cement composites with percolative networks produced by controlling the flow are expected to be utilized as piezoresistive sensors, EMI shielding materials, electrostatic discharge materials, heating elements, etc. Future work will be focused on in-depth study of the origin of the flow effect on the MWNT distribution.

Acknowledgments

This research was supported by the KUSTAR-KAIST Institute, KAIST, Korea and also sponsored by a research project 'Development of fabrication method for floor heating material by use of CNT-cement composites with self-heating capacity of 5W-50 °C' (Project code: 15CTAP-C098086-01) funded by Ministry of Land, Infrastructure and Transport (MOLIT) of Korea Government and Korea Agency for Infrastructure Technology Advancement (KAIA). Authors also deeply thank Prof. Hyeong-Ki Kim in Chosun University and Prof. Sung-Min Choi in KAIST for their important comments on this research and acknowledge June Lee in National NanoFab Center for cooperation in the image analysis.

References

ASTM International. (2013). *ASTM C1437—standard test method for flow of hydraulic cement mortar*. West Conshohocken, PA: ASTM.

Daimon, M., & Roy, D. M. (1979). Rheological properties of cement mixes: II. Zeta potential and preliminary viscosity studies. *Cement and Concrete Research, 9*(1), 103–109.

Fan, M. Z., Bonfield, P. W., Dinwoodie, J. M., & Breese, M. C. (2000). Dimensional instability of cement bonded particleboard: SEM and image analysis. *Journal of Materials Science, 35*(24), 6213–6220.

Kang, S. T., Lee, B. Y., Kim, J. K., & Kim, Y. Y. (2011). The effect of fibre distribution characteristics on the flexural strength of steel fibre-reinforced ultra high strength concrete. *Construction and Building Materials, 25*(5), 2450–2457.

Kim, H. K., Nam, I. W., & Lee, H. K. (2014). Enhanced effect of carbon nanotube on mechanical and electrical properties of cement composites by incorporation of silica fume. *Composite Structures, 107*, 60–69.

Konsta-Gdoutos, M. S., Metaxa, Z. S., & Shah, S. P. (2010). Highly dispersed carbon nanotube reinforced cement based materials. *Cement and Concrete Research, 40*(7), 1052–1059.

Lee, B. Y., Kim, J. K., Kim, J. S., & Kim, Y. Y. (2009). Quantitative evaluation technique of polyvinyl alcohol (PVA) fiber dispersion in engineered cementitious composites. *Cement and Concrete Composites, 31*(6), 408–417.

Lee, Y. H., Lee, S. W., Youn, J. R., Chung, K., & Kang, T. J. (2002). Characterization of fiber orientation in short fiber reinforced composites with an image processing technique. *Materials Research Innovations, 6*, 65–72.

Li, J., Ma, P. C., Chow, W. S., To, C. K., Tang, B. Z., & Kim, J. K. (2007). Correlations between percolation threshold, dispersion state, and aspect ratio of carbon nanotubes. *Advanced Functional Materials, 17*(16), 3207–3215.

Li, H., Xiao, H., & Ou, J. (2006). Effect of compressive strain on electrical resistivity of carbon black-filled cement-based composites. *Cement & Concrete Composites, 28*(9), 824–828.

Liu, J., Li, C., Liu, J., Du, Z., & Cui, G. (2011). Characterization of fiber distribution in steel fiber reinforced cementitious composites with low water-binder ratio. *Indian Journal of Engineering and Materials Sciences, 18*, 449–457.

Nam, I. W., Kim, H. K., & Lee, H. K. (2012). Influence of silica fume additions on electromagnetic interference shielding effectiveness of multi-walled carbon nanotube/cement composites. *Construction and Building Materials, 30*, 480–487.

Nam, I. W., Souri, H., Lee, H. K. (2015). Percolation threshold and piezoresistive response of multi-wall carbon nanotube/cement composites. *Smart Structures and Systems* (in review).

Otsu, N. (1979). A threshold selection method from gray-level histograms. *IEEE Transactions on Systems, Man, and Cybernetics, 9*(1), 62–66.

Ozyurt, N., Woo, L. Y., Mason, T. O., & Shah, S. P. (2006). Monitoring fiber dispersion in fiber-reinforced cementitious materials: Comparison of AC-Impedance Spectroscopy and Image Analysis. *ACI Materials Journal, 103*(5), 340–347.

Redon, C., Chermant, L., Chermant, J. L., & Coster, M. (1999). Automatic image analysis and morphology of fibre reinforced concrete. *Cement & Concrete Composites, 21*(5–6), 403–412.

SEMI MF 43 (2005) Standard test methods for resistivity of semiconductor materials.

Sorensen, C., Berge, E., & Nikolaisen, E. B. (2014). Investigation of fiber distribution in concrete batches discharged from ready-mix truck. *International Journal of Concrete Structures and Materials, 8*(4), 279–287.

Stauffer, D., & Aharony, A. (1994). *Introduction to percolation theory Revised* (2nd ed.). London, UK: Taylor & Francis.

Striolo, A., Chialvo, A. A., Gubbins, K. E., & Cummings, P. T. (2005). Water in carbon nanotubes: Adsorption isotherms and thermodynamic properties from molecular simulation. *The Journal of Chemical Physics, 122*(23), 234712.

Vance, K., Aguayo, M., Dakhane, A., Ravikumar, D., Jain, J., & Neithalath, N. (2014). Microstructural, mechanical, and durability related similarities in concretes based on OPC and alkali-activated slag binders. *International Journal of Concrete Structures and Materials, 8*(4), 289–299.

Wen, S., & Chung, D. D. L. (2007). Double percolation in the electrical conduction in carbon fiber reinforced cement-based materials. *Carbon, 45*(2), 263–267.

Xie, P., Gu, P., & Beaudoin, J. J. (1996). Electrical percolation phenomena in cement composites containing conductive fibres. *Journal of Materials Science, 31*(15), 4093–4097.

13

Monotonic Loading Tests of RC Beam-Column Subassemblage Strengthened to Prevent Progressive Collapse

Jinkoo Kim[1],*, and Hyunhoon Choi[2]

Abstract: In this study the progressive collapse resisting capacity of a RC beam-column subassemblage with and without strengthening was investigated. Total of five specimens were tested; two unreinforced specimens, the one designed as gravity load-resisting system and the other as seismic load-resisting system, and three specimens reinforced with: (i) bonded strand, (ii) unbonded strand, and (iii) side steel plates with stud bolts. The two-span subassemblages were designed as part of an eight-story RC building. Monotonically increasing load was applied at the middle column of the specimens and the force–displacement relationships were plotted. It was observed that the gravity load-resisting specimen failed by fractures of re-bars in the beams. In the other specimens no failure was observed until the maximum displacement capacity of the actuator was reached. Highest strength was observed in the structure with unbonded strand. The test result of the specimen with side steel plates in beam-column joints showed that the force–displacement curve increased without fracture of re-bars. Based on the test results it was concluded that the progressive collapse resisting capacity of a RC frame could be significantly enhanced using unbonded strands or side plates with stud bolts.

Keywords: progressive collapse, beam-column sub-assemblage, catenary action, side plates.

1. Introduction

After collapse of the World Trade Center twin towers in New York, protection of structures against progressive collapse has been an important issue in the field of structural engineering. Researches have been conducted on the collapse behavior of moment-resisting frames caused by sudden loss of columns (Khandelwal and El-Tawil 2005; Tsai and Lin 2008; Kim and An 2009). Milner et al. (2007) and Sasani and Kropelnicki (2008) carried out experiments to investigated the behavior of a scaled model of a continuous perimeter beam in a reinforced concrete frame structure following the removal of a supporting column. Yi et al. (2008) carried out static experimental study of a three-story RC frame structure to investigate progressive failure due to the loss of a lower story column. In those experiments it was observed that after the plastic mechanism formed, the concrete strain in the compression zone at the beam ends reached its ultimate compressive strain, and the compressive re-bars were gradually subject to tension with increasing displacement. Choi and Kim (2011) investigated the

progressive collapse resisting capacity of RC beam-column subassemblages designed with and without seismic load. Qian and Li (2012) carried out experimental study of six RC beam-column substructures with different design detail, span length and span aspect ratio to investigate the dynamic load redistribution performance. Yu and Tan (2013) carried out an experimental program for investigating progressive collapse resistance of reinforced concrete (RC) beam-column sub-assemblages under a middle column removal scenario. Two one-half scaled sub-assemblages were designed with seismic and non-seismic detailing to check the effect of detailing on structural behavior. Song and Sezen (2013) performed a field experiment and numerical simulations to investigate the progressive collapse potential of an existing steel frame building. Four first-story columns were physically removed from the building to understand the subsequent load redistribution within the building. In Qian and Li (2013) experimental study of seven one-third scale RC beam-column substructures were tested to investigate the effect of beam transverse reinforcement ratios, type of design detailing, and beam span aspect ratios. Recently Qian et al. (2015) tested 6 one-quarter scaled specimens to investigate the progressive collapse resisting capacity of RC frames including secondary mechanisms such as membrane actions developed in slabs. Kang et al. (2015) carried out experimental investigation on the behavior of precast concrete beam-column sub-assemblages with engineered cementitious composites (ECC) in structural topping and beam-column joints under middle column removal scenarios to investigate the effectiveness of ECC on mitigating progressive collapse.

[1]Department of Civil and Architectural Engineering, Sungkyunkwan University, Suwon, Korea.
*Corresponding Author; E-mail: jkim12@skku.edu
[2]Research Institute of Technology, Samsung Engineering and Construction Co., Ltd., Seoul, Korea.

In this study monotonic loading tests of five RC beam-column subassemblages, two unstrengthened and three strengthened specimens, were carried out to investigate their progressive collapse resisting capacity and to observe the effect of the reinforcement methods. The two-span sub-assemblage specimens were designed as part of an eight-story RC moment resisting framed building. Three specimens were reinforced with: (i) bonded strands, (ii) unbonded strands, and (iii) side steel plates, and their relative effectiveness in enhancing progressive collapse resisting capacity was addressed. Two unreinforced specimens, the one designed as gravity load-resisting system and the other as seismic load-resisting system, were also tested for comparison.

2. Design of Prototype Structures

As prototype structures of the test specimens two eight-story reinforced concrete structures were designed. The story height of the model structure is 3.5 m, and the design dead and live loads are 5.9 and 2.45 kN/m^2, respectively. The design spectral response acceleration parameters for seismic load, S_{DS} and S_{D1}, are 0.44 and 0.23, respectively in the IBC 2009 (ICC 2009) format. The design compressive strength of concrete (f_c') is 27 MPa and the yield strength of re-bars (f_y) is 392 MPa. Figure 1a shows the structural plan of the prototype structure with core shear walls. In this model all the lateral force is resisted by the shear walls and the ordinary moment frames were designed to resist only gravity loads. Figure 1b shows the structural plan of the other model structure composed of intermediate moment-resisting frames which were designed to resist both gravity and lateral loads.

3. Design of Specimens and Test Setup

To evaluate the progressive collapse resisting capacity of the model structures when subjected to sudden removal of a column, a part of the exterior frame enclosed in the dotted curve in Fig. 1 were manufactured for tests. Total of five specimens, scaled to 37 % of the prototype structure, were constructed for the loading tests: two unstrengthened and three strengthened specimens. The subassembly test specimens are composed of three columns (C2) and two beams (G1) located between the columns. Figure 2 shows the re-bar placements of the specimens for the gravity load-resisting system and the seismic load-resisting system. To take into account the continuation effect of beams in the proto-type structure, the two columns at both sides of a specimen were made to be 1.5 times larger than the center column in size. The longitudinal bars were anchored to the columns with the tail extension of the hook. The re-bar detailing was based on the ACI Detailing Manual (2004). In the case of non-seismically designed specimens, bottom re-bars were extended into the support (the exterior column) without hook, whereas the top and bottom bars of the seismically designed specimen were anchored with standard 90 degree

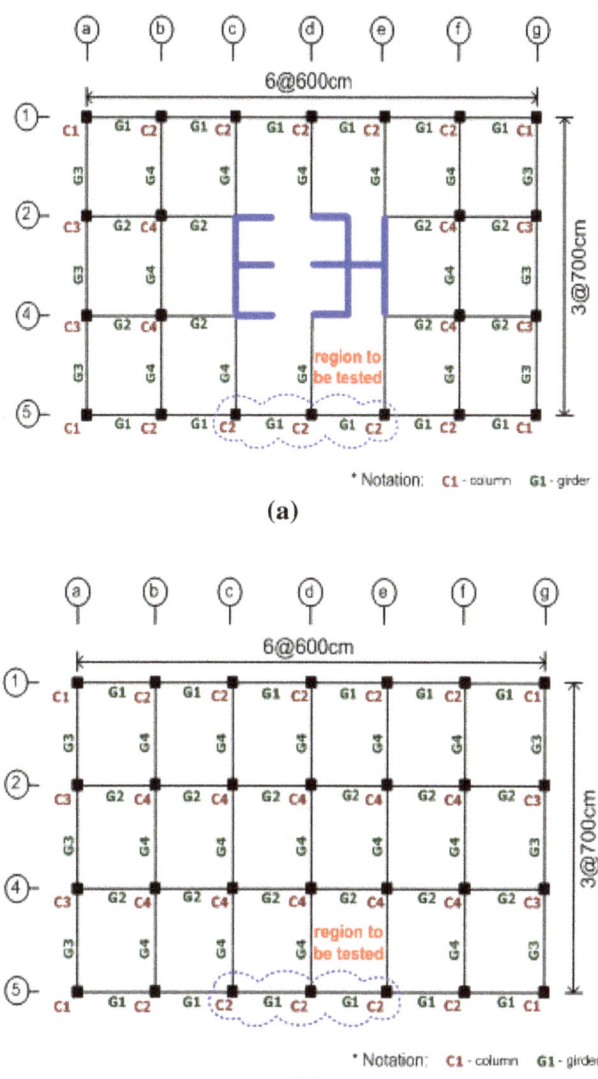

Fig. 1 Structural plans of prototype structures. **a** RC moment resisting frames with shear walls. **b** RC moment resisting frames.

hook into exterior columns. The D10 re-bars with nominal diameter of 9.53 mm were used for main reinforcing steel for beams and columns, and φ 6 steel bars were used for stirrups and tie bars in the specimens. From coupon tests it was observed that the yield strengths of the main re-bars and stirrups/tie bars are 457 and 325 MPa, respectively. The size of the specimens and the number of re-bars are summarized in Table 1.

To increase the progressive collapse resisting capacity, the specimen out of the gravity load-resisting system was strengthened by either high strength strand or steel side plates welded with stud bolts. Figure 3 shows the test specimen for gravity load-resisting system strengthened with a wire strand with diameter of 12.7 mm. The strand was placed along the center of the cross-section of the beams before casting of concrete without prestress, and was anchored at the exterior surfaces of the two end columns using an anchorage as shown in Fig. 4. To compare the effects of bonded and unbonded strands, one specimen was prepared with a bonded strand and another specimen was prepared with an unbonded strand located within a sheath

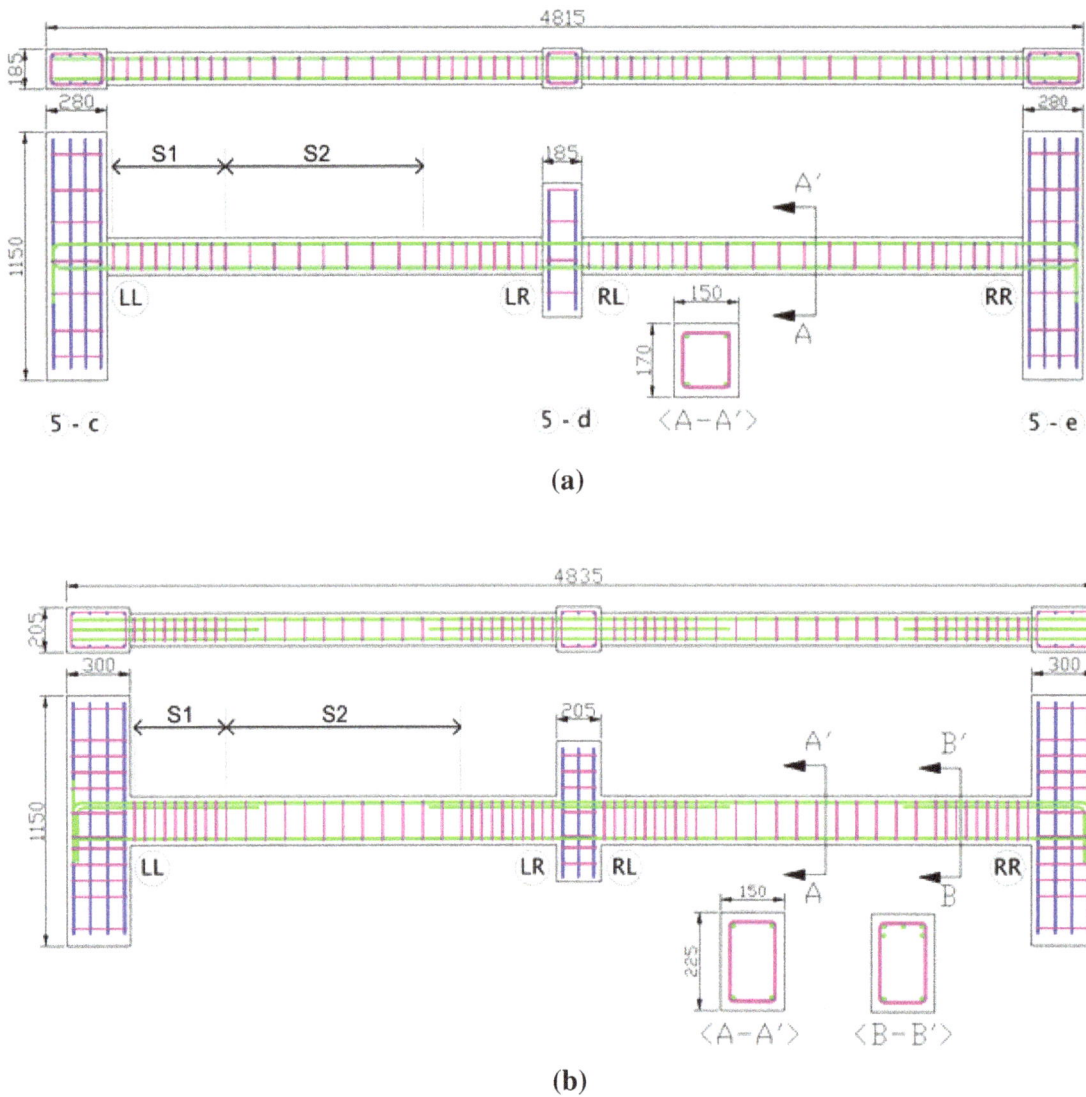

Fig. 2 Re-bar placement in the test specimens. **a** Gravity-load resisting system. **b** Lateral-load resisting system.

pipe. For existing structures the high strength tendons can be installed at both sides of the beams similarly to the external prestressing used to retrofit old structures. External pre-stressing techniques for strengthening beams can be found elsewhere (Harajli 1993; Shin et al. 2013). The other strengthening scheme is to attach steel side plates at both sides of beam ends as shown in Fig. 5. To increase the flexural strength of a beam for gravity load, steel plates are generally attached in the middle of the beam where bending moment is maximized (Ren et al. 2015). In this study, for resisting progressive collapse caused by loss of a column, they were placed at the ends of the beams in the form of side plates. Thirteen high strength bolts with diameter of 8 mm were welded in two rows to the 5 mm-thick side plates as shear connectors to ensure composite action of the plates and the specimen as depicted in Fig. 6. The side plates were installed in position before concrete was poured. However they also can easily be attached to sides of existing beams using chemical anchor bolts.

Figure 7 shows the test setup for the specimens. The right- and left-side-columns were fixed to the jigs and the actuator was connected to the middle column. It was assumed that the column at the location of ⑤-ⓓ of the prototype structure shown in Fig. 1a was suddenly removed by accident, and displacement-controlled monotonic pushdown force was enforced at the middle column of the specimens using a hydraulic actuator with maximum capacity of 2000 kN and maximum stroke of ±250 mm. The tests were carried out horizontally and to prevent vertical deflection of the speci-mens due to self weight, rollers were placed beneath the beam-column joint during the. Strain gages were attached on the longitudinal re-bars and strands located at the ends of girders.

4. Test Results

4.1 Unstrengthened Specimens

Displacement controlled monotonic loading tests were conducted on the five specimens by gradually increasing the displacement at the center column until the displacement capacity of the hydraulic actuator was reached. The force–displacement relationship of the gravity load-resisting specimen not strengthened by strands or side plates is

Table 1 Dimensions and rebar placements of test specimens (unit: mm).

Members		Column (C2)	Girder (G1) (depth × width)		
(a) Gravity load resisting system					
Prototype structure	Size	500 × 500	450 × 400		
	Rebar	4D25	Location	End	Center
			Upper	2D25	2D25
			Lower	2D25	2D25
Test specimen	Size	185 × 185	170 × 150		
	Rebar	4D10	Location	End	Center
			Upper	2D10	2D10
			Lower	2D10	2D10
			Stirrup	S1	φ6@65
				S2	φ6@117
(b) Lateral load resisting system					
Prototype structure	Size	550 × 550	600 × 400		
	Rebar	6D25	Location	End	Center
			Upper	3D25	2D25
				2D25	
			Lower	2D25	2D25
Test specimen	Size	205 × 205	225 × 150		
	Rebar	6D10	Location	End	Center
			Upper	3D10	2D10
				2D10	
			Lower	2D10	2D10
			Stirrup	S1	φ6@48
				S2	φ6@93

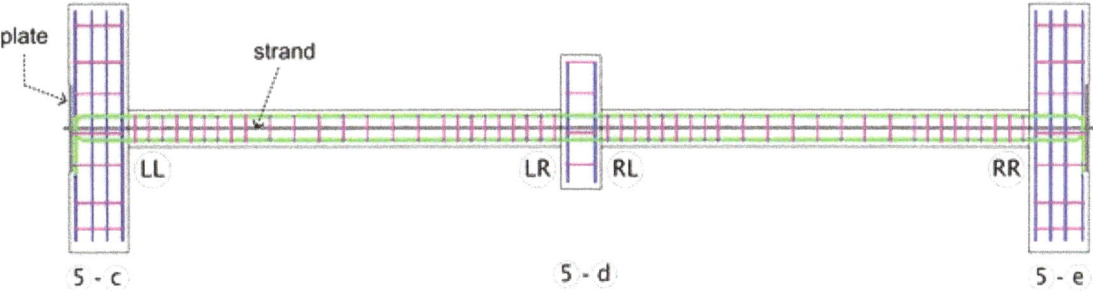

Fig. 3 Specimen strengthened with high-strength strands.

presented in Fig. 8. The arrow mark ╱ represents the fracture of re-bars. The specimen started to yield at the displacement around 40 mm, and full plastic hinge formed at the displacement of about 90 mm. When the displacement reached about 180 mm the strength increased again due to the activation of catenary force of the re-bars. The curve for the specimen dropped slightly at displacement of 293 mm when one of the tension re-bars fractured. The strength

further increased and more re-bar fractured as more displacement was imposed on the specimen, and at displacement of 426 mm one of the interior ends of the beam was completely separated from the column and the specimen failed. The failure criterion for beams recommended in the GSA guidelines is 0.015 radian, which corresponds to vertical displacement of 214 mm in the specimen with clear span of 2035 mm. Therefore the displacement at failure of

Fig. 4 Details of anchorage for strand.

the specimen far exceeded the limit state specified in the guidelines. It also can be noticed that the GSA-specified failure point corresponds to the displacement where the catenary force started to be activated. Figure 9 shows the photograph of the specimen at failure. Major cracks were observed only at both ends of the beams, and relatively few cracks formed in the external columns.

The force–displacement relationship of the lateral load-resisting specimen is presented in Fig. 10. The specimen started to yield at the displacement around 100 mm, which is significantly larger than the first yield point of the gravity load-resisting system. The yield strength of the specimen is also more than twice as high as that of the gravity load-resisting system. After the strength reached 41 kN at the displacement of 144 mm, the flexural strength of the specimen started to decrease as a result of yield of re-bars and

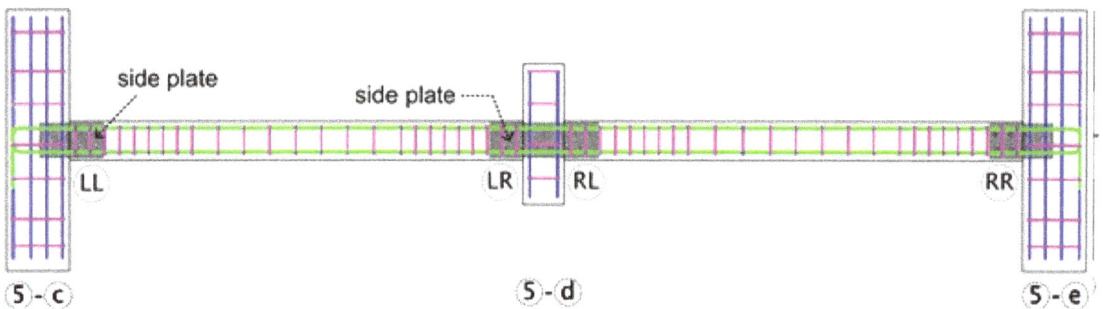

Fig. 5 Specimen strengthened with steel side plates.

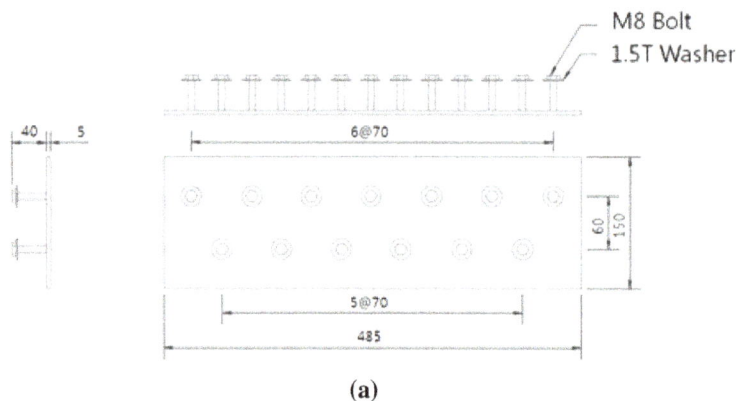

(a)

(b)

Fig. 6 Configuration and detailing of a steel side plate. **a** Dimensions of a side plate. **b** Side plate with stud bolts.

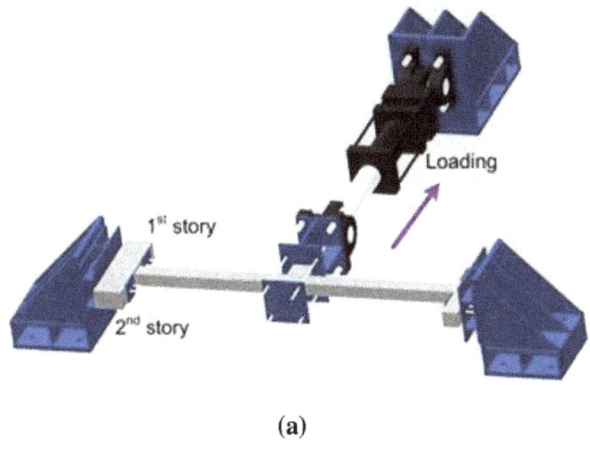

(a)

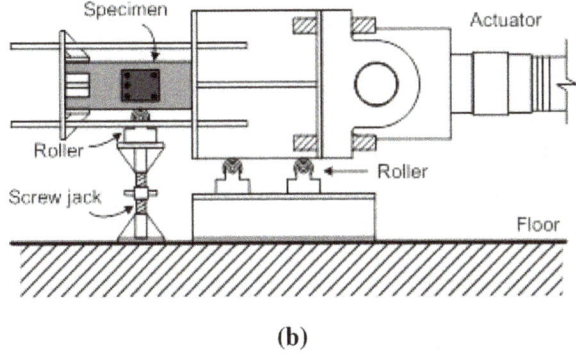

(b)

Fig. 7 Test setup for a subassemblage specimen. **a** Overall view. **b** Roller support at the beam-column joint.

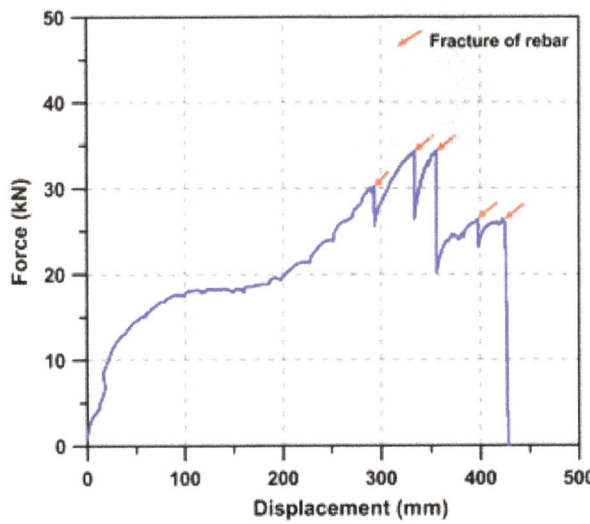

Fig. 8 Load-displacement relationship of the specimen for a gravity-load resisting system.

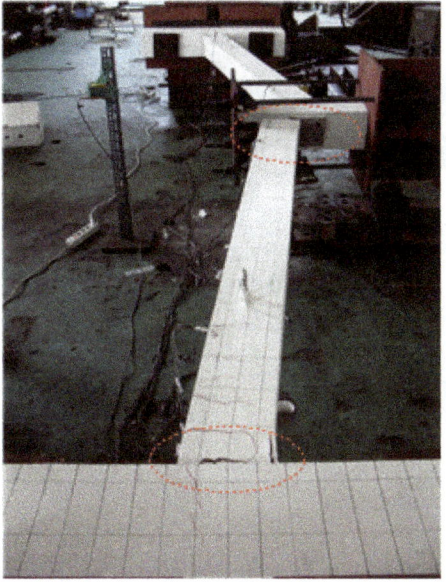

Fig. 9 Failure mode of the gravity-load resisting specimen.

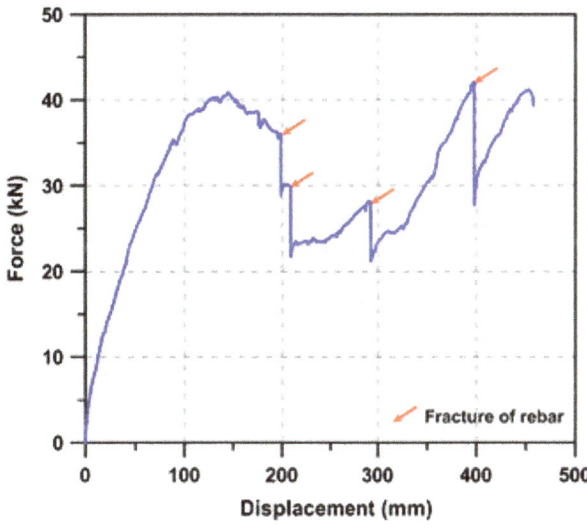

Fig. 10 Load-displacement relationship of the specimen for a lateral-load resisting system.

formation of microcracks. The strength dropped to 54 % of the peak value due to fracture of tension re-bars at the displacements of 198 and 209 mm. The strength increased again up to 27 kN before another re-bar fractured and the strength suddenly dropped. The strength increased again to 44 kN and dropped again at the displacement of about 400 mm due to fracture of another re-bars. Then the strength re-increased due to catenary force of remaining re-bars. The maximum displacement capacity of the actuator was reached

and the test stopped right after the strength re-increased to another peak point. Figure 11 shows the overall view of the damaged specimen after the test was over. It was observed that the plastic hinges formed away from the column faces where the closer stirrup spacing was no longer required. Due to the enhanced shear reinforcement at the ends of the beams and the seismic detailing of re-bars at the beam-column joints including anchoring of bottom re-bars using standard hooks, the effective length of the beam was reduced and consequently the maximum bending strength was increased. The number of fractured re-bars was also reduced compared with the case of gravity load-resisting system. The number of cracks formed in the exterior column was slightly larger than that of the specimen designed only for gravity load. Figure 12 shows the damaged region of the specimen, where it was noticed that the cross-sectional areas of the fractured re-bars were slightly reduced representing typical tension failure.

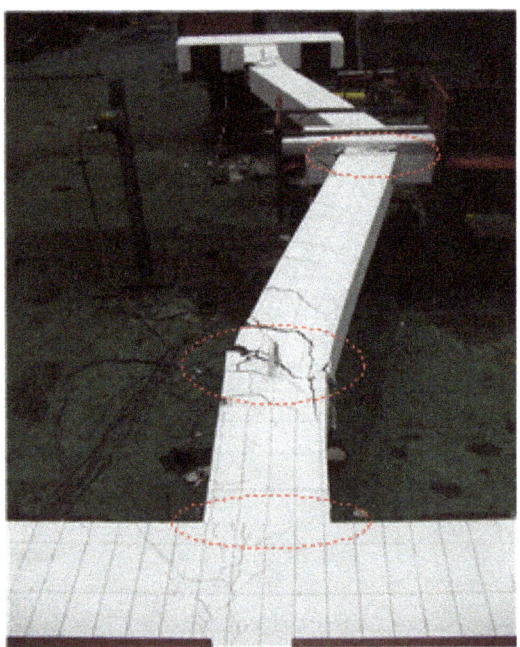

Fig. 11 Failure mode of the lateral-load resisting specimen.

The variation of rebar strain in the gravity-load resisting specimen is shown in Fig. 13. It can be observed in Fig. 13a that the strain of the top rebars of the left end of the right-hand-side beam (location RL in Fig. 2a), which was initially subjected to compression, started to resist tension after the displacement exceeded 40 mm. The tension increased rapidly at the displacement of 330 mm due to the initiation of catenary action. Similar phenomenon was observed in the re-bars located in the right end of the beam (location RR in Fig. 2a) as shown in Fig. 13b. It can be observed that the strain of the bottom bar rapidly increased starting from the displacement of 330 mm due to catenary action. Figure 14 shows the rebar strain in the lateral-load resisting specimen, where it can be noticed that the activation of catenary action is apparent compared with the case of the gravity load resisting specimen. It can be observed that the bottom rebar in the right-end of the right-hand-side beam (location RR in Fig. 2b), which was initially under compression, started to be subjected to tension at the vertical displacement of 400 mm. Therefore if the test had not been terminated early due to the limitation of the displacement capacity of the

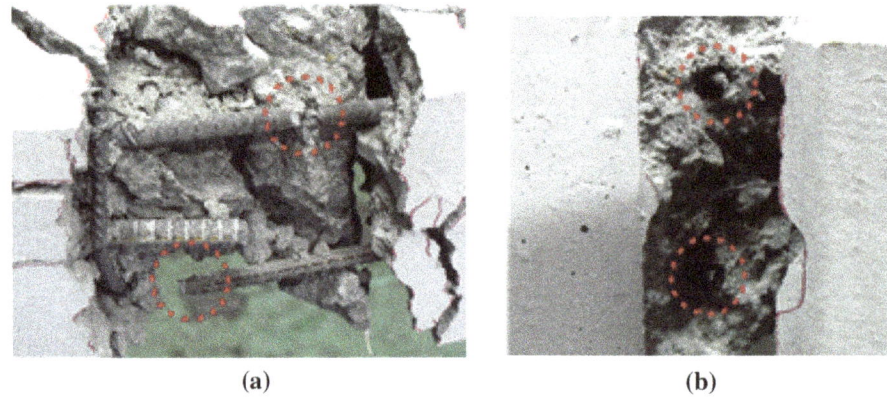

(a)   (b)

Fig. 12 Fracture of beam main rebars in the lateal-load resisting system. **a** Center. **b** End.

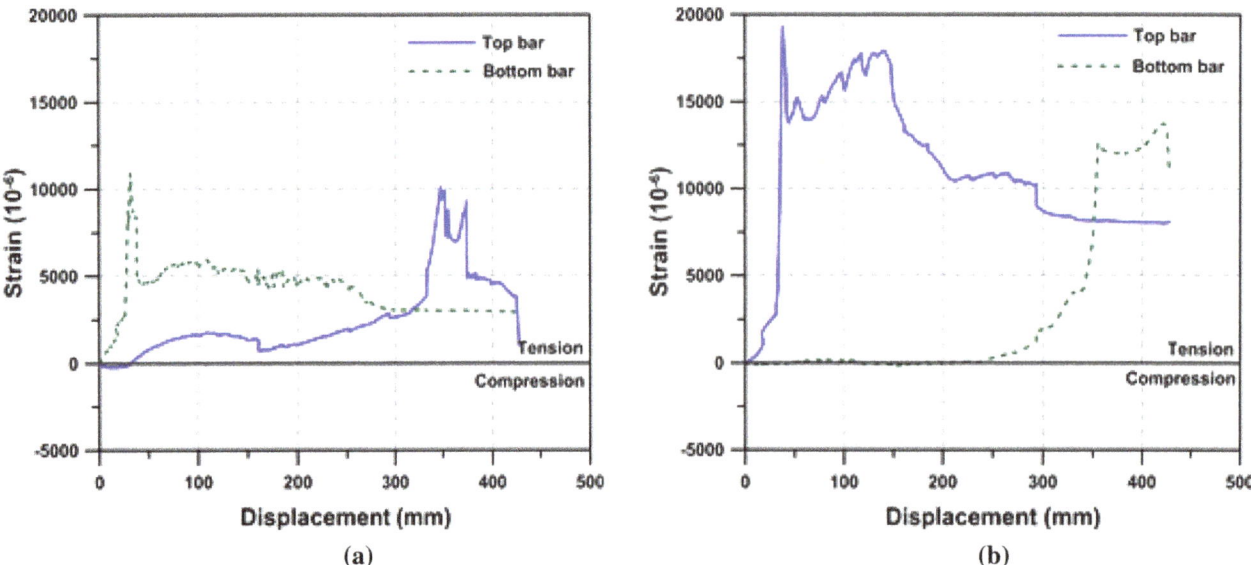

(a)   (b)

Fig. 13 Strain of rebars in the right-hand side beam of the gravity load resisting specimen. **a** Left end (RL). **b** Right end (RR).

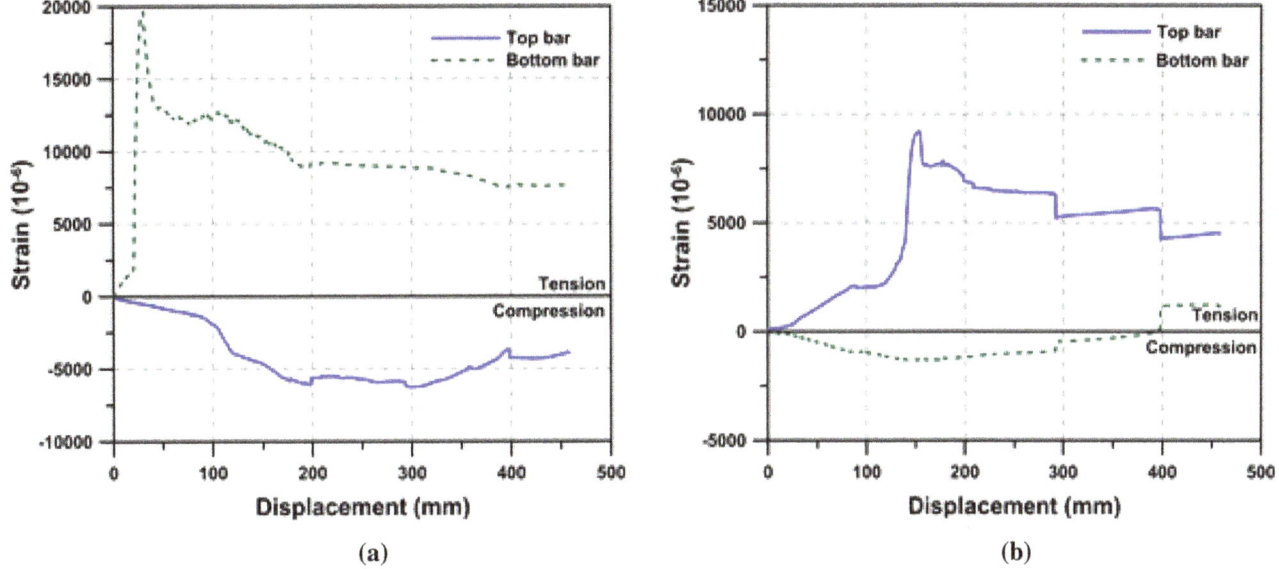

Fig. 14 Strain of rebars in the right-hand side beam of the lateral load resisting specimen. **a** Left end (RL). **b** Right end (RR).

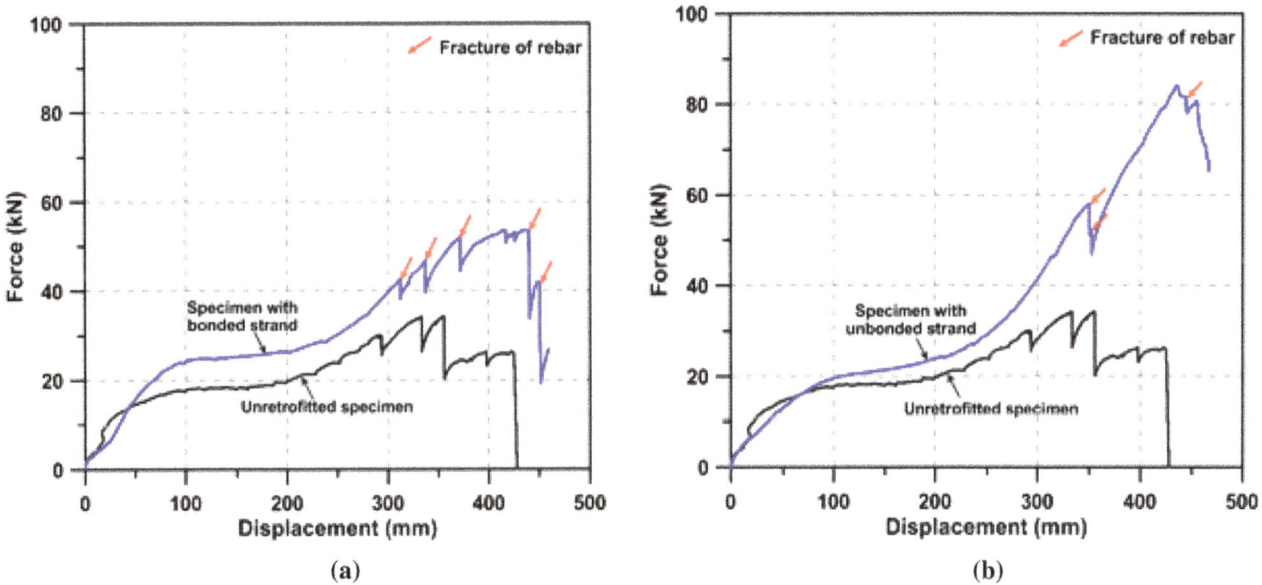

Fig. 15 Load-displacement relationship of the specimen reinforced with high-strength strand. **a** Bonded strand. **b** Unbonded strand.

actuator, more catenary force might have been observed in this specimen. The activation of catenary action in beams with different boundary conditions can be found in Kim and An (2009).

4.2 Strengthened Specimens

The specimen of gravity load system, which showed inferior performance to the specimen designed for seismic load, was reinforced by either high strength strand or side plates to enhance its progressive collapse resisting capacity. The force–displacement relationships of the specimens strengthened by bonded and unbonded strands are shown in Fig. 15. The force–displacement curve for the unstrengthened specimen was also plotted in each figure for comparison. The specimen reinforced with bonded strand showed similar force–displacement relationship to that of the unstrengthened specimen, except that the maximum strength increased by 56 %

and the specimen did not fail completely when the test was over. The number of re-bars fractured was seven, which is the same with that observed in the test of the unstrengthened specimen. In the specimen with unbonded strand, the maximum strength turned out to be 145 % higher than the maximum strength of the specimen without the strand. It was also observed that two re-bars fractured at the displacement of 350 mm, and the force was reduced for about 50 kN. As displacement further increased the force increased again until the strength reached the maximum value of 84 kN at the displacement of 436 mm. The number of fractured re-bars was reduced to three. It can be observed from the figures that the specimen with an unbonded strand showed superior catenary action to that of the specimen with bonded strand.

Figure 16 depicts the strain history of the high strength strand located at the right-end of the left-hand-side beam (LR) and in the right-end of the right-hand-side beam (RR).

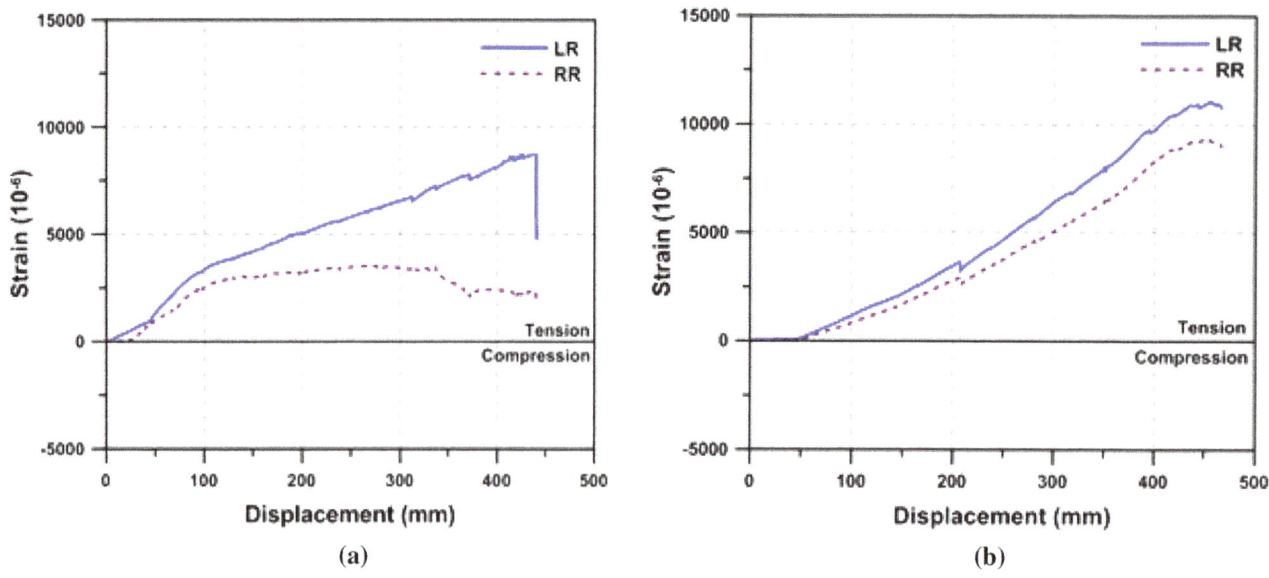

Fig. 16 Strain-displacement relationship of the strand. **a** Bonded strand. **b** Unbonded strand.

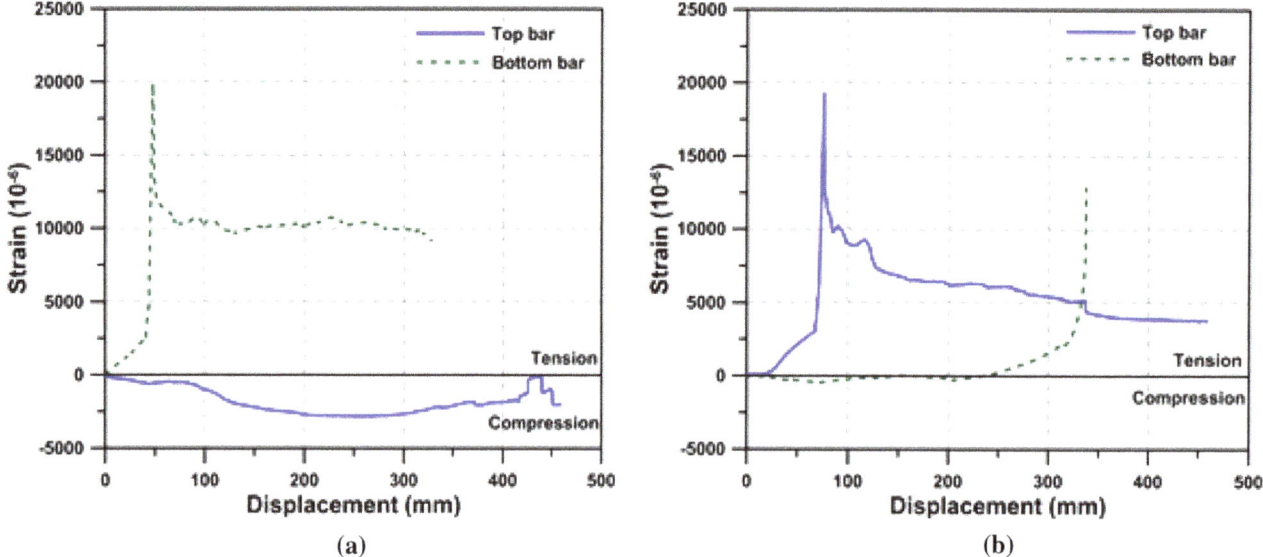

Fig. 17 Strain of rebars in the right-hand side beam of the gravity load resisting specimen strengthened by bonded strand. **a** Left end (RL). **b** Right end (RR).

It can be observed that, except the strain at RR of the bonded specimen, the strains increased almost linearly as the displacement increased. The strain of the strand located at LR of the specimen reinforced with bonded strand is larger than that of the strand at RR, which is due to the larger cracks formed near LR than those formed near RR. However the variations of the strains observed in the specimen with unbonded strand are similar to each other, which implies that damages occurred nearly symmetrically in the two beams.

Figures 17 and 18 depict the variation of the rebar strain in the specimens strengthened with bonded and unbonded strand, respectively. It can be observed that at the vertical displacements around 350 mm the rebars which were initially under compression started to resist large tensile force due to catenary action.

Figure 19 shows the photographs of the damaged specimens reinforced with tendons. It was observed in the force–displacement relationship that the specimen with an unbonded strand showed catenary action superior to that of the specimen with bonded strand. This is due to the fact that in the specimen with a bonded strand the catenary force of the strand was transmitted to the beams evenly along the length, which resulted in separation of the beam from the column face as can be observed in the photograph of the specimen at failure shown in Fig. 19a. However in the specimen strengthened with unbonded strand, where all catenary force in the strand acts on the far face of the exterior column, the beam end was not separated from the column face even at the maximum displacement as can be observed in Fig. 19b. It was observed that, compared with the crack

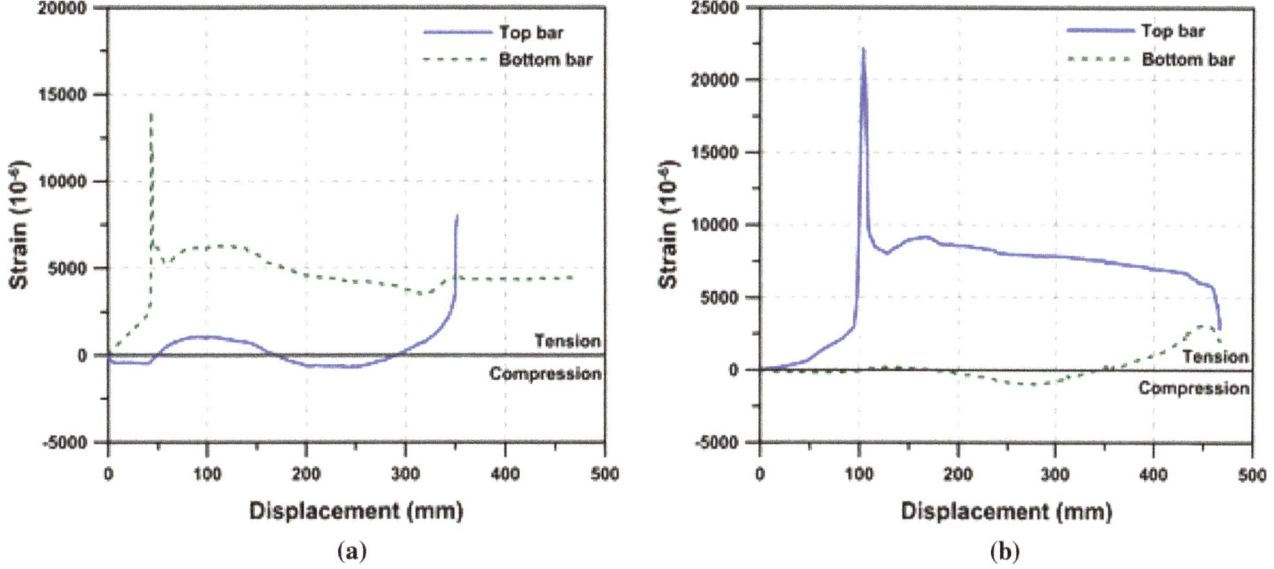

Fig. 18 Strain of rebars in the right-hand side beam of the gravity load resisting specimen strengthened by unbonded strand. **a** Left end (RL). **b** Right end (RR).

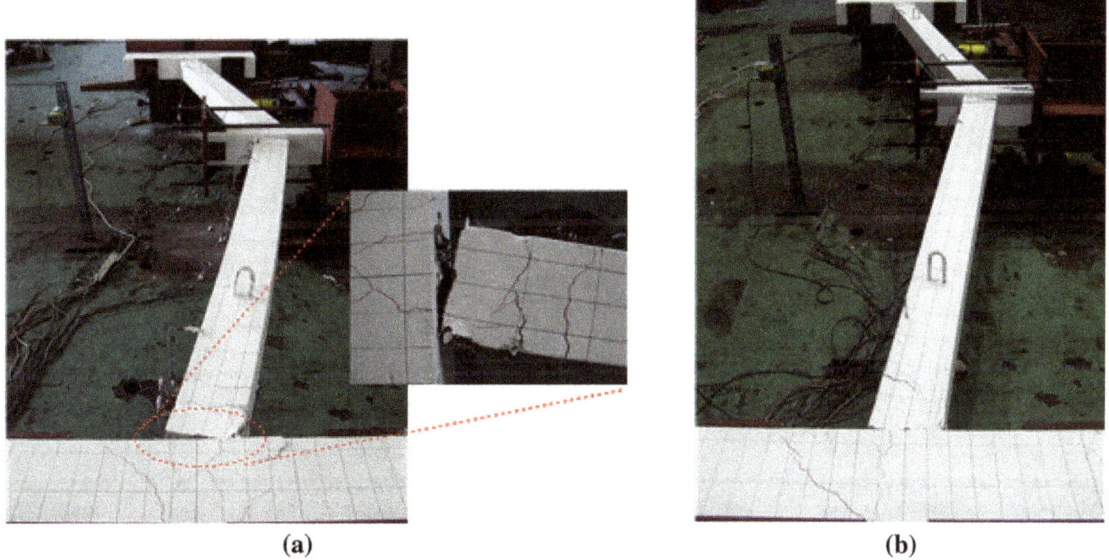

Fig. 19 Failure mode of the gravity-load resisting specimen strengthed by high strength strand. **a** Bonded strand. **b** Unbonded strand.

formation of the specimen without strand (shown in Fig. 9), smaller cracks formed relatively uniformly along the beam length reinforced with high strength strands. The number of cracks formed in the exterior beam-column joint of the specimen with unbonded strand turned out to be smaller than that in the specimen with bonded strand due mainly to the larger confinement effect of the unbonded strand and its anchorage. The cracks formed in the side columns of the specimens, however, may not be found in real buildings because the axial load imposed on the columns and the bending moment of the adjacent beam will compensate for the resultant tensile stress in the column joints.

Figure 20 shows the force–displacement relationship of the specimen reinforced with side plates at both sides of the beam-column joints. Plastic hinges formed at the displacement of about 90 mm, and the specimen showed ductile

behavior until the force increased again at the displacement of around 200 mm due to activation of catenary force. As no rebar was fractured until maximum displacement was reached, the force kept increasing without sudden drop as observed in the other specimens. Compared with the performance of the specimen strengthened with the high strength strands, the specimen reinforced with side plates showed slightly smaller strength but more stable behavior. Moreover, considering the higher expanse involved in the anchoring of strands, the side plate strengthening scheme seems to be more practical means of enhancing progressive collapse resisting capacity of RC moment frames.

The variation of re-bar strain in the specimen reinforced with side plates is shown in Fig. 21. It can be observed in Fig. 21a that the strain of the bottom bars of the left end of the right-hand-side beam, which was initially subjected to

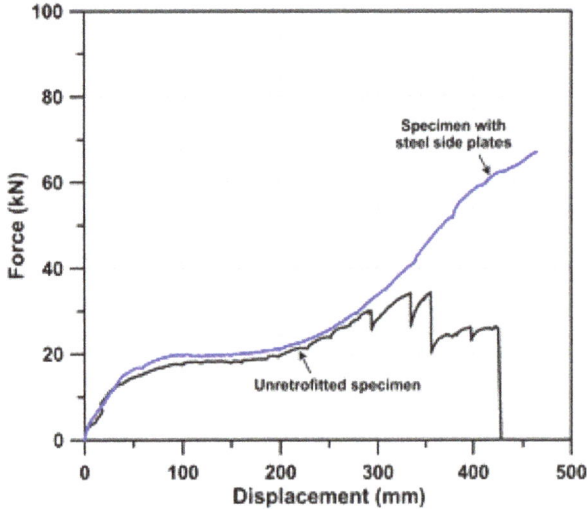

Fig. 20 Load-displacement relationship of the specimen reinforced with steel side plates.

tension, increased until the displacement reached about 180 mm and decreased due to formation of cracks. However the strain of the top bars at the same location, which was initially under compression, started to resist tension after the displacement exceeded 247 mm due to the initiation of catenary action. Similar phenomenon was observed in the rebars located in the right end of the right-hand-side beam as shown in Fig. 21b. In this case the bottom bars were subjected to tension starting from the displacement of 290 mm due to catenary action.

Figure 22 shows the damaged ends of the specimen reinforced with side plates after the test is over. It can be seen that major cracks formed at the end of the side plates, which is 160 mm away from the column face. No major crack was observed within the region covered by the side plates, which is probably due to the confining effect of the plates with stud bolts. It was also observed that due to the catenary force many tension cracks formed along the beam length.

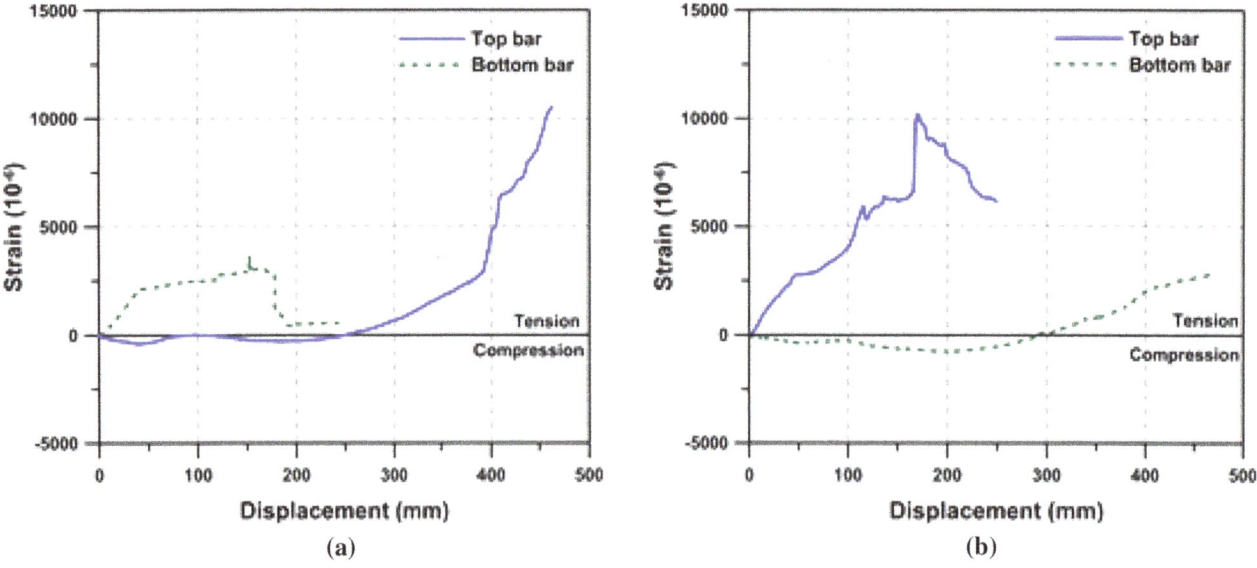

Fig. 21 Strain of rebars in the right-hand side beam of the specimen reinforced with side plates. **a** Left end (LL). **b** Right end (RR).

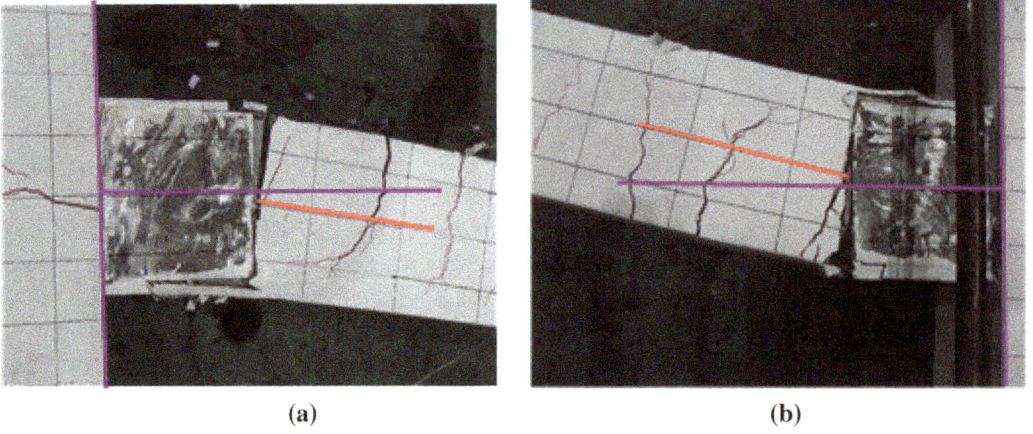

Fig. 22 Failure mode of the specimen strengthed by side plates. **a** Left end (LL). **b** Right end (RR).

5. Conclusion

In this study the progressive collapse resisting capacity of RC beam-column subassemblages with and without strengthening was investigated by a series of displacement controlled static loading tests. The test results showed that the unstrengthened gravity load-resisting specimen failed by fractures of re-bars in the beams, and that due to the activation of the catenary action the maximum displacement at failure turned out to be significantly larger than the limit state for beams recommended in the GSA and DoD guidelines. In the other specimens no failure was observed until the maximum displacement capacity of the actuator was reached. The specimen designed for lateral load showed higher strength and deformation capacity than the specimen designed only for gravity load. Highest strength was observed in the structure with unbonded strand. The test result for the specimen with side steel plates in beam-column joints showed that the force–displacement curve increased without fracture of re-bars. Compared with the performance of the specimen strengthened with the high strength strands, the specimen reinforced with side plates showed slightly smaller strength but more stable behavior. Considering the higher expanse involved in the prestressing of members, the side plate strengthening scheme seems to be more practical means of enhancing progressive collapse resisting capacity of RC moment frames. Based on the test results it was concluded that the progressive collapse resisting capacity of a RC frame could be significantly enhanced using unbonded strands or steel side plates in the beam-column subassemblages exposed to abnormal loads.

Finally it should be pointed out that, as the tests were carried out using 2D beam-column subassemblages, the 3D effects of transverse frames and floor slabs could not be considered in this study, which can provide significant resistance against progressive collapse as pointed out by Qian et al. (2015). Also the dynamic effect caused by sudden removal of a column could not be considered in this study of static loading tests.

Acknowledgments

This research was supported by a grant (13AUDP-B066083-01) from Architecture & Urban Development Research Program funded by Ministry of Land, Infrastructure and Transport of Korean government.

References

ACI Committee 315. (2004). *ACI Detailing Manual-2004*. Farmington Hills, MI: Publication SP-66(04), American Concrete Institute.

Choi, H., & Kim, J. (2011). Progressive collapse-resisting capacity of reinforced concrete beam-column subassemblage. *Magazine of Concrete Research, 63*(4), 297–310.

Ellingwood, B. R., Smilowitz, R., Dusenberry, D. O., Duthinh, D., Lew, H. S., & Carino, N. J. (2007). Best practices for reducing the potential for progressive collapse in buildings, Report No. NISTIR 7396, National Institute of Standards for Technology.

General Services Administration. (2003). *Progressive collapse analysis and design guidelines for new federal office buildings and major modernization projects*. Washington, DC: General Services Administration.

Harajli, M. H. (1993). Strengthening of concrete beams by external prestressing. *PCI Journal, 38*(6), 76–88.

ICC. (2009). *International building code*. Falls Church, VA: International Code Council.

Kang, S. B., Tan, K. H., & Yang, E. H. (2015). Progressive collapse resistance of precast beam–column sub-assemblages with engineered cementitious composites. *Engineering Structures, 98*(1), 186–200.

Khandelwal, K., & El-Tawil, S. (2005). Progressive collapse of moment resisting steel frame buildings, Proceedings of the 2005 Structures Congress and the 2005 Forensic Engineering Symposium, New York, NY.

Kim, J., & An, D. (2009). Evaluation of progressive collapse potential of steel moment frames considering catenary action. *The Structural Design of Tall and Special Buildings, 18*(4), 455–465.

Milner, D., Gran, J., Lawver, D., Vaughan, D., Vanadit-Ellis, W., & Levine, H. (2007). *FLEX analysis and scaled testing for prediction of progressive collapse, first international workshop on performance, protection & strengthening of structures under extreme loading (PROTECT 2007)*. Canada: Whistler.

Powell, G. (2005). Progressive collapse: Case studies using nonlinear analysis: Proceedings of the 2005 Structures Congress and the 2005 Forensic Engineering Symposium, New York, NY.

Qian, K., & Li, B. (2012). Dynamic performance of RC beam-column substructures under the scenario of the loss of a corner column—experimental results. *Engineering Structures, 42*, 154–167.

Qian, K., & Li, B. (2013). Performance of three-dimensional reinforced concrete beam-column substructures under loss of a corner column scenario. *ASCE Journal of Structural Engineering, 139*(4), 584–594.

Qian, K., Li, B., & Ma, J. X. (2015). Load carrying mechanism to resist progressive collapse of RC buildings. *ASCE Journal of Structural Engineering, 141*(2), 04014107.

Ren, W., Sneed, L. H., Gai, Y., & Kang, X. (2015). Test results and nonlinear analysis of RC T-beams strengthened by bonded steel plates. *International Journal of Concrete Structures and Materials, 9*(2), 133–143.

Sasani, M., & Kropelnicki, J. (2008). Progressive collapse analysis of an RC structure. *The Structural Design of Tall and Special Buildings, 17*(4), 757–771.

Shin, K.-J., Lee, S.-H., & Kang, T. H.-K. (2013). External post-tensioning of reinforced concrete beams using a V-shaped steel rod system. *ASCE Journal of Structural Engineering, 140*(3), 04013067.

Song, B. I., & Sezen, H. (2013). Experimental and analytical progressive collapse assessment of a steel frame building. *Engineering Structures, 56*, 664–672.

Tsai, M. H., & Lin, B. H. (2008). Investigation of progressive collapse resistance and inelastic response for an earthquake-resistant RC building subjected to column failure. *Engineering Structures, 30*(12), 3619–3628.

Yi, W. J., He, Q. F., Xiao, Y., & Kunnath, S. K. (2008). Experimental study on progressive collapse-resistant behavior of reinforced concrete frame structures. *ACI Structural Journal, 105*(4), 433–439.

Yu, J., & Tan, K. H. (2013). Experimental and numerical investigation on progressive collapse resistance of reinforced concrete beam column sub-assemblages. *Engineering Structures, 55*, 90–106.

14

Effect of Wet Curing Duration on Long-Term Performance of Concrete in Tidal Zone of Marine Environment

Mehdi Khanzadeh-Moradllo[1,2),*], Mohammad H. Meshkini[1)], Ehsan Eslamdoost[1)], Seyedhamed Sadati[3)], and Mohammad Shekarchi[1)]

Abstract: A proper initial curing is a very simple and inexpensive alternative to improve concrete cover quality and accordingly extend the service life of reinforced concrete structures exposed to aggressive species. A current study investigates the effect of wet curing duration on chloride penetration in plain and blended cement concretes which subjected to tidal exposure condition in south of Iran for 5 years. The results show that wet curing extension preserves concrete against high rate of chloride penetration at early ages and decreases the difference between initial and long-term diffusion coefficients due to improvement of concrete cover quality. But, as the length of exposure period to marine environment increased the effects of initial wet curing became less pronounced. Furthermore, a relationship is developed between wet curing time and diffusion coefficient at early ages and the effect of curing length on time-to-corrosion initiation of concrete is addressed.

Keywords: curing, diffusion, chloride, silica fume, service life, durability.

1. Introduction

The chloride-induced corrosion of the embedded steel has become the most common cause of loss of integrity and failure in concrete structures and infrastructures placed in the marine environment (Swamy 1988; Neville 2000; Radlińska et al. 2014; Pritzl et al. 2014; Ghassemzadeh et al. 2011). Hence, the chloride permeability has been recognized to be a critical intrinsic property of the concrete (Guneyisi et al. 2005, 2009), and a lot of research has been conducted to enhance concrete resistance to chloride permeability (Shekarchi et al. 2009).

From durability point of view, concrete cover quality plays significant role in blocking of aggressive substance ingress such as chloride ions into the reinforced concrete (Thomas 1991; Bonavetti et al. 2000). There are several methods to improve the quality of the concrete cover such as use of supplementary cementitious materials, reduction in water-to-cementitious materials ratio (w/cm), and appropriate initial curing regimes (Neville and Brooks 1990; Ghassemzadeh et al. 2010). Although it is a very simple and inexpensive procedure, proper initial curing, prior to exposure to marine environment, has an important influence on improving concrete cover quality so that the concrete acting as a fine barrier to the access of aggressive species and accordingly extend the service life of reinforced concrete structures exposed to chloride (Alizadeh et al. 2008; Khatib and Mangat 2002; Khatib 2014; Radlinski and Olek 2015).

The objective of curing is considered by the duration of providing concrete with sufficient humidity and appropriate temperature conditions to reduce the loss of moisture to ensure the progress of hydration reactions causing the filling and segmentation of capillary voids by hydrated compounds (Guneyisi et al. 2005, 2009). On the contrary, drying of concrete particularly at the concrete surface, caused by a poor curing regime, leads to a restricted hydration and thus higher porosity and permeability in the surface layers which form covers for the reinforcement protections (Mangat and Limbachiya 1999; Khanzadeh-Moradllo et al. 2009).

The matter would be more critical in the case of concrete containing silica fume replacement because the pozzolanic reaction is, in general, very sensitive to the curing procedure (Toutanji and Bayasi 1999; Atis et al. 2005). According to the ACI 308 Recommended Practice (ACI Committee 308 1998), the curing period should be extended to 14 days when the cement contains supplementary cementitious materials, owing to the slow hydration reactions between supplementary cementitious materials and the calcium hydroxide. In addition, curing condition also could be an important parameter in controlling durability of the reinforced concrete in a harsh condition of the marine tidal zone

---

[1)]School of Civil Engineering, Construction Materials Institute, University of Tehran, Tehran, Iran.

[2)]Department of Civil and Environmental Engineering, Oklahoma State University, Stillwater, OK 74078, USA.
*Corresponding Author;
E-mail: mehdi.khanzadeh_moradllo@okstate.edu

[3)]Civil, Architecture, and Environmental Engineering Department, Missouri University of Science and Technology, Rolla, MO 65401, USA.

prior to exposing to sea water, where the concrete cover is subjected to wetting–drying cycles.

A considerable volume of research has been conducted on different curing regimes and related effects on concrete properties. However, the effect of curing conditions on the chloride penetration into the concrete in real field condition at long term has not been well studied. Also, despite the importance of this object in Persian Gulf region, which is one of the high aggressive environments in the world, a few investigations were conducted in this region in long-term (Neville 2000; Shekarchi et al. 2009; Khanzadeh-Moradllo et al. 2012). In this regard, a comprehensive effort is accomplished in Construction Material Institute (CMI) to examine the short and long-term effect of curing regimes on durability of concretes located in Persian Gulf marine environment. The objective of this study is to investigate the effect of wet curing duration on chloride penetration in plain and blended cement concretes with 7.5 % silica fume which subjected to tidal exposure condition in Persian Gulf for 5 years.

2. Experimental Program

2.1 Materials and Mixture Proportions

The cementitious materials used in this study were Portland cement (PC) equivalent to ASTM Type II with a specific gravity of 3.14 and a fineness of 290 m^2/kg, and silica fume (SF) obtained from Azna ferro-silicon alloy manufacture with a specific gravity of 2.20 and a specific surface area of 20,000 m^2/kg. The chemical and physical properties of these materials are given in Table 1. The aggregates used were crushed limestone from Metosak plant and were graded according to ASTM C 33. The coarse aggregate had maximum size of 12.5 mm and specific gravity and absorption values of 2.79 and 1.9 %, respectively. The fine aggregate had specific gravity and absorption values of 2.59 and 3.2 %, respectively. The fineness modulus of fine aggregates was 3.2. Polycarboxylate ether polymer superplasticizer was used for the mixes in order to improve the workability of fresh concrete.

The concrete mixture proportions detailed in Table 2 were used to study the effect of curing on both normal Portland cement and silica fume concrete (labeled as NPC and SFC, respectively); w/cm ratio is 0.5 and cementitious material content 400 kg/m^3.

The concrete mix proportions used in this study were not specifically chosen to meet the durability requirements given in Iranian National Code for concrete durability in Persian Gulf for the conditions of exposure used, but to provide concretes which would undergo a measurable amount of chloride penetration and deterioration in the short exposure periods used. Therefore, the results of this study may not be generalized for concrete made with a low w/cm. In addition, silica fume content of 7.5 % by weight of cement was used, because a previous study (Shekarchi et al. 2009) showed that there is an optimum silica fume content of 7.5 % by weight of cement beyond which additional silica fume does not produce additional benefits in line with the additional costs.

2.2 Casting and Curing of Concrete Specimens

The concrete mixture was prepared in the laboratory using a 0.1 m^3 countercurrent pan mixer. The fresh concrete was tested for air content (ASTM C 231), slump (ASTM C 143) and unit weight (ASTM C 138). Cubes of 150 × 150 × 150 mm and prisms of 150 × 150 × 600 mm in dimension were casted in steel molds and were compacted using a vibrating table. The 150 mm cubes were used for the determination of the compressive strength in accordance with DIN 1048, while the prisms were used to be tested for chloride penetration in the field. Properties of fresh and hardened concrete are summarized in Table 3. Five different curing procedures were applied on both cube and prism concrete specimens. All molds were covered with wet burlap for the first 24 h after casting. Concrete specimens were demolded after 24 h. Four of the five concrete prisms and their cube specimens were cured in water for 1, 3, 6, and 27 days labeled 1-D, 3-D, 6-D, and 27-D, respectively. The other one remained in ambient conditions of laboratory without moist curing at 40–50 % RH and 19–23 °C until exposure to seawater called 0-D for no-curing.

Table 1 Chemical properties of binders.

Oxide composition % by mass	Cement	Silica fume
CaO	62.25	–
SiO$_2$	21.22	93.16
Al$_2$O$_3$	4.68	1.13
Fe$_2$O$_3$	3.68	0.72
MgO	3.63	1.6
Na$_2$O	0.25	–
K$_2$O	0.75	–
SO$_3$	1.74	0.05
L.O.I.	1.37	1.58

Table 2 Details of the concrete mixtures.

code	Cement (kg/m^3)	Silica fume (kg/m^3)	w/cm	Water (kg/m^3)	Coarse aggregate (kg/m^3)	Fine aggregate (kg/m^3)	Plasticizer (%)
NPC	400	0	0.5	200	956	778	1.2
SFC	370	30	0.5	200	959	784	1.2

Table 3 Properties of fresh and hardened concrete.

Code	Density (kg/m^3)	Air content (%)	Slump (mm)	Curing	Compressive strength (MPa)	
					7 days	28 days
NPC	2370	2.7	80	0-D	26.0	31.3
				1-D	30.0	36.6
				3-D	30.8	37.9
				6-D	33.5	39.0
				27-D	33.5	39.8
SFC	2355	1.6	65	0-D	26.0	37.1
				1-D	29.9	45.3
				3-D	32.9	51.4
				6-D	34.4	53.6
				27-D	34.4	55.2

2.3 Exposure Condition

Next, the investigated prism specimens which were located in Bandar-Abbas coast in south of Iran are sealed on four sides using epoxy polyurethane coating to ensure one-dimensional diffusion. The performance of this type of coating has been confirmed by previous studies (Khanzadeh-Moradllo et al. 2012). Specimens were then subjected to tidal zone exposure condition in Persian Gulf for the entire period of investigation (60 months). Tidal exposure was situated at the about 2.2 m from sea level, so that concrete specimens were in contact with sea water for 12 h per day then they were exposed to dry condition (air) for rest of the day, simulating the tidal zone condition.

The Persian Gulf water is highly saline (Table 4) due to its enclosed condition (mostly surrounded by lands of Iran and Arabian Peninsula) and the high evaporation rate. Also, there are large fluctuations in daily and seasonal temperature and humidity regimes. Temperature can vary by as much as 30 °C during a typical summer day and relative humidity can range from 40 to 100 % within 24 h (Al-Amoudi and Bader 2001).

2.4 Sampling and Testing

Sampling is carried out at the ages of 3, 9, 36, and 60 months of exposure in tidal zone. Each time, 100 mm long prisms are cut from the end of the prism specimens (Fig. 1). The cut surface of the remaining part is coated and moved back to the exposure condition for future sampling. The 150 × 150 × 100 mm slices are taken to the laboratory in order to determinate the chloride penetration.

In the laboratory, a nominal 45 mm diameter core is taken from each slice to provide chloride concentration profiles. Each core is grinded in eight increments from the finished surface to an estimated depth of chloride penetration. The method used to estimate the chloride penetration depth was according to the procedure described by NordTest NT Build 492, which involves measuring the depth of color change of a freshly cut concrete surface in the direction of the chloride flow using 0.1 M AgNO$_3$ aqueous solution. Fine particles for chloride analysis are collected using a profile grinder parallel to the exposed surface according to NordTest NT Build 443 method with the accuracy of 0.5 mm at eight different depths. The first 1 mm fine particles are not included in calculations as it might be affected by actions such as washout, etc. The profile grinder and a grind hole are cleaned between depth increments to reduce the possibility of cross-contamination of samples from different depths. For each sample of concrete, fine particles are collected, the depth below the exposed surface is calculated as the average of six uniformly distributed measurements using a slide

Table 4 Concentration of various ions in seawater at Persian Gulf.

Ion type	K$^+$	Ca^{++}	Mg^{++}	SO$_4^{--}$	Na$^+$	Cl$^-$	Total salt
Concentration (ppm)	470	480	1600	3300	12600	23,400	41,850

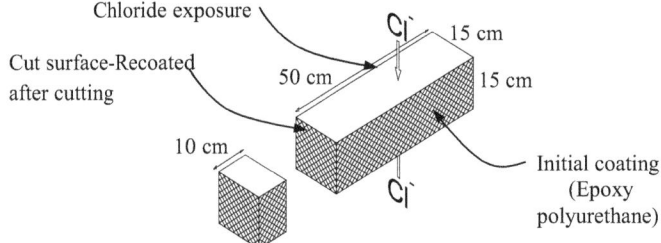

Fig. 1 Initial concrete prism specimens + cuts for laboratory tests.

caliper. The fine particles from each layer is collected and pulverized so that all the material will pass a 850-μm (No. 20) sieve. At each depths, a sample having a mass of approximately 10 g is selected to the nearest 0.01 g and then analyzed for acid-soluble chloride content by the potentiometric titration of chloride with silver nitrate according with ASTM C 1152, and ASTM C 114, part 19. The cross-sectional area of a 45 mm diameter core is large enough to represent the concrete so that there is no need to be concerned about variations from sample to sample due to varying aggregate contents.

2.5 Chloride Diffusion Coefficient (D_c) and Surface Chloride Content (C_s) Calculation

The chloride penetration rate as a function of depth from the concrete surface and time can reasonably be represented by Fick's second law of diffusion according to following expression (Crank 1975):

$$\frac{\partial C}{\partial t} = D_c \frac{\partial^2 C}{\partial x^2}. \tag{1}$$

The solution for Eq. (1):

$$C_{(x,t)} = C_s \left(1 - \mathrm{erf}\left(\frac{x}{2\sqrt{D_c t}}\right)\right) \tag{2}$$

$$C_{(x,0)} = 0 \quad x > 0 \quad C_{(0,t)} = C_s \quad t \geq 0$$

where x is distance from concrete surface; t denotes time; D_c is diffusion coefficient; C_s is surface chloride concentration; $C_{(x,t)}$ represents chloride concentration at the depth of x from the surface after time t; and erf is the error function.

Fick's second law for one-dimensional diffusion, as shown in Eq. (1), is a special case of a more generalized model of diffusion where concrete is assumed to be a homogenous material; chloride concentration at the exposure surface is considered constant; no chemical or physical binding between the diffusing species and material occurs; and the effect of co-existing ions is constant. In other words, these limitations of analysis may be neglected as measured data are used for comparison purposes within the same set of exposure conditions. Also, the effect of other mechanisms of chloride ion penetration such as a capillary suction or sorption mechanism is not considered in this study.

Using a computer statistical analysis program, the non-linear regression is carried out on the experimental data and by curve fitting of solutions of Fick's second law of diffusion, the values of D_c and C_s in the Eq. (2) are determined. The curve fitting has been done in such a way that the chloride profiles are fitted where the correlation between the measured and fitted profiles has a maximum. Curve fitting has been performed in accordance with a procedure described in NordTest NT Build 443 and resulted in two regression parameters; Namely a diffusion coefficient and surface chloride content. For each specimen, at the time of testing, a single measurement of chloride concentration at each specified depth has been done and the diffusion coefficient and surface chloride build-up have been calculated accordingly.

3. Results and Discussion

3.1 Chloride Profiles at Varying Exposure Time

Chloride concentration profiles of the concrete specimens with different curing regimes which were placed in tidal zone at varying exposure time (3, 9, 36, and 60 months) are presented in Fig. 2. Based on Fig. 2a, it can be concluded that the 27-D and 6-D curing regimes present better performance to prevent chloride transmission into the concrete for NPC sample after 3 months exposure. While by a visual judgment it seems that extending the wet curing period till 3 days have not significantly improved the NPC sample performance against the chloride uptake. It is obvious from chloride profiles that there is a noticeable improvement in reducing chloride concentration is seen in 27 days wet cured SFC sample in comparison with the other curing regimes, especially in early ages. The less pronounced effect of wet curing in SFC until 6 days could be due to the micro filling or particle packing effect of silica fume which conceal the wet curing influence on chloride pentration. From Fig. 2b, d, it is noticeable that wet cured samples of SFC show a higher surface chloride concentration than the uncured sample at early ages which could be due to the higher level of chloride binding and sorptivity. Because, it is suggested that the higher hydration rate for the the pozzolanic action of silica fume results in the formation of a higher content of C–S–H phases. This, in turn, increases the physical chloride binding due to the relatively high surface area of the C–S–H (Beaudoin et al. 1990; Luping and Nilsson 1993; Dousti et al. 2011).

It is obvious from chloride concentration profiles during exposure time (Fig. 2) that the wet curing effect is time-dependent and its influence on chloride resistance diminishes in long-term irrespective of the concrete mixture. Further

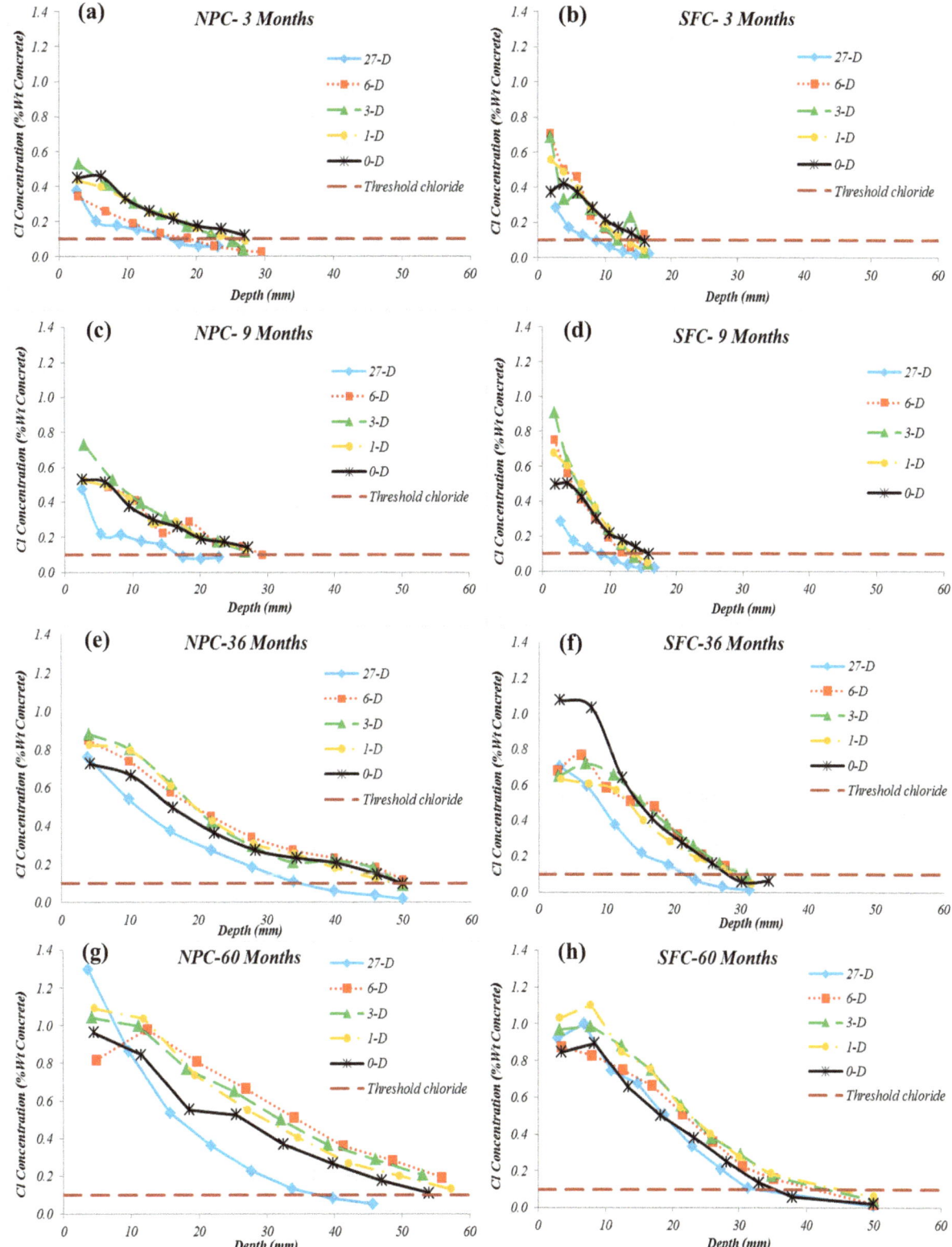

Fig. 2 Chloride concentration profiles of concrete specimens with different curing regimes placed in tidal zone of Persian Gulf region at varying exposure time. *Note* chloride threshold value is assumed 0.1 % by weight of concrete for corrosion initiation (Thomas 1996; Pargar et al. 2007).

analysis is provided in following sections based on calculated diffusion coefficient and surface chloride content.

3.2 Chloride Diffusion Coefficient (D_c)

Diffusion coefficient changes, over time for different curing periods in tidal zone are presented in Fig. 3. From results, it can be seen that extending the wet curing period has reduced the diffusion coefficient in comparison with no-cured sample in plain and blended specimens at early ages (3 and 9 months) as the same as previous studies (Guneyisi et al. 2007) that indicated long-term curing results in higher resistance to chloride permeability. But as the time goes on,

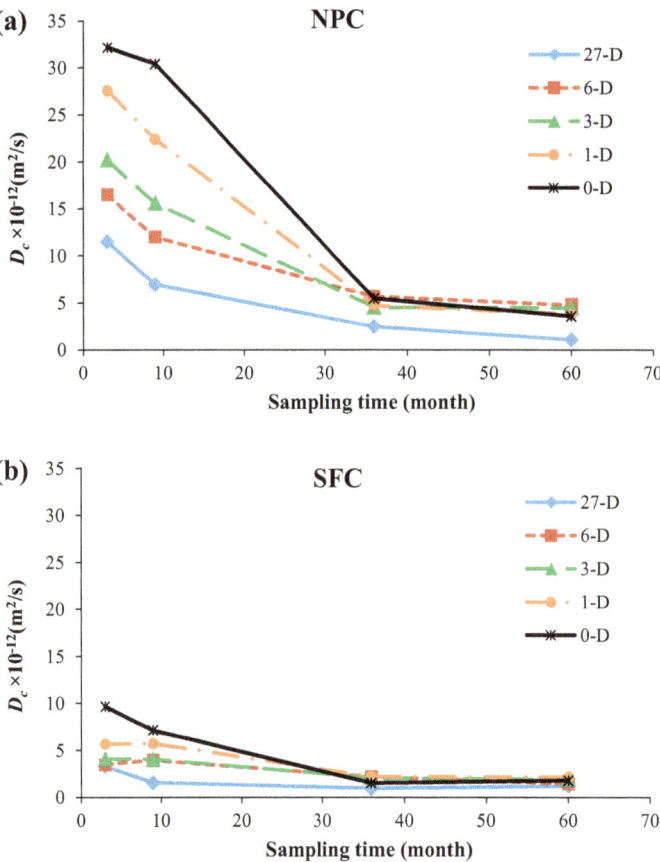

Fig. 3 Diffusion coefficient values versus sampling time for different curing regimes in tidal zone: **a** NPC specimens, **b** SFC specimens.

this discrepancy reduces and the rate of diffusion coefficient diminishes too. This is thought to be due to the curing effects of the seawater masking initial differences (Thomas and Matthews 1990). Mohammed et al. (Mohammed et al. 2002, 2004) found that seawater causes a reduction in the pore volume or shifts the pore size to smaller pores at concrete surface.

According to Fig. 3, the differences in value of diffusion coefficient in various ages, especially between early and long-term ages, were notable in short-term curing regimes and non-curing, but this is not the case in 27 days curing. Indeed, as the curing time increases, the decreasing rate of diffusion coefficient upon time reduces while the microstructure of concrete improves and the ingress of chloride ions into the concrete diminishes in early ages. This reduction in early age diffusion preserves the concrete against high rate of chloride penetration at early ages. So the initiation time of reinforcement corrosion will delay and the service life of the structure will increase.

As it is shown in Fig. 3, with comparison of silica fume and plain specimens, it is found that the silica fume specimens have lower diffusion coefficient in all curing times which confirms that concretes containing silica fume exhibit improved chloride penetration resistance compared to those of plain Portland cement concretes.

With regard to the time dependent variation of chloride diffusion coefficient, following equation was employed to express the D_c as a function of exposure time:

$$D_c = at^b \tag{3}$$

where D_c is the diffusion coefficient (mm²/s), t is the exposure time (s), "a" and "b" are the regression parameters presented at Table 5.

Based on Table 5, good correlation between D_c and exposure time is observed for all NPC and SFC samples with regression coefficients varying from 0.78 to 0.99. The model introduced in Eq. (3) can be employed to estimate the variation of chloride diffusion coefficient with time for different curing regimes. The fitted equation has also been incorporated in estimating the time-to-corrosion initiation which will be discussed in following sections.

3.3 Relationship Between Diffusion Coefficient and Curing Time

From the modeling point of view, the effect of curing conditions on the chloride diffusion coefficient has not been studied well (Alizadeh et al. 2008). To understand better the relationship between the curing time and a diffusion coefficient in short-term and long-term exposure periods, $k_{curing} = D_t/D_0$ versus curing times are plotted in Fig. 4 for NPC and SFC samples in tidal zone, where D_0 is the diffusion coefficient of no-cured concrete, D_t is the diffusion coefficient of wet cured specimen, and k_{curing} is the curing factor. As it is seen from Fig. 4, there is a power functional relationship between curing factor (k_{curing}) and time of wet curing (t_{curing}) at early ages:

Table 5 Modeling the chloride ion diffusivity versus exposure time.

Curing	NPC			SFC		
	a	b	R^2	a	b	R^2
27-D	2.2991	−0.760	0.95	0.0008	−0.357	0.83
6-D	0.0165	−0.432	0.99	0.0005	−0.296	0.78
3-D	0.1863	−0.568	0.94	0.0004	−0.286	0.89
1-D	3.5321	−0.728	0.93	0.0026	−0.378	0.87
0-D	13.688	−0.797	0.90	0.3849	−0.659	0.91

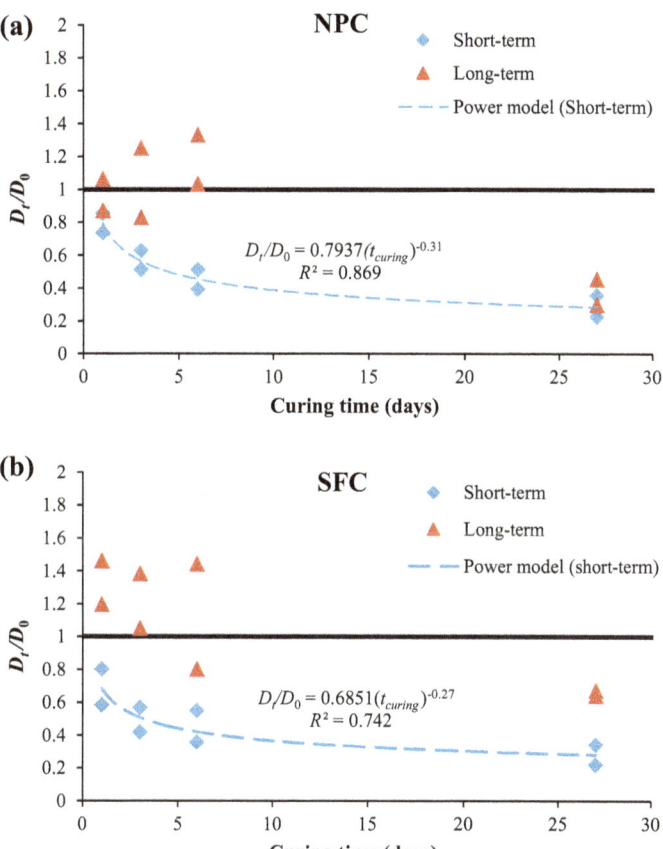

Fig. 4 Curing factor values versus curing time for different concrete mixtures in tidal zone: **a** NPC specimens, **b** SFC specimens.

$$k_{curing} = \alpha \times t_{curing}^{-m} \quad (4)$$

where "α" and "m" are coefficients varying with concrete mixture design, which are determined for different concrete mixtures with implementing the nonlinear regression on the experimental data. The curing factor decreases with increasing the wet curing time for NPC and SFC samples in short-term ages (3 and 9 months), which represents the efficiency of wet curing period in reducing the diffusion coefficient. This decrease in diffusion coefficient is very sharp until curing time of 6 days, especially for SFC, while the curing factor decreasing rate significantly diminishes from 6 days to 27 days curing time.

Based on Fig. 4, there is no distinct relationship between k_{curing} and t_{curing} in long-term (36 and 60 months) and some of the curing regimes are not any more effective in reducing

the diffusion coefficient. A "$k_{curing} = 1$" is considered as a efficiency boundary of curing regimes, where the wet cured sample acts similar to no-cured sample. According to results in long-term ages, a 27 days wet curing is the only curing regime which preserves its efficiency in reducing diffusion coefficient in both of NPC and SFC mixtures. As mentioned, this might be due to the curing effects of the seawater masking initial differences.

The results of long-term ages also imply that there is a slight increase in diffusion coefficient from a 0-D curing condition to the 6-D and 3-D in NPC and SFC samples ($k_{curing} > 1$). William F. Perenchio observed the same trend between initial curing period and long-term drying shrinkage (Perenchio 1997). According to Perenchio's suggestion (Perenchio 1997), it is possible that there is a pessimum initial curing time with respect to drying shrinkage or other parameters which produces the greatest value for that parameter. This higher drying shrinkage might cause microcracks and respectively an increase in diffusion coefficient. More work is needed to further understand this behavior.

According to the European Union RC structures durability design guidline (The European Union 2000), it has been suggested to include a coefficient for curing in RC structures service life design in order to take into the account the effect of curing regime on concrete diffusion coefficient and service life. The European Union suggested values for curing factor are compared to observed values from this study in early age in Table 6. Based on Table 6, the European Union suggested values are about 20 % higher than the calculated values for NPC samples. This difference can be due to the variations in concrete mixture proportions and environmental factors.

3.4 The Influence of the Curing Time on Surface Chloride Content (C_s)

The time-dependent characteristic of the chloride content at a concrete surface is another significant parameter in predicting the chloride ingress at the depth of steel and the concrete structures service life (Ann et al. 2009). Therefore, the influence of curing conditions on surface chloride content is addressed in current study.

The surface chloride content versus an exposure time is plotted in Fig. 5 for different curing regimes in tidal zone. Surface chloride content data were scattered with increment of the curing time, there is no regular trend observed. The general trend of surface chloride shows its increment as the time goes on. The SFC specimens have more surface chloride content in comparison with NPC specimens in early ages as shown in Fig. 5, presumably due to the higher level of chloride binding and sorptivity. With increased chloride binding capacity, total chloride contents increase nearer the surface of the concrete, but decrease deeper in the concrete (Glass and Buenfeld 2000; Song et al. 2008). As discussed earlier, it is suggested that the higher hydration rate for the the pozzolanic action of silica fume results in the formation of a higher content of C–S–H phases. This, in turn, increases the physical chloride binding due to the relatively high surface area of the C–S–H (Beaudoin et al. 1990; Luping and Nilsson 1993; Dousti et al. 2011). As the time goes, the rate of surface chloride content increment is lower in SFC specimens in contrast to NPC specimens.

In addition, it has been observed that the linear build-up model can express the time dependent nature of the C_s

(Khanzadeh-Moradllo et al. 2012; Sadati et al. 2015). The following equation was employed to express the C_s as a function of exposure time:

$$C_s = k \times t + C_0 \tag{5}$$

where C_s is the surface chloride content (% weight of concrete), t is the time (s), "k" is the regression coefficient, and C_0 is the earliest available measurement on surface chloride concentration; i.e. the measurement at 3 months.

Table 7 summarizes the regression coefficients. Based on Table 7, good correlation between C_s and exposure time is observed for all NPC and SFC samples with regression coefficients varying from 0.76 to 0.99. This indicates that a linear regression with the initial value is good representatives of the surface chloride ion build-up in tidal zone. This equation is proposed for further investigating the time-to-corrosion initiation of structures at following section.

3.5 The Influence of the Wet Curing Time on Time-to-Corrosion Initiation of Concrete Structures

The efficiency of the curing condition in preventing the ingress of chloride can be further emphasized by estimating time-to-corrosion initiation of concrete structures which subjected to different curing regimes. It is worth to mention that the concrete structure corrosion occurs in two steps, corrosion initiation time and corrosion propogation time (Morga and Marano 2015). The latter step is not considered in this study. The time-to-corrosion initiation is considered to be the time required for chloride ion concentration to reach to a certain threshold value at reinforcement cover depth. The regression results obtained for predicting the chloride ion diffusion and surface chloride build-up (Eqs. (3) and (5)) were incorporated to calculate the corrosion initiation time based on Fick's second law of diffusion (Crank 1975):

$$C_{(x,t)} = (C_0 + kt)\left(1 - \mathrm{erf}\left(\frac{x}{2 \times \sqrt{(at^b) \cdot t}}\right)\right) \tag{6}$$

where "a", "b", "C_0", and "k" are the regression parameters presented at Tables 5 and 7. In this study the chloride threshold value is supposed to be 0.1 % weight of the concrete, based on previous studies (Thomas 1996; Pargar et al. 2007), and the thickness of concrete reinforcement cover is considered to be 50 mm. Therefore, Eq. (6) was solved for

Table 6 Curing factor for different curing conditions.

Curing condition	$k_{c,cl}$* (NPC)	$k_{c,cl}$ (SFC)	$k_{c,cl}$ (The European Union 2000)
1 day wet curing	1.77	1.54	2.08
3 days wet curing	1.26	1.10	1.50
7 days wet curing**	1.00	1.00	1.00
28 days wet curing**	0.64	0.68	0.79

* $k_{c,cl}$ (curing factor for different curing regimes) = D_c (for correspondence curing)/D_c (7 days wet curing).
** To compare curing factors, 6-D and 27-D curings are assumed equal to 7 and 28 days wet curings, respectively.

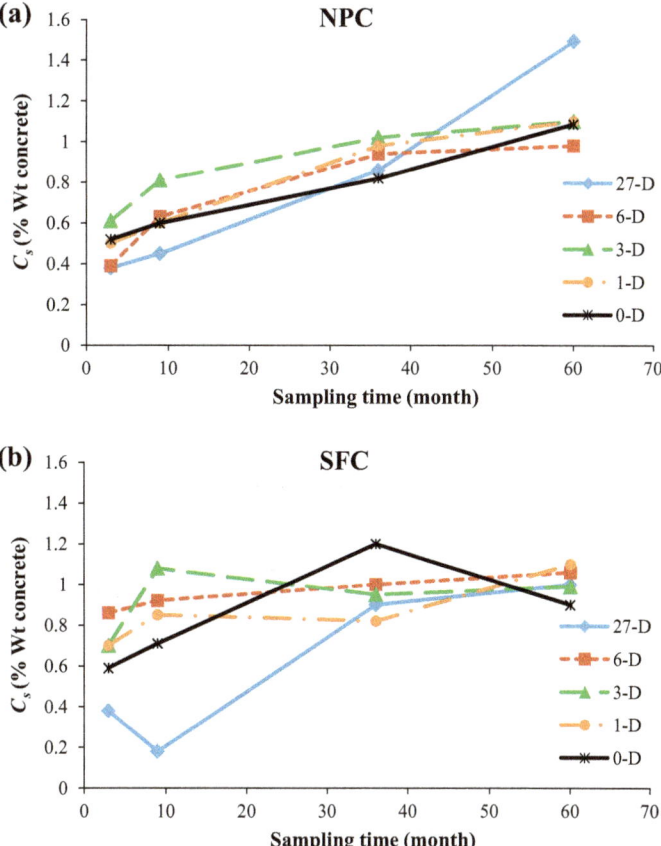

Fig. 5 Surface chloride content values versus sampling time for different curing regimes in tidal zone: **a** NPC specimens, **b** SFC specimens.

Table 7 Modeling the surface chloride concentration versus exposure time.

Sample	Curing	k	R^2
NPC	27-D	7.00E−09	0.95
	6-D	4.00E−09	0.79
	3-D	4.00E−09	0.83
	1-D	4.00E−09	0.95
	0-D	4.00E−09	0.99
SFC	27-D	4.00E−09	0.77
	6-D	1.00E−09	0.95
	3-D	2.00E−09	0.91
	1-D	2.00E−09	0.76
	0-D	6.00E−09	0.98

finding time-to-corrosion initiation (t) corresponding to $C_{(x,t)} = 0.1$ at $x = 50$ mm for different samples.

An estimated time-to-corrosion initiation versus curing time for different mixtures are plotted in Fig. 6. Both plain and silica fume specimens show that 27 days wet curing causes tangible increase in time-to-corrosion initiation. It seems that no difference is observable between the time-to-corrosion initiation values of the curing times less than 6 days in both silica fume and plain specimens based on time dependent results from field. It might happen because of the availability of continuous capillary pores in concrete

specimens with less amount of wet curing (Shamim Khan et al. 1993; Zhang et al. 1999). Based on Powers et al. (1959), the approximate time required to produce maturity at which capillaries become discontinuous is approximately 14 days for concrete with w/c of 0.50. As explained earlier, microcracks due to drying shrinkage can also cause this phenomenon.

According to results, it seems that longer wet curing (27 days) provides high quality skin layer in concrete surface, and accordingly it plays significant role as a barrier to concrete inner depths in controlling chloride ions penetration

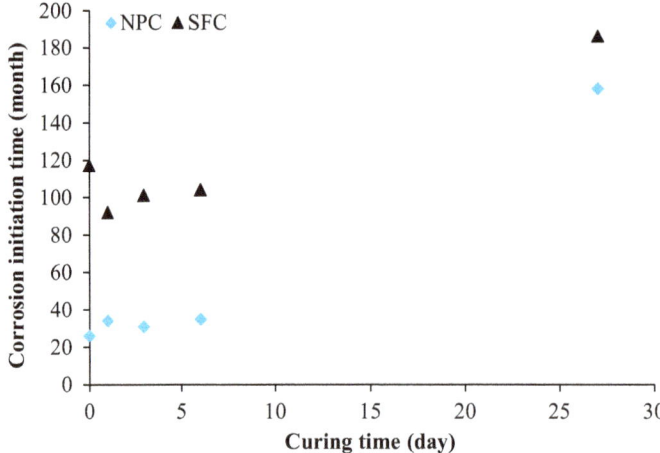

Fig. 6 An estimated corrosion initiation time vs curing time for different concrete mixtures.

into the concrete specimens. These results are for the materials, mixtures, environmental conditions, and specifications used on this study.

4. Conclusion

This study was conducted to investigate the effect of wet curing duration on chloride penetration in plain (NPC) and blended cement concretes with 7.5 % silica fume (SFC) and water-to-cement ratio of 0.5 which were subjected to tidal exposure condition in Persian Gulf for five years. Based on the test results, the major conclusions of this study could be summarized as follows:

- A wet curing extension decreases difference between initial and long-term diffusion coefficients due to improvement of concrete cover quality and blocking the ingress of aggressive substance in initial ages. This reduction in early age diffusion preserves concrete against high rate of chloride penetration at early ages.
- As the length of exposure period to marine environment increased the effects of initial wet curing became less pronounced. This might be due to the curing effects of the seawater which compensate for the differences observed in early age diffusion coefficient due to the duration of initial wet curing. In long-term ages, a 27 days wet curing is the only curing regime which preserves its efficiency in reducing diffusion coefficient in both of NPC and SFC mixtures.
- A power functional relationship is derived between curing factor ($k_{curing} = D_t/D_0$, where D_0 is the diffusion coefficient of no-cured concrete, D_t is the diffusion coefficient of wet cured specimen) and time of wet curing (t_{curing}) at early ages.
- The general trend of surface chloride shows its increment as the time goes on. The SFC specimens have more surface chloride content in comparison with NPC specimens in early ages presumably due to the higher level of chloride binding and sorptivity. But as the time

goes, the rate of surface chloride content increment is lower in SFC specimens in contrast to NPC specimens.
- Both plain and silica fume specimens show that 27 days wet curing causes tangible increase in time-to-corrosion initiation and service life of concrete structures. It seems that no difference is observable between the time-to-corrosion initiation values of the curing times less than 6 days in both silica fume and plain specimens based on time dependent results from field. It might happen because of the availability of continuous capillary pores in concrete specimens (w/c of 0.50) with less amount of wet curing.

References

ACI Committee 308. (1998). *Recommended practice for curing concrete*. Farmington Hills, MI: MCP, American Concrete Institute.

Al-Amoudi, O. S. B., Maslehuddin, M., & Bader, M. A. (2001). Characteristics of silica fume and its impacts on concrete in the Arabian Gulf. *Concrete, 35*(2), 45–50 (London, UK).

Alizadeh, R., Ghods, P., Chini, M., Hoseini, M., Ghalibafian, M., & Shekarchi, M. (2008). Effect of curing conditions on the service life design of RC structures in the Persian Gulf Region. *Journal of Materials in Civil Engineering, 20*, 2–8.

Ann, K. Y., Ahn, J. H., & Ryou, J. S. (2009). The importance of chloride content at the concrete surface in assessing the time to corrosion of steel in concrete structures. *Journal of Construction and Building Materials, 23*(1), 239–245.

Atis, C. D., Ozcan, F., Kilic, A., Karahan, O., Bilim, C., & Severcan, M. H. (2005). Influence of dry and wet curing conditions on compressive strength of silica fume concrete. *Building and Environment Journal, 40*, 1678–1683.

Beaudoin, J. J., Ramachandran, V. S., & Feldman, R. F. (1990). Interaction of chloride and CSH. *Cement and Concrete Research, 20*(6), 875–883.

Bonavetti, V., Donza, H., Rahhal, V., & Irassar, E. (2000). Influence of initial curing on the properties of concrete containing limestone blended cement. *Journal of Cement and Concrete Research, 30*, 703–708.

Crank, J. (1975). *The mathematics of diffusion* (2nd ed.). London, UK: Oxford.

Dousti, A., Shekarchi, M., Alizadeh, R., & Taheri-Motlagh, A. (2011). Binding of externally supplied chlorides in micro silica concrete under field exposure conditions. *Cement & Concrete Composites, 33*(10), 1071–1079.

Ghassemzadeh, F., Shekarchi, M., Sajedi, S., Khanzadeh-Moradllo, M., & Sadati, H. (2010). Effect of silica fume and GGBS on shrinkage in the high performance concrete. In *Proceedings of the sixth international conference on concrete under severe conditions: environment and loading*, Yucatan, Mexico (pp. 1007–1012).

Ghassemzadeh, F., Shekarchi, M., Sajedi, S., Mohebbi, M. J. & Khanzadeh-Moradllo, M. (2011). Performance of repair concretes in marine environments. In *Proceedings of the annual conference canadian society for civil Engineering* (pp. 1654–1664). Ottawa, Canada.

Glass, G. K., & Buenfeld, N. R. (2000). The influence of chloride binding on the chloride induced corrosion risk in reinforced concrete. *Journal of Corrosion Science., 42*, 329–344.

Guneyisi, E., Gesoglu, M., Ozturan, T., & Ozbay, E. (2009). Estimation of chloride permeability of concretes by empirical modeling: Considering effects of cement type, curing condition and age. *Journal of Construction and Building Materials, 23*, 469–481.

Guneyisi, E., Ozturan, T., & Gesoglu, M. (2005). A study on reinforcement corrosion and related properties of plain and blended cement concretes under different curing conditions. *Journal of Cement and Concrete Composites, 27*, 449–461.

Guneyisi, E., Ozturan, T., & Gesoglu, M. (2007). Effect of initial curing on chloride ingress and corrosion resistance characteristics of concretes made with plain and blended cements. *Journal of Building and Environment, 42*, 2676–2685.

Khanzadeh-Moradllo, M., Ghassemzadeh, F., Shekarchi, M., Mosadarolom, A., & Roujei, H. (2009). Effect of curing conditions on chloride diffusion into the concrete structures in Persian Gulf region. In *Proceedings of the 4th international conference on construction materials: performance, innovations and structural implications*, Nagoya, Japan (pp. 703–708).

Khanzadeh-Moradllo, M., Shekarchi, M., & Hoseini, M. (2012). Time-dependent performance of concrete surface coatings in tidal zone of marine environment. *Journal of Construction and Building Materials, 30*, 198–205.

Khatib, J. M. (2014). Effect of initial curing on absorption and pore size distribution of paste and concrete containing slag. *KSCE Journal of Civil Engineering, 18*(1), 264–272.

Khatib, J. M., & Mangat, P. S. (2002). Influence of high-temperature and low-humidity curing on chloride penetration in blended cement concrete. *Journal of Cement and Concrete Research, 32*, 1743–1753.

Luping, T., & Nilsson, L. O. (1993). Chloride binding capacity and binding isotherms of OPC pastes and mortars. *Cement and Concrete Research, 23*(2), 247–253.

Mangat, P. S., & Limbachiya, M. C. (1999). Effect of initial curing on chloride diffusion in concrete repair materials. *Journal of Cement and Concrete Research, 29*, 1475–1485.

Mohammed, T. U., Hamada, H., & Yamaji, T. (2004). Concrete after 30 years of exposure? Part II: Chloride ingress and corrosion of steel bars. *ACI Materials Journal, 101*(1), 13–18.

Mohammed, T. U., Yamaji, T., & Hamada, H. (2002). Chloride diffusion, microstructure, and mineralogy of concrete after 15 years of exposure in tidal environment. *ACI Materials Journal, 99*(3), 256–263.

Morga, M., & Marano, G. C. (2015). Chloride penetration in circular concrete columns. *International Journal of Concrete Structures and Materials, 9*(2), 1–11.

Neville, A. (2000). Good reinforced concrete in the Arabian Gulf. *Journal of Materials and Structures, 33*, 655–664.

Neville, A. M., & Brooks, J. (1990). *Concrete technology.* Singapore, Singapore: Longman Scientific and Technical.

Pargar, F., Layssi, H., & Shekarchi, M. (2007). Investigation on chloride threshold value in an old concrete structure. In *Proceeding of the fifth international conference on concrete under severe conditions, CONSEC'07*, Tours, France (pp. 175 182).

Perenchio, W. F. (1997). The drying shrinkage dilemma. *Concrete Construction, 42*, 379–383.

Powers, T. C., Copeland, L. E., & Mann, H. M. (1959). Capillary continuity or discontinuity in cement pastes. *Journal of Portland Cement Association, Research and Development Laboratories, 1*(2), 38–48.

Pritzl, M. D., Tabatabai, H., & Ghorbanpoor, A. (2014). Laboratory evaluation of select methods of corrosion prevention in reinforced concrete bridges. *International Journal of Concrete Structures and Materials, 8*(3), 201–212.

Radlińska, A., McCarthy, L. M., Matzke, J., & Nagel, F. (2014). Synthesis of DOT use of beam end protection for extending the life of bridges. *International Journal of Concrete Structures and Materials, 8*(3), 185–199.

Radlinski, M., & Olek, J. (2015). Effects of curing conditions on the properties of ternary (ordinary portland cement/fly ash/silica fume) concrete. *ACI Materials Journal, 112*(1), 49–58.

Sadati, S., Arezoumandi, M., & Shekarchi, M. (2015). Long-term performance of concrete surface coatings in soil exposure of marine environments. *Construction and Building Materials, 94*, 656–663.

Shamim Khan, M., & Ayers, M. E. (1993). Curing requirements of silica fume and fly ash mortars. *Journal of Cement and Concrete Research, 23*, 1480–1490.

Shekarchi, M., Rafiee, A., & Layssi, H. (2009). Long-term chloride diffusion in silica fumes concrete in harsh marine climates. *Journal of Cement and Concrete Composites, 31*(10), 769–775.

Song, H. W., Lee, C. H., & Ann, K. Y. (2008). Factors influencing chloride transport in concrete structures exposed to marine environments. *Journal of Cement and Concrete Composites, 30*(1), 113–121.

Swamy, R. N. (1988). Durability of steel reinforcement in marine environment. In *Proceedings of the 2nd international conference on concrete in marine environment*, St. Andrews By-The-Sea, Canada (pp. 147–161).

The European Union, Brite EuRam III, DuraCrete R17. (2000). Probabilistic performance based durability design of concrete structures, includes general guidelines for durability design and redesign. DuraCrete Final Technical Report, Document BE95-1347/R17.

Thomas, M. D. A. (1991). Marine performance of PFA concrete. *Magazine of Concrete Research, 43*(151), 171–185.

Thomas, M. (1996). Chloride thresholds in marine concrete. *Cement and Concrete Research, 26*(4), 513–519.

Thomas, M. D. A., Matthews, J. D., & Haynes C. A. (1990). Chloride diffusion and reinforcement corrosion in marine exposed concretes containing pulverized-fuel ash. In C. L. Page, K. W. J. Treadaway & P. B. Bamforth (Eds.), *Proceedings of the third international symposium on corrosion of reinforcement in concrete construction*, Garston, UK (pp. 198–212).

Toutanji, H. A., & Bayasi, Z. (1999). Effect of curing procedures on properties of silica fume concrete. *Journal of Cement and Concrete Research, 29*, 497–501.

Zhang, M. H., Bilodeau, A., Malhotra, V. M., Kim, K. S., & Kim, J. C. (1999). Concrete incorporating supplementary cementing materials: Effect on compressive strength and resistance to chloride-ion penetration. *ACI Materials Journal, 96*(2), 181–189.

15

Effects of Isolation Period Difference and Beam-Column Stiffness Ratio on the Dynamic Response of Reinforced Concrete Buildings

Young-Soo Chun[1], and Moo-Won Hur[2],*

Abstract: This study analyzed the isolation effect for a 15-story reinforced concrete (RC) building with regard to changes in the beam-column stiffness ratio and the difference in the vibration period between the superstructure and an isolation layer in order to provide basic data that are needed to devise a framework for the design of isolated RC buildings. First, this analytical study proposes to design RC building frames by securing an isolation period that is at least 2.5 times longer than the natural vibration period of a superstructure and configuring a target isolation period that is 3.0 s or longer. To verify the proposed plan, shaking table tests were conducted on a scaled-down model of 15-story RC building installed with laminated rubber bearings. The experimental results indicate that the tested isolated structure, which complied with the proposed conditions, exhibited an almost constant response distribution, verifying that the behavior of the structure improved in terms of usability. The RC building's response to inter-story drift (which causes structural damage) was reduced by about one-third that of a non-isolated structure, thereby confirming that the safety of such a superstructure can be achieved through the building's improved seismic performance.

Keywords: seismic isolation, period ratio, isolation effect, shaking table test, acceleration.

1. Introduction

Isolation is a method that reduces the seismic response of a building by extending its vibration period, which is achieved by inserting a special device between the building and its foundation or into a middle story. It is an effective technology that brings about great improvement in the seismic behavior of a superstructure. This technology has been verified empirically and commercialized in countries such as Japan, China, the United States, and New Zealand, all of which have had extensive experience with earthquakes, and it has been acknowledged for its excellent results. The main targets of early isolation methods were low-story buildings and high stiffness buildings, and the outstanding and successful effects of these early methods already have been demonstrated. However, the latest architectural trends are high-story, lightweight, slender buildings, and so, isolation technology is now being developed to address these gradually increasing building trends. Although most apartment housing structures in Asian countries were built originally as slab-wall structures, recent changes include more flexible structural systems such as reinforced concrete (RC) beam-column structures and flat-plate structures that can be remodeled easily. The effects of isolation technology for such building structures have not yet been confirmed (Chun et al. 2007).

Several researchers have investigated isolated building systems. For example, Ariga et al. (2006) studied the resonant behavior of base-isolated high-rise buildings in terms of long-term ground motion, and Olsen et al. (2008) also investigated long-term building responses. Deb (2004), Dicleli and Buddaram (2007), Casciati and Hamdaoui (2008), Di Egidio and Contento (2010) also have made progress in the study of isolated building systems. Komodromos et al. (2007) and Kilar and Koren (2009) focused on seismic behavior and responses through the dynamic analyses of isolated buildings. Nonetheless, further work is needed for the practical application of an isolation device. Low to medium earthquake risk regions in Asian countries are prone to seismic hazards. Thus, for building construction in these zones, seismic base isolation can be a suitable strategy as it ensures the flexibility of the building and reduces the lateral forces in a drastic manner. Although the application of an isolator is similar all over the world, currently, proper research is lacking that could implement such a device practically. So, this concern is an urgent matter for this study.

Building codes of various countries (Japan, United States, etc.) that provide for isolation standards point out that these effects may differ according to the difference in the vibration period between the superstructure and the isolation layer, and so, many codes include relevant regulations for limitations. Feng (2007) and Feng et al. (2012) presented a comparative report of the building codes of Japan for 2000,

[1] Land and Housing Institute, Korea Land & Housing Corporation, Daejeon 305-731, Korea.

[2] Department of Architectural Engineering, Dankook University, Gyeonggi-do 448-701, Korea.

*Corresponding Author; E-mail: hmwsyh@gmail.com

China for 2001, USA IBC 2012, Italy for 2008, and Taiwan for 2002, which was updated in 2010. The important point that these reports made is that regulations for limits are different in each code. Likewise, the isolation effect is expected to differ based on how the target isolation period is configured according to the characteristics of a particular building. However, until now, no definite research records have been presented on this subject.

In response to this need, this study aims to analyze isolation effects according to the difference in the vibration period between the superstructure and the isolation layer in order to provide information about how the isolation effect changes according to the configuration of the superstructure period and the target isolation period and the characteristics of the superstructure, and to compare the isolation response of a model 15-story RC flat-plate apartment building to the seismic response of a model non-isolated building.

2. Analytical Study

2.1 Studied Building and Modeling

As suggested by Roehl (1972), in this study, the dynamic characteristics of an RC frame are defined as follows based on the fundamental vibration period (T) and the beam-column stiffness ratio (ρ) of a building. The RC frame is modeled as a single-span frame with a story height of h and span of $2h$. It is assumed that all the members of the RC building have identical cross-sections and that the distribution of the mass and stiffness for each story is constant. Figure 1 is a diagram of the study model. The major variables considered in this study and its scope are as follows:

(1)   Beam-column stiffness ratio (ρ)

The beam-column stiffness ratio can be defined as shown in Eq. (1). This ratio represents the different characteristics of the frame as the value changes from 0 to ∞.

$$\rho = \sum_{beam} \frac{EI_b}{L_b} / \sum_{column} \frac{EI_c}{L_c} \qquad (1)$$

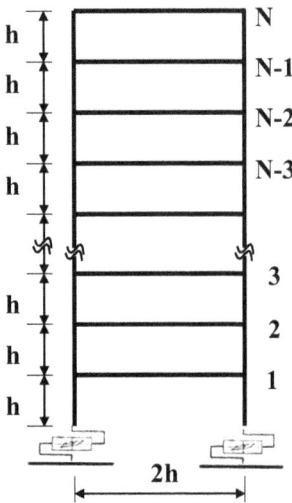

Fig. 1 Modeling for the object of study.

Here, E is the elastic modulus, I_b is the geometrical moment of the inertia of the beam, I_c is the geometrical moment of the inertia of the column, L_b is the length of the beam, and L_c is the length of the column. The value of ρ is taken from the middle story of the RC building. The limit value $\rho = 0$ represents a cantilever that consists of a beam with no restraint on its nodal rotation, and the limit value $\rho = \infty$ represents a shear building in which the nodal rotation is completely restrained. The value between the two limit values represents a frame that induces bending deflection depending on the degree of the nodal rotation in the beam and column. In this paper, the range of the ρ value is defined as between 0.05 and 2.0, which is sufficient to show the characteristics of building frames in Asia, based on existing studies. Analysis was carried out on seven values: 0.05, 0.1, 0.15, 0.2, 0.5, 1.0 and 2.0.

(2)   Fundamental vibration period of building (T_1)

The fundamental vibration period of an RC building has the range shown in Eq. (2), which approximates the number of stories in the building.

$$0.6T \leq T_1 \leq 1.4T \qquad (2)$$

Here, $T = 0.1N$ and $N =$ number of stories in the building.

In this study, the number of stories in the RC building is 15 stories. Seven stiffness ratio models and six period models were developed with consideration of the above variables and the scope of the study, thereby interpreting and analyzing a total of 42 models. For the hysteretic characteristics of the isolation device, the characteristics (bilinear model) of a hybrid-type isolation system applied to an actual isolation design were used. For seismic motion, time history analysis was performed using the most widely used earthquake data for the El Centro (1942), Taft (1952), and Hachinohe (1968) earthquakes (refer to Fig. 2). ETABS v8.48 was the program used for analysis. The modeling of the superstructure was designed as for a frame element, and Isolator1 element was used to perform local nonlinear modeling of the isolation device. Also, five percent viscous damping was assumed for the damping of the superstructure. For the isolation layer, any viscous damping other than hysteretic damping of the isolation device was not taken into account.

2.2 Analytical Results and Comments

(1)   Isolation effect according to period difference between superstructure and isolation layer

Figures 3 and 4 show the distribution of the maximum response accelerations for each story according to the change in the period of the isolated RC building when $\rho = 0.05 \sim \rho = 2.0$, which are used to examine the isolation effect according to the period difference between the superstructure and the isolation layer of the isolated building. In this study, the isolation effect according to the change in the stiffness ratio between the superstructure and isolation

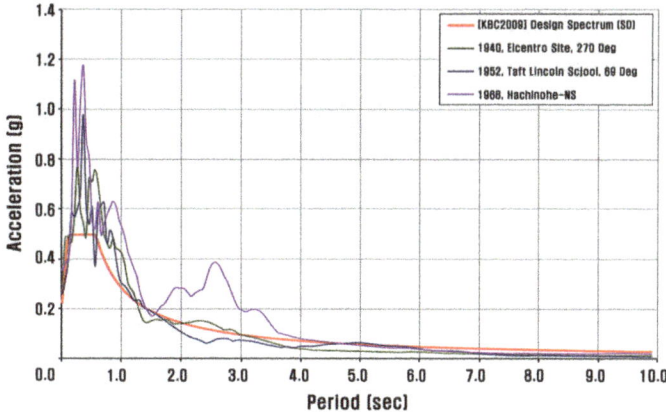

Fig. 2 Scaled response spectra of earthquake data.

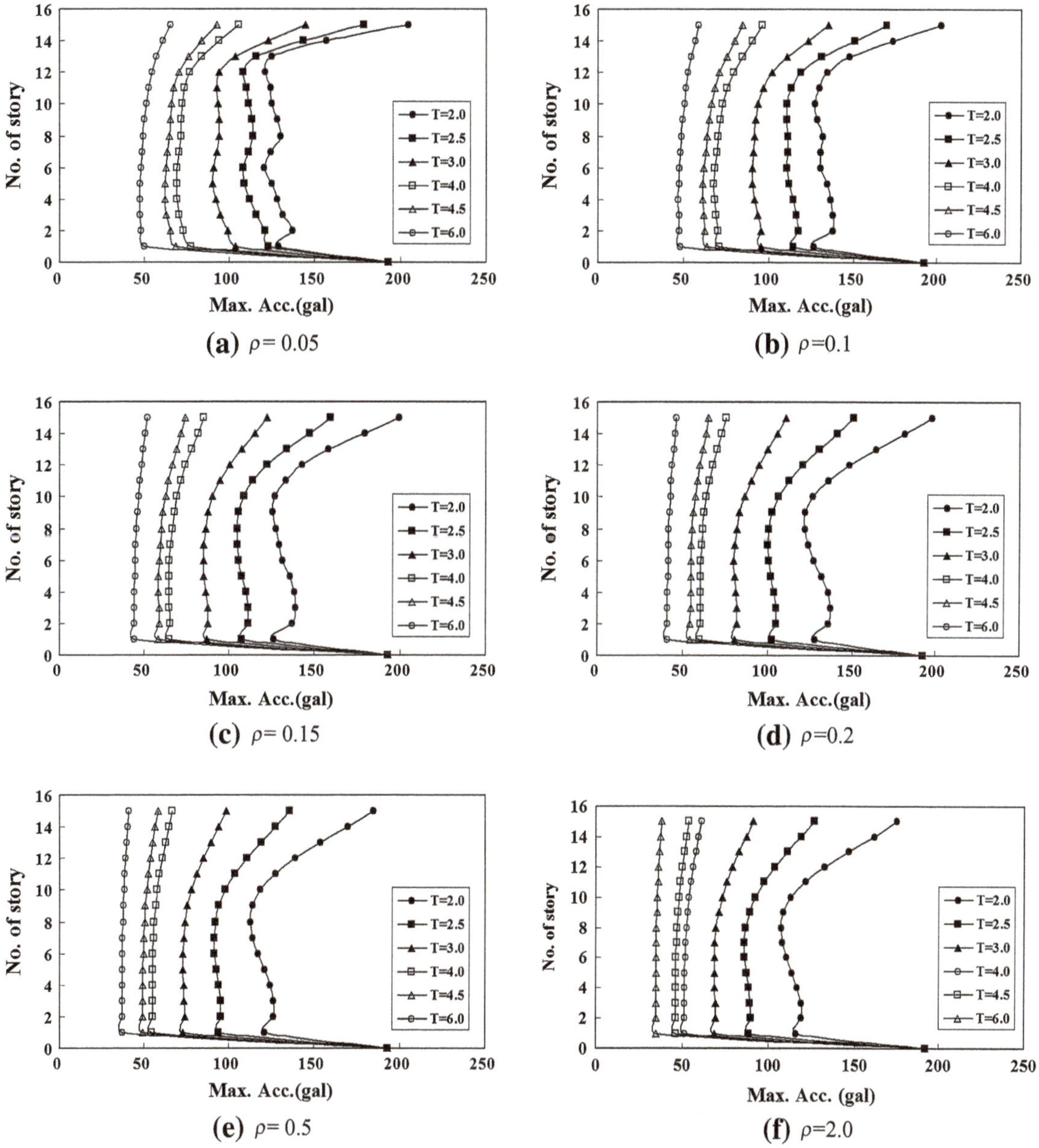

(a) $\rho = 0.05$

(b) $\rho = 0.1$

(c) $\rho = 0.15$

(d) $\rho = 0.2$

(e) $\rho = 0.5$

(f) $\rho = 2.0$

Fig. 3 Distribution of the acceleration of superstructure due to variation in period ratio.

layer was examined by changing the vibration period of the isolated building under the fixed vibration period of the superstructure.

As shown in Figs. 3 and 4, with an increase in the period of the isolated building when $\rho = 0.05 \sim \rho = 2.0$, the distribution of the response acceleration in each story of the

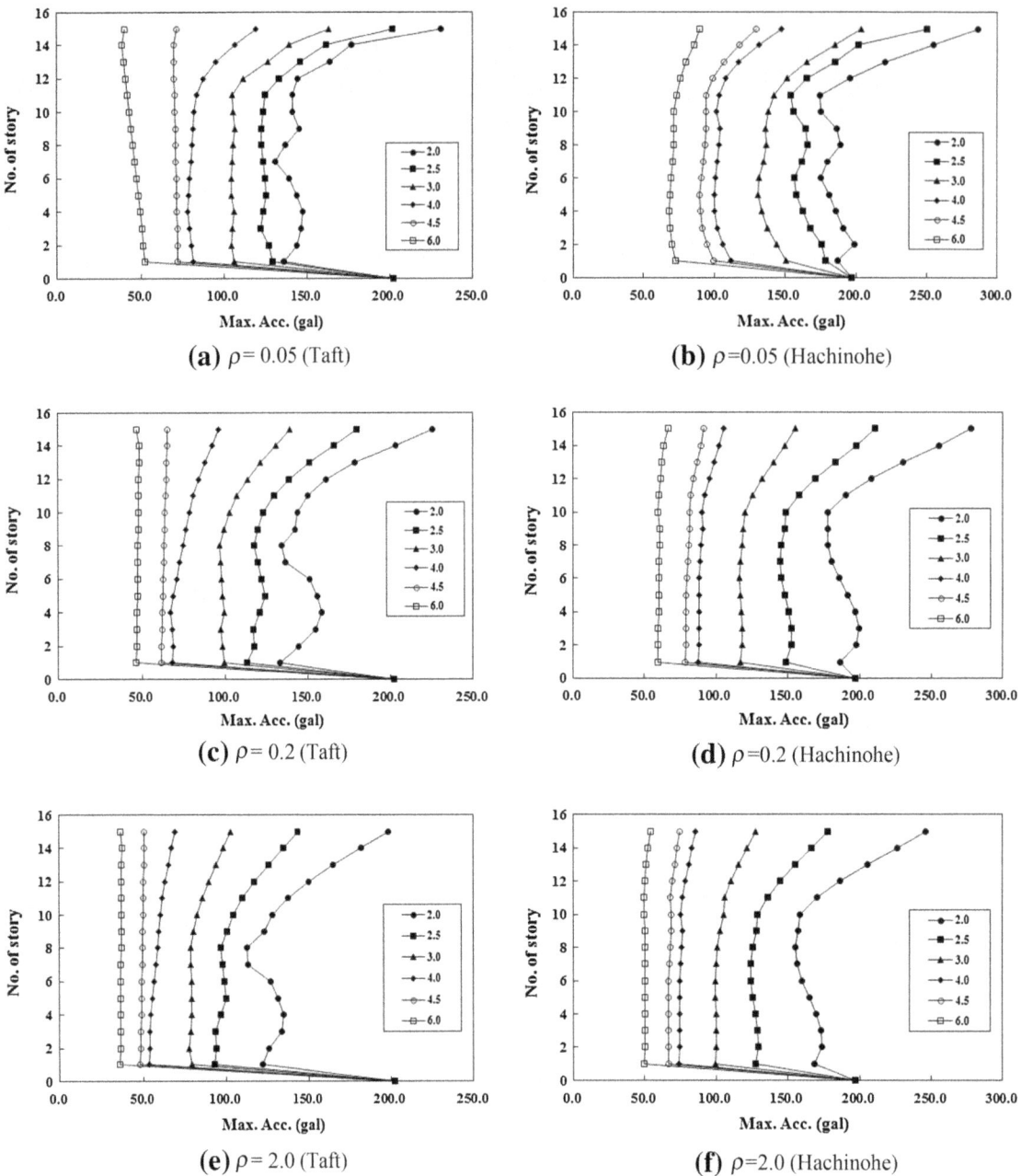

(a) $\rho = 0.05$ (Taft)

(b) $\rho = 0.05$ (Hachinohe)

(c) $\rho = 0.2$ (Taft)

(d) $\rho = 0.2$ (Hachinohe)

(e) $\rho = 2.0$ (Taft)

(f) $\rho = 2.0$ (Hachinohe)

Fig. 4 Distribution of the acceleration of superstructure due to variation in period ratio (Taft and Hachinohe).

building changes from the form that includes the effect of a higher order mode to a constant distribution that is close to rigid body behavior. In terms of reduced acceleration, both buildings showed a reduction in acceleration, but it was difficult to anticipate the effective isolation, as acceleration tends to increase at the uppermost story of a building when the difference of the vibration period between the superstructure and the isolation layer is small. Considering the fact that the reduction in the acceleration of the superstructure refers to the reduced load in the member design, it is necessary to design the difference of the vibration period between the superstructure and the isolation layer to be above a certain level in order to achieve an effective isolation effect.

When the vibration period ratio between the superstructure and the isolation layer is low, the distribution of the response

acceleration becomes inconsistent due to the effect of the higher order mode, which makes it difficult to achieve very much improved usability of the building. Also, a large load can act on a specific unpredicted layer to form a soft layer, which also causes difficulty in predicting the overall behavior of the RC building and assessing hazards. Because such careful design is required, it is desirable to design the model for above a certain level of the vibration period ratio between the superstructure and the isolation layer. From a practical viewpoint, it is necessary to set forth a quantitative standard for the level of the period ratio that must be achieved in order to obtain an effective isolation effect. However, no accurately established quantitative standards are currently available that can determine the isolation effect from the form of a specific response. Nevertheless, based on the comparison results shown in Figs. 3 and 4, and considering the fact that the aim of an isolation design is to achieve a consistent response acceleration distribution in order to decrease the response acceleration and increase usability, it is recommended that the subject RC building should be designed with a isolation period that is at least 2.5 times longer than that of superstructure in order to obtain an effective isolation effect for the building.

(2) Effects according to beam-column stiffness ratio

Another interesting point that can be observed from Figs. 3 and 4 is that no large difference is evident in the responses to the beam-column stiffness ratio (ρ) of the superstructure when the period ratios of the superstructure and isolation layer are the same. This finding implies that the target isolation period does not need to be considered differently in terms of the stiffness (or the frame characteristics) of the superstructure for the isolation design. It shows that the isolation effect can be differentiated only in terms of the vibration period ratio of the superstructure and isolation layer.

3. Experimental Study

3.1 Experimental Program and Test Specimens

For the experiments in this study, an isolated RC building model and a non-isolated RC building model scaled to one-tenth of an actual building size were designed with consideration of the capacity of a shaking table. The effects of the model building foundations and site on the seismic behavior of the RC buildings were considered using seismic waves with different characteristics. The ground motion records used for the tests include those for the Central Chile earthquake S2A059 (059), San Fernando earthquake S3A103 (103), Eureka earthquake S2A105 (105), and an artificial earthquake. Earthquake 105 and the artificial earthquake represent a harmonic type of motion that is critical for long period structures. The spectrum characteristics of these seismic waves are compatible with the Korean Building Code (KBC 2009) and the US Building Code (ASCE/SEI7-10, 2010) design spectrum. The fabrication of the specimens and experimentation were carried out at the China Academy of Building Research (CABR) in Beijing, China. Figure 5 shows a typical floor plan of the subject buildings and the facade of the specimen installed on top of a shaking table. A full-scale model of the specimen, shown in Fig. 5a, is a flat-type medium sized apartment building with 15 stories and four units in each story.

The total height of the RC building is 45,800 mm including the water tank room. The plane dimensions are 50,720 mm × 12,270 mm. The story height of each story is 2800 mm, and the height of the water tank room is 1400 mm. The target isolation period of the full-scale model originally was to be configured as 4.0 s based on the criteria described in Sect. 2.2, but the isolation period was changed to 3.0 s based on the manufacturing limits of the scaled model isolation device and the effectiveness of the isolation effect. Eighteen lead rubber bearings (LRB 80) and 23 laminated rubber bearings (RB 80) were used to make the isolation layer in order to attain the target isolation period (JSSI 2006; SIVIC 2009). Table 1 presents the characteristics of each isolation device. The scaled specimen was modeled using a maximum manufacturing size that is one-tenth an actual building size to accommodate the capacity of the shaking table (6 m × 6 m, 800 KN, 6 DOF). It is a 15-story model with a long side and a short side of 5.07 m × 1.23 m respectively, and an overall height of 4.6 m. Table 2 summarizes the law of similarity for the mixed similarity model applied in this study.

3.2 Manufacture of Specimens

The concrete strength values of the full-scale model were C30 (30 MPa), C35 (35 MPa) and C40 (40 MPa), and the corresponding concrete strength values of the scaled model were M5.5 (5.5 MPa), M6.0 (6.0 MPa) and M6.5 (6.5 MPa). For the rebar, annealed fine-drawn steel bars with diameters of $0.9 \sim 2.2$ mm were used based on similarity conditions. Table 3 shows the test results for the small aggregate concrete and rebar. The specimens were manufactured by first making a base-plate of reinforced concrete, then installing and curing the scaled isolation device on top of the base-plate, and creating a base-plate and superstructure for manufacture of the superstructure. The concrete placement was performed story by story, and the concrete was cured for about 3 days after placement before sequentially carrying out the construction of the upper stories.

3.3 Test Setup and Loading Procedure

Figure 6 shows a complete view of the specimen installed on top of the shaking table. After fixing the specimen to the shaking table, the insufficient additional mass was determined according to the law of similarity, as shown in Fig. 6b.

For additional artificial mass, 0.1 kN iron ingot was distributed as uniformly as possible as a single layer on top of the floor slab of each story. Seismic motion was applied sequentially in the x, y, and $x + y$ directions of the model, and the peak ground acceleration (PGA) was increased gradually to 0.07, 0.1, 0.22, 0.4 and 0.9 g. In this study, the tests were conducted with 0.22 g as the design seismic motion in order to assess the residential performance and 0.9 g as the seismic motion used to verify the safety performance. The time of the seismic motion was scaled down to the ratio of 1/4.472 for the recorded time according to the law of similarity.

The story acceleration and inter-story drift were measured to compare the capacity of the test buildings. Based on the test conditions, the story acceleration was measured every two floors, measuring four points in each direction and the center point, to find the lateral movement and warping histories of each floor, whereas the inter-story drift was measured every five floors by connecting the corners of the upper and lower sides of each floor diagonally. The ARJ-50A (Tokyo Sokki, Japan) was employed as the accelerometer, and a cable-extension displacement sensor, CDS-30 (Vishay Precision Group, Inc.), was employed as the displacement meter.

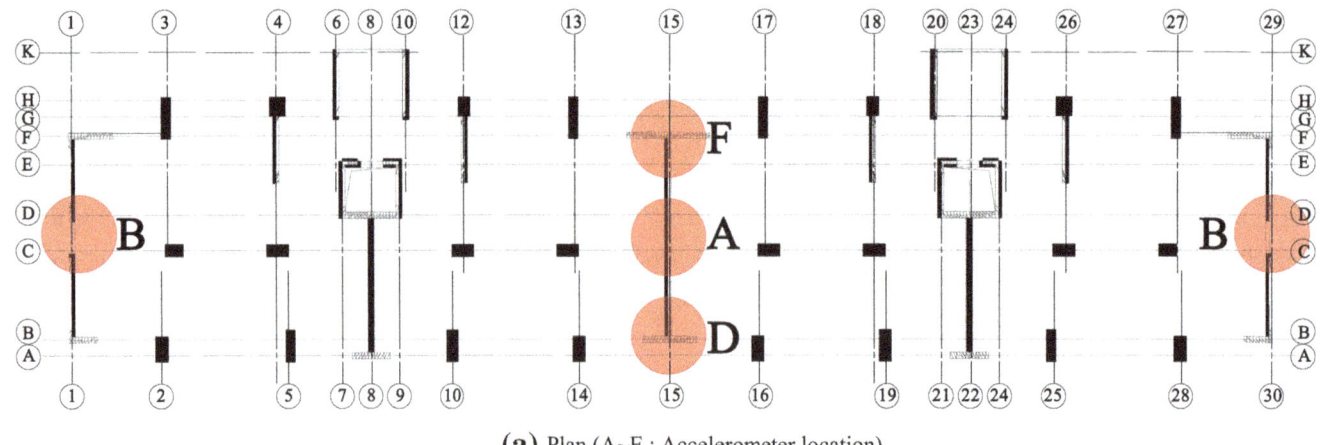

(a) Plan (A~E : Accelerometer location)

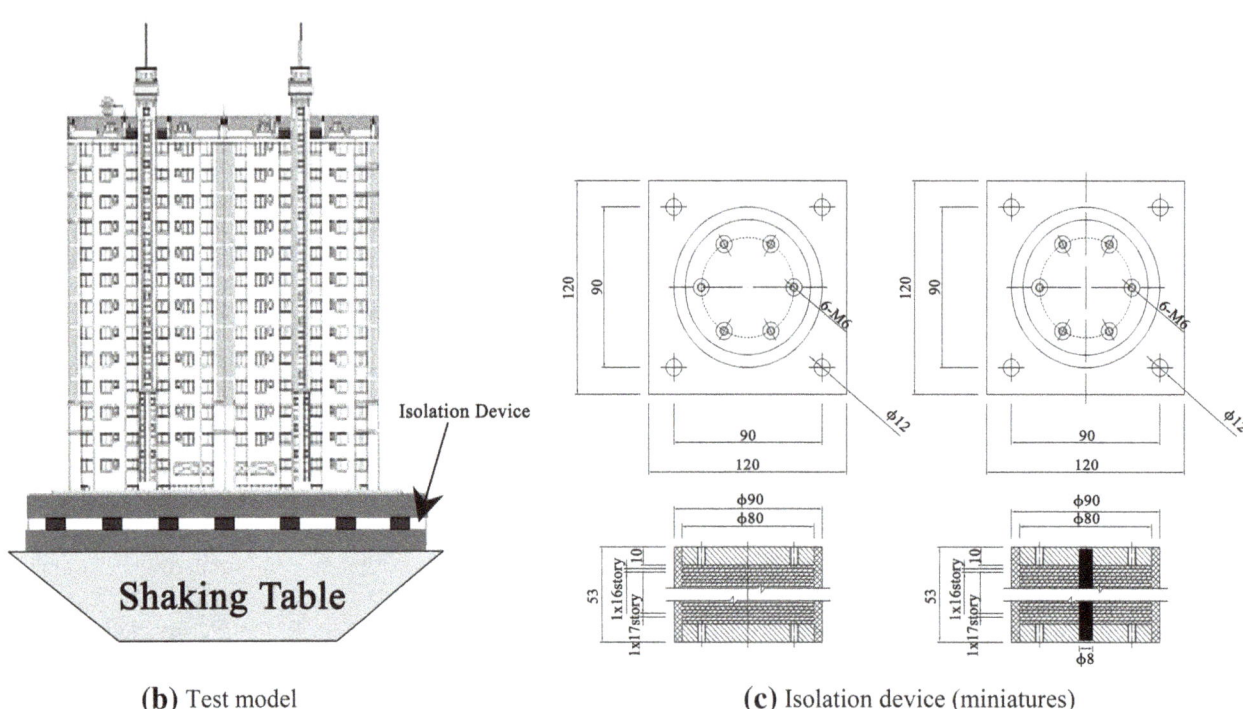

(b) Test model   **(c)** Isolation device (miniatures)

Fig. 5 Plan of study model, test model and isolation device.

3.4 Test Results

(1) Crack pattern and mode of failure

The isolated structure and non-isolated structure showed substantial differences in terms of scale and the type of seismic motion. First, with regard to cracking and the status of the failure, the isolated RC structure did not show any visually observed cracks in the superstructure until 0.9 g. No destructive symptom appeared until the end of the experiment. In contrast, the non-isolated RC structure was found to maintain an elastic state under small seismic loads, but cracks occurred in a portion of a structural member at under 0.22 g. Also, the structure clearly reached an inelastic state after experiencing 0.9 g, and many cracks and the destruction of some members were observed (refer to Fig. 7).

(2) Maximum drift and inter-story drift

Figure 8 shows the response to inter-story drift, which is a phenomenon that causes structural damage. In comparison to the non-isolated structure, the isolated structure shows that the almost constant response reduced by about one-third, depending on the height. Considering the fact that inter-story drift is limited as a means to determine the seismic performance of buildings, and the reduced inter-story drift in an isolated structure implies improved seismic performance, then the safety of a superstructure can be achieved. Also, Fig. 9 shows that the isolated RC structure experienced a very large drift in the isolation layer, but the maximum response drift of the isolation layer under 0.22 g was 14.76 mm, which is much lower than the allowable drift, i.e., smaller value between 0.55 d (d = diameter of isolation device) and 3 t_r

Table 1 Details of isolation devices used in 1/10-scaled model.

Group	LRB 80	RB 80
Number	18	23
Diameter × height (mm)[a]	80 × 79	80 × 79
Inner steel plate (mm)	19@1.0	16@1.0
Inner rubber plate (mm)	20@1.0	17@1.0
1st shape coefficient	17.5	17.5
2nd shape coefficient	3.5	4.12
Vertical stiffness (N/mm)	7830	3880
Effective stiffness at 50 % horizontal strain (N/mm)[b]	78	103
Damping ratio at 50 % horizontal strain[b]	0.12	0.05
Effective stiffness at 100 % horizontal strain (N/mm)[b]	62	100
Damping ratio at 100 % horizontal strain[b]	0.1	0.05
1st Stiffness (N/mm)	140	–
2nd Stiffness (N/mm)	52	–
Bucking load (N)	200	–

[a] Height: included the thickness of top and bottom plates.
[b] Effective stiffness and damping ratio in this table are average value. Difference between effective stiffness of isolation devices is no greater than ten percent.

Table 2 Scale factor of reduction model.

Quantity	Dimension	Ratio	Quantity	Dimension	Ratio
Length	L	1/10	E. Modulus	FL^{-2}	1/2.5
Stress	FL^{-2}	1/2.5	Frequency	T^{-1}	1/0.2236
Mass	FT^2L^{-1}	1/500	Time	T	1/4.472
Stiffness	FL^{-1}	1/25	Acceleration	LT^{-2}	2/1

Table 3 Test results of small aggregate concrete and characteristics of steel.

Strength	Average strength (MPa)	E. Modulus ($\times 10^4$ MPa)	Variety	Diameter	Steel no.	E. Modulus (N/mm^2)	Yield strength (N/mm^2)	Max. strength (N/mm^2)	Elongation
M5.5	7.5	1.12	14#	2.2	2#	0.8×10^5	310	400	0.26
					6#	0.8×10^5	300	400	0.30
M6.0	8.5	1.21	16#	1.6	3#	0.7×10^5	360	420	0.13
					4#	1.0×10^5	360	420	0.12
M6.5	10.0	1.40	18#	1.2	5#	0.65×10^5	360	480	0.32
					7#	0.65×10^5	300	420	0.26

(t_r = total thickness of rubber layer in isolation device), in the Chinese standard GB50011-2010 (2010) and CECS126 (2001). The maximum response drift under 0.9 g was found to be 26.46 mm, confirming that the design is appropriate for the target performance.

(3) Acceleration response

Figure 10 presents a comparison of the peak acceleration response of each story for the isolated RC structure and non-isolated RC structure. While the response acceleration of the

Fig. 6 Specimens and added mass setting.

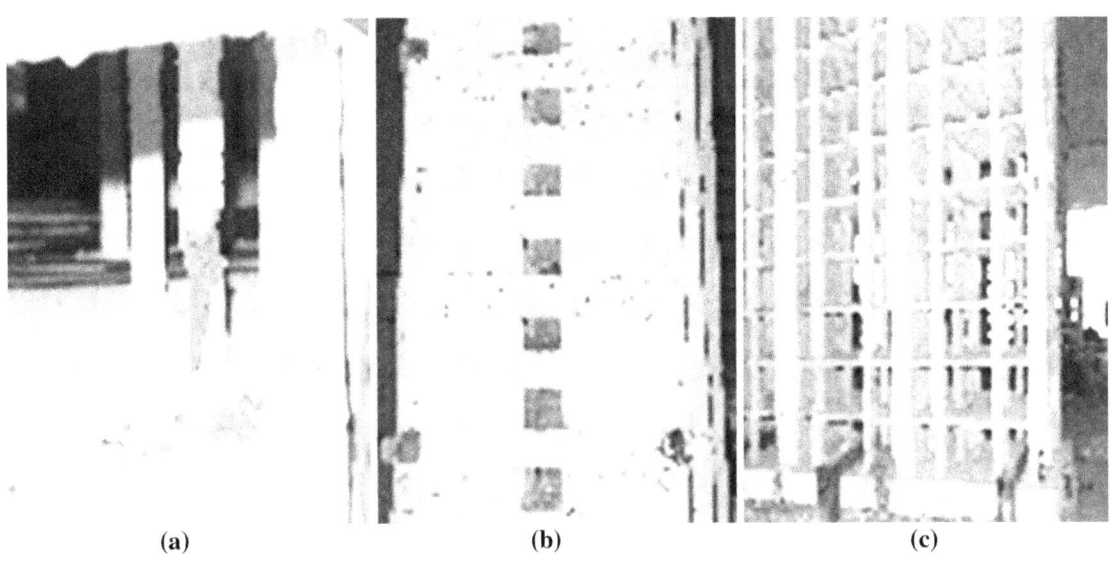

Fig. 7 Building cracks and destruction of 0.9 g.

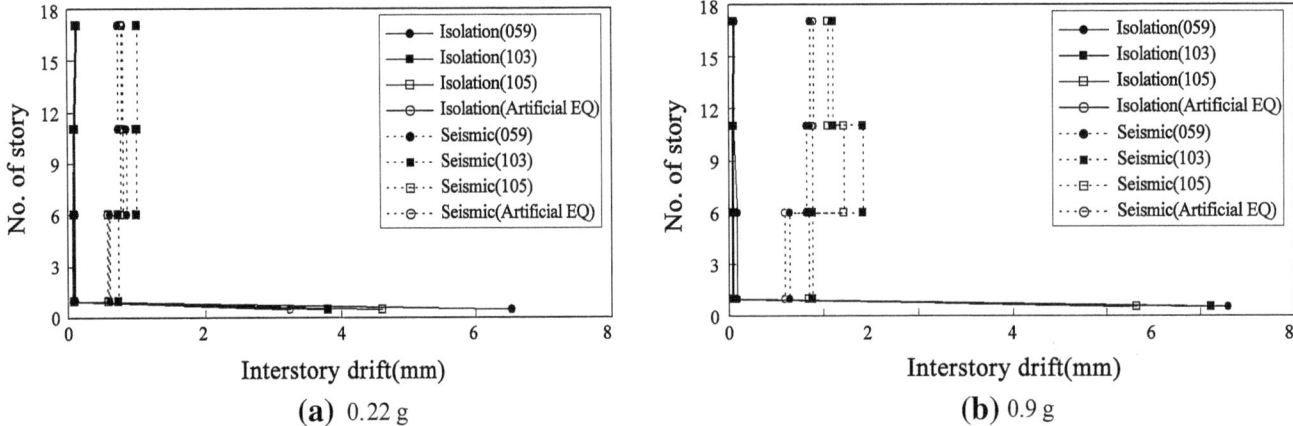

Fig. 8 Distribution of inter-story drift (Y-direction).

non-isolated structure shows a response distribution of a typical fixed-based structure in which the acceleration gradually increases as the response moves closer to the upper stories, the isolated structure shows that the almost constant response distribution and acceleration at the uppermost story is reduced by about two-thirds compared to the non-isolated structure. This improved behavior can be expected from the isolated RC structure in terms of usability, and amplification of

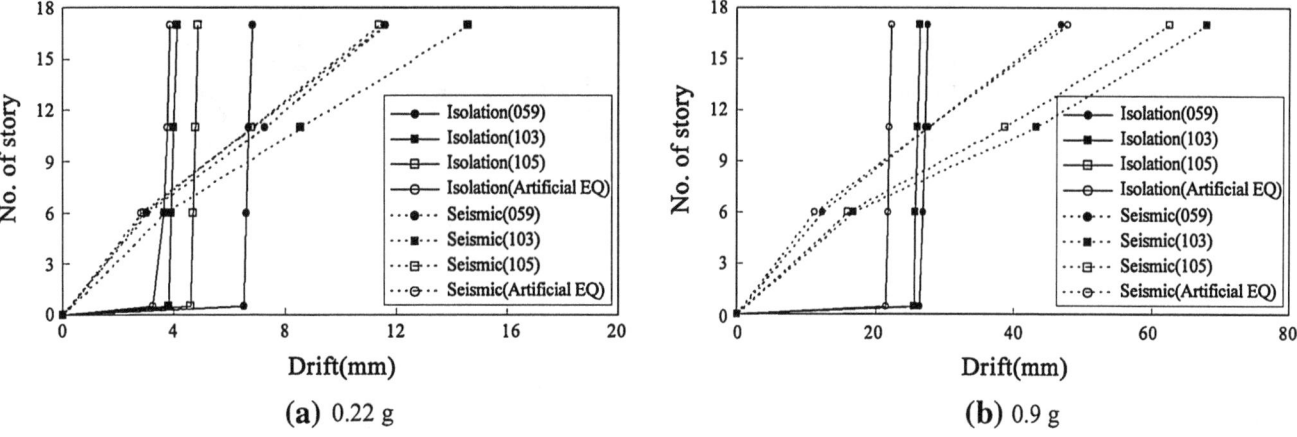

Fig. 9 Distribution of maximum drift response (Y-direction).

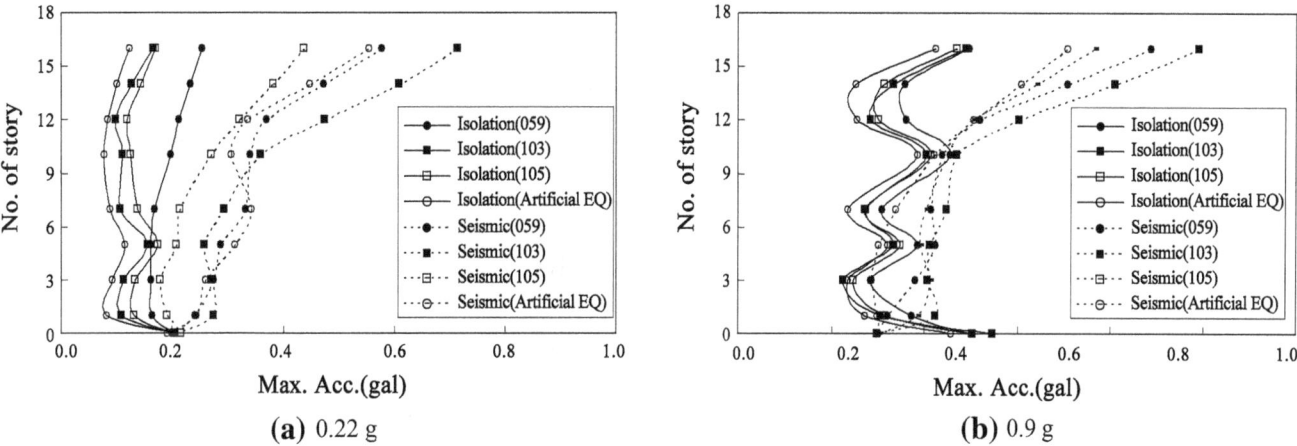

Fig. 10 Distribution of maximum response acceleration (Y-direction) on each story.

the response acceleration in the superstructure did not appear until 0.9 g. This phenomenon clearly explains the isolation effect. In addition to increased usability, greater reduction of the shear force in the superstructure can be obtained in comparison to that of the non-isolated RC structure.

However, because the period ratio between the super-structure and the isolation layer in this case did not configure as larger than 2.5 times, the distribution of the response acceleration in the isolated structure is closer to a K-shape than complete rigid body behavior. The effect of the higher order mode seems to be included, and the acceleration value did not decrease as much as the input value.

3.5 Comparison of Test Results with Response of Full-Scaled Reinforced Concrete Building

The acceleration response of the full-scale RC building could be obtained from the following relational equation with consideration of the similarity relationship presented in Table 2.

$$a_i = K_i a_g \qquad (3)$$

Here, a_i is the maximum acceleration response (g) of the i story of the full-scale building, a_g is the maximum

acceleration (g) of the input seismic motion, and K_i is the dynamic magnification factor of the i story of the scaled model that corresponds to the input seismic motion of a full-scaled RC structure.

Figure 11 presents a comparison of the maximum acceleration response of each story that was obtained from the test results and analytical results according to different seismic waves.

As illustrated in the figure, the experimental values are slightly higher than the values obtained from the analysis of seismic motion to assess residential performance. The figure shows an extremely large difference of about two to three times in the seismic motion values, which confirms the model's safety performance. Also, in terms of the mode shape, although the analytical results are similar to those for rigid body behavior, the experimental results include some effects of the higher order mode. The main cause of this difference is probably a technical error associated with the fabrication of the scaled model. As this problem occurred not only in the scaled model experiments but also can occur in practice, strict experimental tests and a review of the characteristics of the isolation device are essential for on-site application. The problem becomes even greater in terms

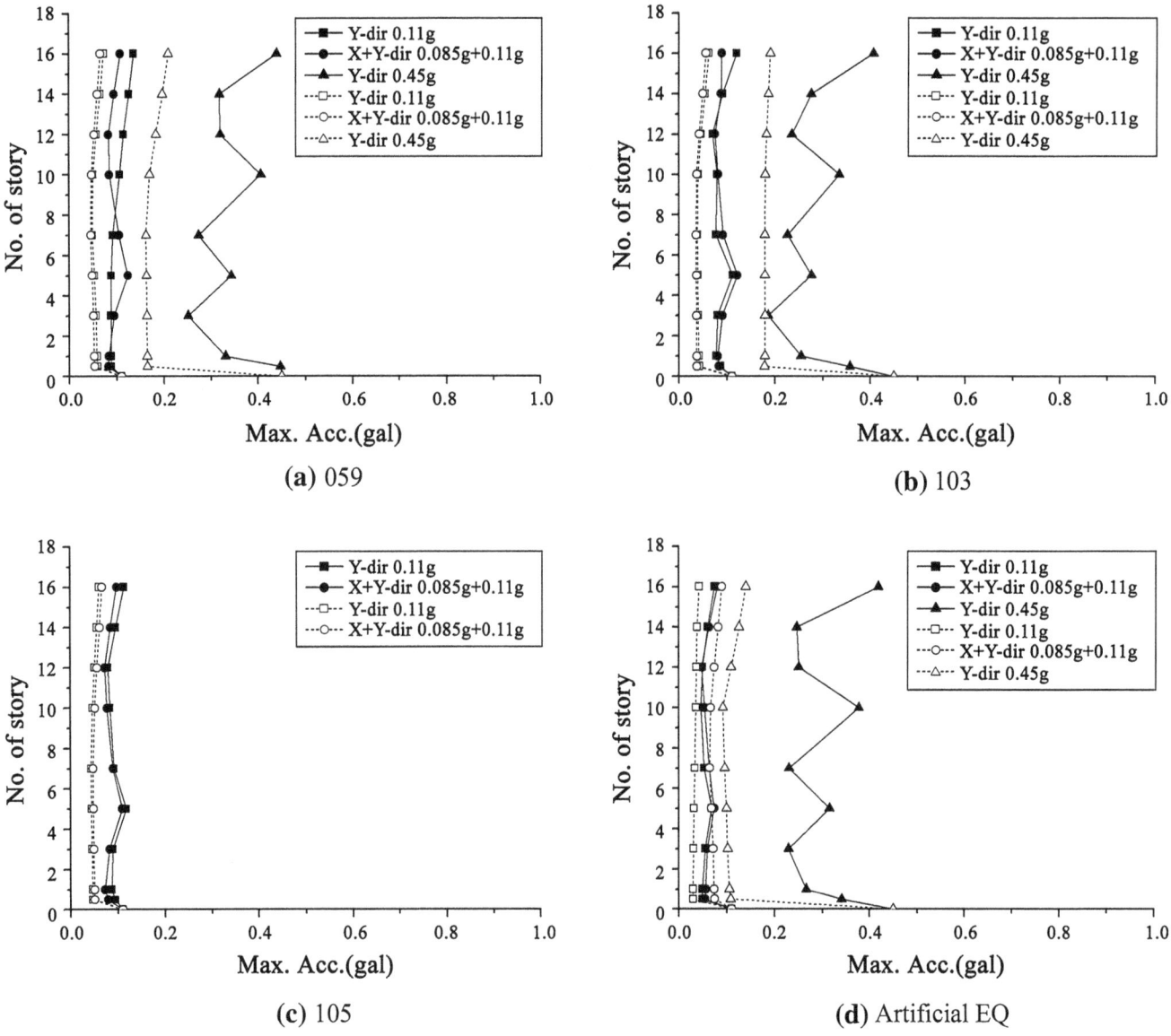

Fig. 11 Comparison of maximum acceleration response.

of the seismic motion that is used to confirm safety performance. The large difference in responses probably resulted from the large reduction in the isolation effect on the superstructure due to the smaller stiffness value difference between the superstructure and the isolation layer than the target value and from the deepening effects of the higher order mode and torsion under strong vibration due to the increased eccentricity between the superstructure and the isolation layer. However, as shown by the responses of the seismic motion that are used for the assessment of residential performance, a very similar response is evident in the one-way seismic motion, except for the slight difference in the response acceleration that is caused by the difference in stiffness value between the test results and analysis results. Based on these findings, a sufficiently reliable response can be obtained for the interpretive isolation model by paying close attention to the performance of the isolation device and appropriately modeling the stiffness of the isolation layer.

The drift response of a full-scale RC building can be obtained from the following relational equation with consideration of the similarity relationship presented in Table 2.

$$D_i = \frac{a_{mg}D_{mi}}{a_{tg}S_d} \tag{4}$$

Here, D_i is the drift response of the i story in a full-scale building (mm), D_{mi} is the drift response of the i story in the scaled model (mm), a_{mg} is the maximum input acceleration (g) of the scaled model shaking table (g), a_{tg} is the maximum input acceleration (g) of the scaled model shaking table that corresponds to D_{mi}, and S_d is the drift similarity factor of the scaled model.

Figures 12 and 13 respectively show the maximum acceleration response and maximum story displacement response that correspond to the design seismic motion level in order to assess residential performance by comparing the experimental and analytical results. As shown in the figures, the drift responses are extremely similar and constant except

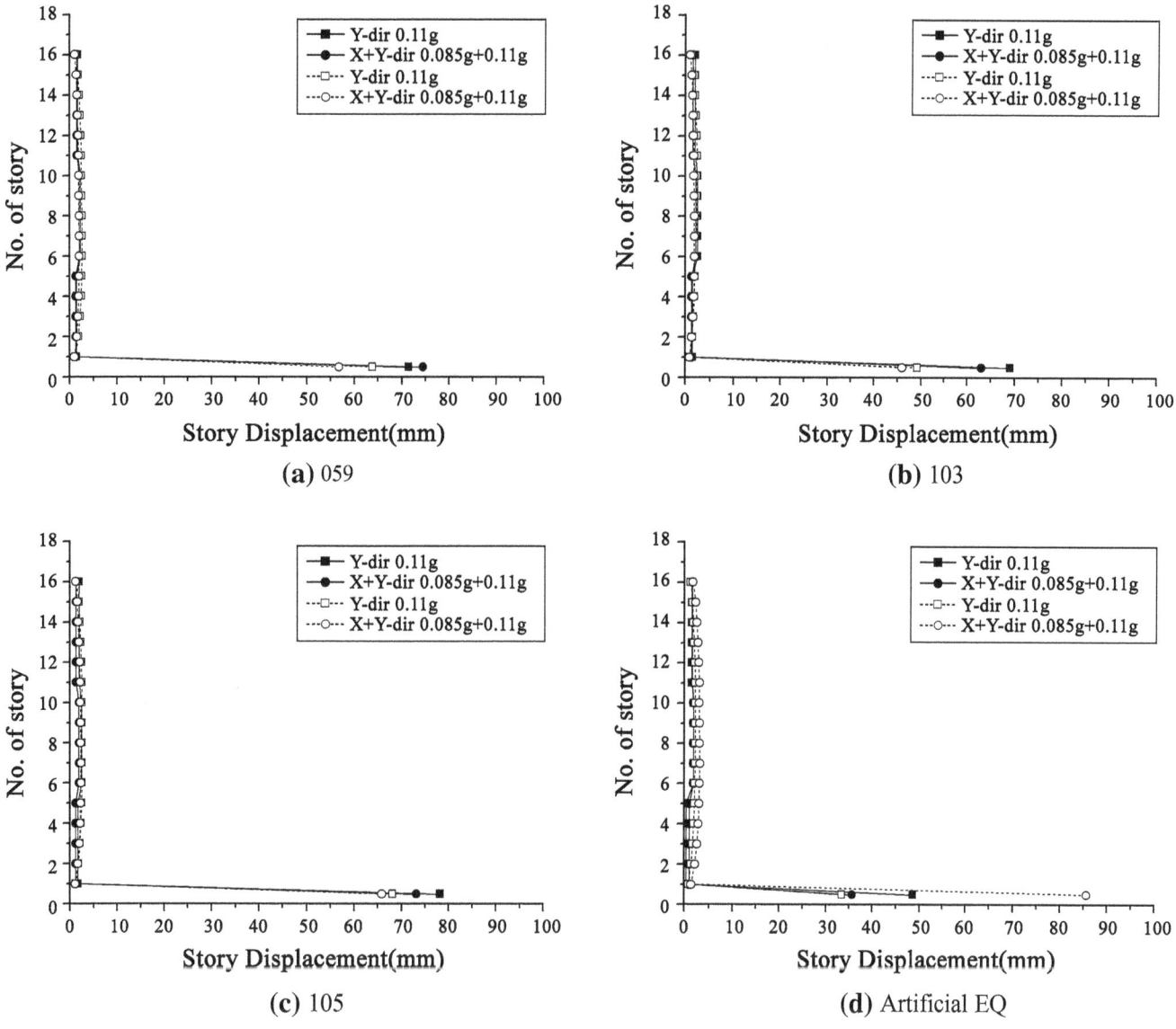

Fig. 12 Comparison of maximum acceleration response.

that the analytical results for the artificial seismic wave are slightly larger than the experimental results. The established analytical model can be used appropriately to confirm the limitation of the isolation device and to review the seismic performance of an isolated building.

4. Conclusion

In this paper, the isolation effect of a 15-story medium-rise RC building in terms of the vibration period ratio (difference in stiffness values) between a superstructure and an isolation layer and and the diverse characteristics of a frame was analyzed. Based on this analysis, the seismic behavior and performance of an isolated RC flat-plate structure were assessed by conducting shaking table tests on a model that scaled down an apartment building to one-tenth size. The conclusions of this study are as follows:

(1) In order to obtain valid seismic isolation effects for a 15-story medium-rise RC building, the isolation period must be over two and half times the fundamental vibration period of the upper structure, and the target isolation period must be more than 3 s.

(2) Based on the test results, the isolated RC structure showed that the acceleration of the uppermost story reduced by about two-thirds compared to the non-isolated RC structure. The seismic motion that is assessed to confirm the safety performance decreased by over two-thirds. Thus, a RC flat-plate structure can achieve excellent seismic performance through isolation.

(3) In terms of the inter-story drift response that causes structural damage, the response of the isolated structure decreased by about one-third compared to the non-isolated structure. This decrease in the inter-story drift can also help to achieve the safety of a superstructure.

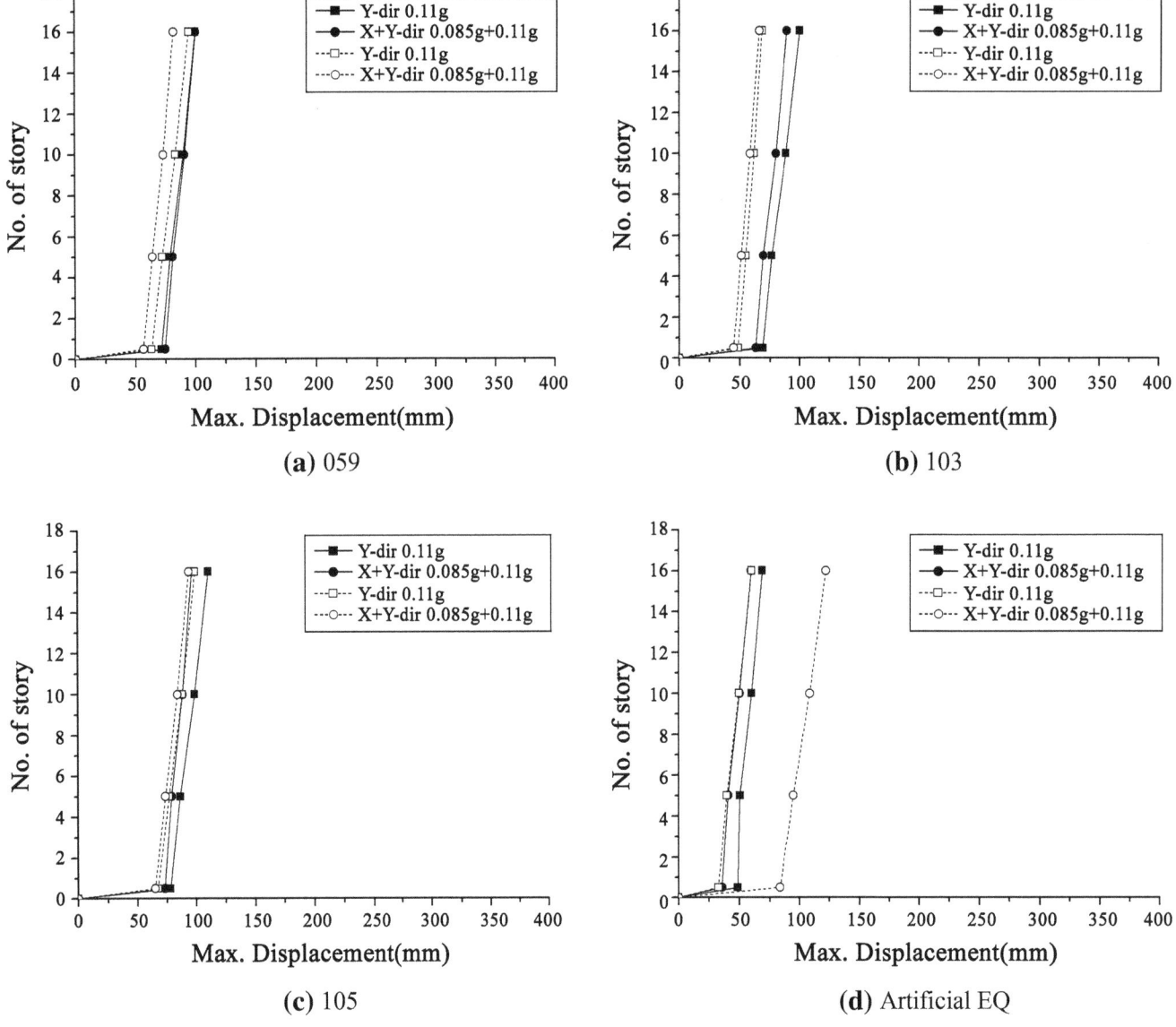

Fig. 13 Comparison of maximum displacement response.

(4) By comparing the experimental response acceleration with the analytical values, very similar responses were obtained for one-way seismic motion, except for a slight difference between the experimental results and the analysis results that was caused by the difference in stiffness value. Thus, the analytical isolation model can obtain sufficiently reliable responses through the careful confirmation of the performance of the isolation device and appropriate modeling of the stiffness in the isolation layer.

(5) As a part of the methodology to achieve effective seismic isolation behavior for a medium-rise RC building, this study presents ways to set the isolation period against the period of the upper structure and the beam-column stiffness ratio through a limited analytical study and tests on a fifteen-story reinforced concrete building. However, seismic isolation behavior is affected by a variety of variables, such as seismic waves, ground conditions, and the damping ratio, so additional research is needed to investigate these additional factors.

Acknowledgments

This research was supported by the Research Program funded by the Land and Housing Institute. Also, the China Academy of Building Research (CABR) in Beijing, China, contributed to the success of the tests and provided helpful advice. This support is gratefully acknowledged.

References

American Society of Civil Engineers (ASCE). (2010). Minimum design loads for buildings and other structures. ASCE/SEI 7-10.

Architectural Institute of Korea. (2009). Korean Building Code and Commentary (KBC) (in Korean).

Ariga, T., Kanno, Y., & Takewaki, I. (2006). Resonant behaviour of base isolated high-rise buildings under long-period ground motions. *Structural Design Tall Special Buildings, 15*, 325–338.

Casciati, F., & Hamdaoui, K. (2008). Modelling the uncertainty in the response of a base isolator. *Probabilistic Engineering Mechanics, 23*, 427–437.

China Association for Engineering Construction Standardization. (2001). Technical specification for seismic-isolation with laminated rubber bearing isolators. CECS126 (in Chinese).

Chun, Y. S., Son, C. H., Joo, I. D., Ahn, K. S., Kim, J. P., & Choi, K. Y. (2007). Development of flat plate structure with base isolation system and analysis of economical efficiency. Research Report of Housing & Urban Research Institute, pp. 30–36.

Deb, S. K. (2004). Seismic base isolation-an overview. *Current Science, 87*(10), 1426–1430.

Di Egidio, A., & Contento, A. (2010). Seismic response of a non-symmetric rigid block on a constrained oscillating base. *Engineering Structures, 32*, 3028–3039.

Dicleli, M., & Buddaram, S. (2007). Comprehensive evaluation of equivalent linear analysis method for seismic-isolated structures represented by SDOF systems. *Engineering Structures, 29*, 1653–1663.

Feng, D. (2007). A comparative study of seismic isolation codes worldwide. *Proceedings of SIViC international seminar* (pp. 1–28).

Feng, D., et al. (2012). A new design procedure for seismically isolated buildings based on seismic isolation codes worldwide. *Proceedings of 15WCEE*: Paper No. 0435. Lisbon, Portugal.

Japan Society of Seismic Isolation (JSSI). (2006). Response control and seismic isolation of buildings. Tokyo, Japan: JSSI (in Japanese).

Kilar, V., & Koren, D. (2009). Seismic behaviour of asymmetric base isolated structures with various distributions of isolators. *Engineering Structures, 31*, 910–921.

Komodromos, P., Polycarpou, P., Papaloizou, L., & Phocas, M. (2007). Response of seismically isolated buildings considering poundings. *Earthquake Engineering and Structural Dynamics, 36*, 1605–1622.

Korea Society of Seismic Isolation and Vibration Control (SIVIC). (2009). Isolation design manual and collection of practical examples. Technical Report. Seoul, Korea (in Korean).

Ministry of Construction, P. R. China. (2010). Code for Seismic Design of Buildings. GB50011-2010 (in Chinese).

Olsen, A., Aagaard, B., & Heaton, T. (2008). Long-period building response to earthquakes in the San Francisco Bay area. *Bulletin of the Seismological Society of America, 98*(2), 1047–1065.

Roehl, J. L. (1972). *Dynamic response of ground-excited building frames*. Houston, TX: Rice University.

Permissions

All chapters in this book were first published in IJCSM, by Springer; hereby published with permission under the Creative Commons Attribution License or equivalent. Every chapter published in this book has been scrutinized by our experts. Their significance has been extensively debated. The topics covered herein carry significant findings which will fuel the growth of the discipline. They may even be implemented as practical applications or may be referred to as a beginning point for another development.

The contributors of this book come from diverse backgrounds, making this book a truly international effort. This book will bring forth new frontiers with its revolutionizing research information and detailed analysis of the nascent developments around the world.

We would like to thank all the contributing authors for lending their expertise to make the book truly unique. They have played a crucial role in the development of this book. Without their invaluable contributions this book wouldn't have been possible. They have made vital efforts to compile up to date information on the varied aspects of this subject to make this book a valuable addition to the collection of many professionals and students.

This book was conceptualized with the vision of imparting up-to-date information and advanced data in this field. To ensure the same, a matchless editorial board was set up. Every individual on the board went through rigorous rounds of assessment to prove their worth. After which they invested a large part of their time researching and compiling the most relevant data for our readers.

The editorial board has been involved in producing this book since its inception. They have spent rigorous hours researching and exploring the diverse topics which have resulted in the successful publishing of this book. They have passed on their knowledge of decades through this book. To expedite this challenging task, the publisher supported the team at every step. A small team of assistant editors was also appointed to further simplify the editing procedure and attain best results for the readers.

Apart from the editorial board, the designing team has also invested a significant amount of their time in understanding the subject and creating the most relevant covers. They scrutinized every image to scout for the most suitable representation of the subject and create an appropriate cover for the book.

The publishing team has been an ardent support to the editorial, designing and production team. Their endless efforts to recruit the best for this project, has resulted in the accomplishment of this book. They are a veteran in the field of academics and their pool of knowledge is as vast as their experience in printing. Their expertise and guidance has proved useful at every step. Their uncompromising quality standards have made this book an exceptional effort. Their encouragement from time to time has been an inspiration for everyone.

The publisher and the editorial board hope that this book will prove to be a valuable piece of knowledge for researchers, students, practitioners and scholars across the globe.

List of Contributors

Sherif Yehia, Kareem Helal, Anaam Abusharkh, Amani Zaher and Hiba Istaitiyeh
Department of Civil Engineering, American University of Sharjah, Sharjah, United Arab Emirates

Arezou Babaahmadi and Luping Tang
Division of Building Technology (Building Materials), Chalmers University of Technology, Gothenburg, Sweden

Zareen Abbas
Department of Chemistry, University of Gothenburg, Gothenburg, Sweden

Per Mårtensson
Division of Low and Intermediate Level Nuclear Waste, Swedish Nuclear Fuel and Waste Management Company, Stockholm, Sweden

Seong-Hoon Kee
Department of Architectural Engineering, Dong-A University, Busan 604-714, Korea

Boohyun Nam
Department of Civil, Environmental and Construction Engineering, University of Central Florida, Orlando, FL 32816, USA

Mostafa Fakharifar
Department of Civil, Architectural and Environmental Engineering, Missouri University of Science and Technology, Rolla, MO 65409, USA

Genda Chen
Missouri University of Science and Technology, Rolla, MO 65409, USA

Ahmad Dalvand
Department of Engineering, Lorestan University, Khorramabad, Iran

Anoosh Shamsabadi
Office of Earthquake Engineering, California Department of Transportation, Sacramento, CA 95816, USA

Mustafa Kaya
Faculty of Engineering, Aksaray University, Ankara, Turkey

Zeynel Çağdaş Kankal
Republic of Turkey Ministry of Health, Ankara, Turkey

Hyeonggil Choi
Graduate School of Engineering, Muroran Institute of Technology, Hokkaido, Japan

Takafumi Noguchi
Graduate School of Engineering, The University of Tokyo, Tokyo, Japan

Se-Jin Jeon and Sang-Hyun Kim
Department of Civil Systems Engineering, Ajou University, Suwon-si, Gyeonggi-do 443-749, Korea

Sung Yong Park, Sung Tae Kim and Young Hwan Park
Structural Engineering Research Institute, Korea Institute of Civil Engineering and Building Technology, Goyang-si, Gyeonggi-do 411-712, Korea

G. M. Kim, J. G. Jang, Faizan Naeem and H. K. Lee
Department of Civil and Environmental Engineering, Korea Advanced Institute of Science and Technology, Daejeon 305-701, South Korea

H. S. S. Abou El-Mal
Civil Engineering Department, Menofia University, Shibin El-Kom 32511, Egypt

A. S. Sherbini
Civil Engineering Department, Suez Canal University, Ismailya 41522, Egypt

H. E. M. Sallam
Department of Civil Engineering, Faculty of Engineering, Jazan University, Jazan 82822-6694, Saudi Arabia

Hailong Ye
Department of Civil and Environmental Engineering, The Pennsylvania State University, State College, PA 16801, USA

Padmanabha Rao Tadepalli
American Global Maritime, Houston, TX 77079, USA

Hemant B. Dhonde
Civil Engineering Department, Vishwakarma Institute of Information Technology, Pune 411048, India

Y. L. Mo and Thomas T. C. Hsu
Department of Civil and Environmental Engineering, University of Houston, Houston, TX 77204, USA

I. W. Nam and H. K. Lee
Department of Civil and Environmental Engineering, Korea Advanced Institute of Science and Technology, Daejeon 34141, South Korea

Jinkoo Kim
Department of Civil and Architectural Engineering, Sungkyunkwan University, Suwon, Korea

Hyunhoon Choi
Research Institute of Technology, Samsung Engineering
and Construction Co., Ltd., Seoul, Korea

Mehdi Khanzadeh-Moradllo
School of Civil Engineering, Construction Materials
Institute, University of Tehran, Tehran, Iran
Department of Civil and Environmental Engineering,
Oklahoma State University, Stillwater, OK 74078, USA

**Mohammad H. Meshkini, Ehsan Eslamdoost and
Mohammad Shekarchi**
School of Civil Engineering, Construction Materials
Institute, University of Tehran, Tehran, Iran

Seyedhamed Sadati
Civil, Architecture, and Environmental Engineering
Department, Missouri University of Science and
Technology, Rolla, MO 65401, USA

Young-Soo Chun
Land and Housing Institute, Korea Land & Housing
Corporation, Daejeon 305-731, Korea

Moo-Won Hur
Department of Architectural Engineering, Dankook
University, Gyeonggi-do 448-701, Korea